Pragmatism and Embodied Cognitive Science

# HUMANPROJEKT
# Interdisziplinäre Anthropologie

---

Im Auftrag der Berlin-Brandenburgischen
Akademie der Wissenschaften
herausgegeben von Detlev Ganten, Volker Gerhardt,
Jan-Christoph Heilinger und Julian Nida-Rümelin

**Band 14**

# Pragmatism and Embodied Cognitive Science

From Bodily Intersubjectivity to Symbolic Articulation

Edited by
Roman Madzia and Matthias Jung

DE GRUYTER

ISBN 978-3-11-047889-1
e-ISBN (PDF) 978-3-11-048023-8
e-ISBN (EPUB) 978-3-11-047893-8
ISSN 1868-8144

**Library of Congress Cataloging-in-Publication Data**
A CIP catalog record for this book has been applied for at the Library of Congress.

**Bibliografische Information der Deutschen Nationalbibliothek**
Die Deutsche Nationalbibliothek verzeichnet diese Publikation in der Deutschen Nationalbibliografie; detaillierte bibliografische Daten sind im Internet über http://dnb.dnb.de abrufbar.


Einbandgestaltung: Martin Zech, Bremen
Druck und Bindung: Hubert & Co. GmbH & Co. KG, Göttingen
♾ Gedruckt auf säurefreiem Papier
Printed in Germany

www.degruyter.com

# Contents

Roman Madzia and Matthias Jung

# Introduction: What a Pragmatist Cognitive Science Is and What It Should Be

## 1 The paradigm shift to embodied cognition

Over the last two decades, cognitive sciences, as well as human sciences in general, have witnessed a considerable paradigm shift with regard to their outlook on the nature of the human mind, cognition, perception and action. Traditional cognitive sciences have, without adequate critical examination, inherited their conceptual framework and theoretical outlook on what cognition is from representatives of the early modern philosophical tradition such as René Descartes, John Locke and/or David Hume. Within the early modern framework, the mind was understood as a peculiar sort of an object that could be characterized by specific properties. From the Cartesian perspective, the mind was perceived as an *organ* for thinking (Rowlands 2003). In a similar manner as the stomach digests and eyes see, according to early Modern thinkers—"the mind thinks". Within that philosophical perspective, the mind, being a specific kind of organ, was allegedly contained in the brain (internalism). Unlike other bodily organs that are situated entirely in the physical world and are, hence, governed by purely mechanical principles, the mind was taken to be governed primarily by the principles of *reason* (intellectualism).

In a very similar manner to the early Modern outlook on the mind as a capacity governed by the principles of reason, the so-called "first-generation cognitive science" (cognitivism and computationalism [see Lakoff & Johnson 1999]) understands the human mind as some sort of an abstract program of the brain, governed by the rules of formal logic. Within such a conceptual framework, cognition is viewed exclusively as a process of computation over formal symbols (in the early Modern rendition—assembling and re-assembling of ideas), taking place in our heads. Hence, seen from the perspective of the first-generation cognitive science (but also first-generation artificial intelligence and robotics), the human mind and cognition are:

1. *Disembodied*—the structure of the body, its sensorimotor capacities and bodily skills add nothing to the nature of the processes in question.
2. *Formal*—cognitive processes are instantiations of formal sets of rules (programs). Mental representations are, in themselves, meaningless formal symbols that acquire content either a) on the basis of their mutual relations, or b) by virtue of representing something outside the formal system.

3. *Internal*—cognition is something that happens in our brains by means of representing the external environmental objects and events. Hence, human beings are, in principle, cut off from the world—they can only be acquainted with its "copy" but can never achieve a direct contact with that world as such.

Since its formation as a discipline that aims for a naturalistic account of the mental, cognitive science has been dominated by a view of cognition as represented in 1–3. This classical paradigm has been partly fruitful and has stimulated important research in the early decades of cognitive science. However, starting two decades ago, significant criticisms have been raised with regard to various insufficiencies of the first-generation cognitive science in providing a viable and empirically responsible account of the nature of cognition, mentality and human experience in general (e.g. Varela et al. 1991; Dreyfus 1992; Clark 1997; Lakoff & Johnson 1999, etc.). As an alternative, the so-called "embodied" (Lakoff & Johnson), "embedded" (Rupert), "enactive" (Varela), "extended" (Clark) approaches to cognition were introduced. Although these approaches sometimes differ from one another quite significantly, their common denominator is the belief that cognition is not to be understood as a formal capacity for deriving world-models but rather, it has been emphasized that cognitive processes are so closely intertwined with *action* that they would be best understood as the exercise of skillful bodily know-how of organisms engaging with their environments. According to the proponents of "4EA" (embodied, extended, embedded, enactive, affective) approach, cognition is not a process of creating representational models of the world, but rather *models bodily interaction* with the world. In the 4EA approach, the mind is not a disembodied and self-sufficient computing device but is understood as a never fully formalizable set of practical skills (strategies) by means of which physically embodied and environmentally embedded agents negotiate their worlds. The criterion for the success of cognitive operations is not to recover pre-existing features or to construct a veridical *representation* of the environment—as the classical, first-generation cognitive science would have had it. According to 4EA, cognition is not a process that unfolds exclusively in our heads. The cognitive feedback loops between organisms and their environments are so dense and continuous that, for scientists studying the nature of cognitive activity, the isolated brain is not a meaningful unit of analysis. Cognition, as more and more cognitive scientists are ready to admit, takes place on the entire brain-body-world chain without the possibility to indicate which one of these elements constitutes its substance. As Andreas K. Engel et al. (2013) write in their recent article "Where's the action? The pragmatic turn in cognitive science": "Cognition is a form of practice". The function of the mind is to guide action, and cognitive mechanisms such as perception and memory must be understood in terms of their ultimate contribution to situation-appropriate behavior (Wilson 2002).Situation-appropriateness, however, would be misunderstood if taken to refer only to the immediate physical surrounding of the human organism. Human beings develop long-term plans, engage in various activities decoupled from immediate response to situations (cognitive science itself being a compelling example), and tend to frame their coping behavior by developing and adhering to comprehensive world views. The ultimate situation in which the embodied mind shapes action is thus nothing less than the *conditio humana* in general. This is one important reason, among others, why American pragmatism ought to enter the picture sketched so far. Pragmatism, keenly aware of the fact that cognition is for action, but action transcends specific situations, is embedded in a culture and ultimately refers to the place of human beings in the world, is a potent antidote against a too narrow identification of action with situated coping.

## 2 Pragmatism as the philosophical foundation of embodied cognition

Many contemporary representatives of the "4EA" develop their theories mostly by elaborating on works by phenomenologists like Maurice Merleau-Ponty (Gallagher 2005; Dreyfus 1979; 1991; 2005; 2013; Noë 2004; 2009; 2012) and/or Martin Heidegger (Wheeler 2005). Although phenomenology and even hermeneutics (Kurthen 1994) have been highly inspiring as a conceptual source in the debate on embodiment both in philosophy and cognitive science, its American counterpart, the tradition of American pragmatism (for which the notion of cognition as an ongoing organism-environment coupling was the founding idea), by far still has not received the attention it undoubtedly deserves. Although especially the views on consciousness by John Dewey have significantly entered the debate in the work of certain authors (Johnson 2007; Rockwell 2005; Solymosi 2011), the important voices of other pragmatist philosophers, such as George H. Mead, William James, and Charles S. Peirce, have yet to be heard. The editors hope that this volume, largely the outcome of a conference on the topic at the University of Koblenz-Landau in 2015, will help to change this unsatisfying situation. More specifically, our wish is to encourage discussions not only about the appropriate philosophical foundations of cognitive science, but also about the 'greater picture', the relation of science to the pressing questions of social life and human self-understanding. Antidualism, as the pragmatists saw it, not only questions the separation of the mind from the physical world but it is also highly skeptical about severing the normative und evaluative questions of human life from the scientific understanding of the embodied mind.

Let us now start with sketching some of the most important aspects in which embodied cognitive science may profit from classical pragmatism. First and foremost, Mead's naturalistic philosophy of action provides the right conceptual tools for introducing a naturalized notion of intentionality. In the classical analytic rendition, intentionality is characterized by two basic properties: i) directedness (aboutness) and ii) conditions of satisfaction. Within condition i) consciousness is always directed at something other than itself, most of the time—at worldly objects. As for the condition ii), a certain mental state is intentional if we can indicate specified conditions under which the mental state in question is true. If we accept such a definition (especially property [ii]), however, we face a serious problem as to how to account for conscious awareness and clear directedness toward objects in pre- and non-linguistic creatures (infants, animals, etc.). On account of their consciousness, non-linguistic creatures cannot propositionally reflect on the content of their thought, given the classical definition, as they are not intentional beings. As experience tells us, however, this is, quite evidently, not the case. Mead's theory of the act, and especially his concept of attitudes, provides the right tools to solve this pressing problem of accounting for continuity between human (reflective) consciousness and consciousness of non-linguistic creatures. The key to this solution is Mead's conviction that all bodily coping, mediated by attitudes (neurologically realized bodily dispositions encoding perceived objects in terms of possible reactions toward them), is inherently goal-directed (quasi-intentional) but, in most cases, does not involve mental content. Hence, elaborating on Mead's concept of attitudes, the notion of intentionality can be divided into "bodily-interactive" and "social-symbolic" intentionality. The former shows that intentionality is not primarily a property of the mind but of the active body and may serve as a conceptual ground for social (second-order) intentionality. It is on the level of social intentionality on which we can properly talk about mental content and the division between the subject and the object. Here Mead's theory of intersubjectivity (along with his well-known social-psychological concepts like taking the attitude/role of the other, consciousness of meaning, and significant symbol) will play a crucial role.

Such a concept of intentionality will bring about strong consequences also for social cognition. As opposed to classical theories of empathy (theory-theory and simulation theory) that presuppose that mental states are in principle inner to every subject and accessible only from the first-person perspective, it claims that, most of the time, mental states of others are directly given to us in perception. Mead's theory of social cognition and intersubjectivity can therefore successfully avoid dualism about the self and the other, and consequently also skepticism about other minds.

Since in existential phenomenology (Heidegger and Merleau-Ponty) we can find a very similar approach toward empathy and intersubjectivity, some authors (e.g. Gallagher 2005, 2012) have advocated the so-called "interactionist theory of social cognition" that is supposed to undermine the prevailing intellectualist paradigm in contemporary philosophy of psychology. In that respect, Mead's empirically well-founded pragmatist theory of intersubjectivity (see Franks 2010) will be an inspiring and much needed ally.

Given the above-mentioned affinities between pragmatism and phenomenology, the contributions to this volume will further examine the approach of both these traditions toward the problem of the consciousness in relation to embodied experience and skillful expertise. As opposed to overly intellectualist theories in today's philosophy of action and traditional cognitive science, second-wave phenomenologists like Heidegger or Merleau-Ponty hold that our basic experience of the world is not that of detachment, disengagement and intellectual contemplation but, on the contrary, that of a physically embodied and environmentally embedded agent skillfully coping with his practical world. Exactly the same, for that matter, counts for pragmatism. However, virtually no research has been done on mutual relations, affinities and tensions between these two influential traditions especially with respect to the two above-mentioned problem areas.

The importance of such a research is directly tied to the basic controversy between first- and second-generation cognitive science. The second-generation cognitive science contests the intellectualist claim that the basic experiential mode of cognitive agents is *theoretical*, contemplative and disengaged, that is to say, that it primarily takes up the form of creating and computing representations. Pragmatism and the later phenomenology, likewise, focus on action over contemplation, practice over theory, and body over mind. However, whereas certain phenomenologists (e.g. H. Dreyfus) tend to underestimate the role of what might be called "inner information processing"—pragmatism offers a more balanced vision of the role of conscious cognitive processes in all sorts of cognitive tasks. According to Ch. S. Peirce, for instance, the situated brain is indeed a sort of an 'engine' for producing predictive beliefs about events and structures that the acting organism is likely to be confronted with in the environment: "perhaps all belief might involve expectation as its essence" (CP 5.542). Peirce, however, did not think of 'beliefs' as necessarily having a representational (propositional) structure. The problem about how to conceive of 'beliefs' or 'predictions' in a non-representational way has been central to the ongoing debate in the embodied approaches to otherwise very promising "prediction minimization theory" (Clark 2013; 2015).

The authors of this volume directly confront the classical rendition of cognition as *always* mediated by mental content (an approach which can be best characterized by words of Fred Dretske: "the manipulation and use of representations is the primary job of the mind" [1995]). Their contributions show that according to pragmatism, phenomenology and 4EA, not all mentality requires individuals to construct representations of their world. On the embodied level of intelligent, practical coping with our experience of the world is not mediated by mental representations but by our sensorimotor bodily skills through which we negotiate the world "directly" as a space of "affordances" (Gibson 1979). Although this notion is well known in the 4EA debate, the ontological status of affordances is still seen as problematic. In order to elucidate this ontological status, certain authors analyze this notion from the point of view of W. James' radical empiricism, according to which not only objects but also the relations between them are of the same ontological importance. In that respect, some of the contributors link insights of pragmatism with the most radical theories of embodied cognition, such as the extended mind theory. To that end, answers are necessary to questions such as: Is pragmatism fully compatible with extended mind theory? What new insights can pragmatism bring to support the notion of extended cognition and *vice versa?* Can Mead's theory of the act, Dewey's concept of qualitative thought and other insights from classical pragmatism help extended cognition overcome the so-called "Hard Problem of Content" (Hutto & Myin 2013)? If so, how does that alter the classical notion of mental content (mental representation)?

In accordance with John Dewey's basic philosophical conviction, the contributors to this volume all share the conviction that a good portion of intelligent action is best understood without adopting notions of mental representation, which term is, for that matter, also difficult to reconcile with explanatory naturalism. In that respect, the overall goal of the current volume is to systematically argue in favor of the anti-intellectualist claim that our most elementary ways of engaging the world and understanding others are mindful in the sense of being purposive and phenomenally charged, despite being non-representational and contentless. That is, of course, not to say that there is no place for mental content in our investigations into the nature of human cognition. The challenge is to understand how elementary and reflective cognition interact with each other. Symbolic, higher-order thinking remains crucial for making us the kind of beings we are, but it always comes structurally coupled with, so to speak, bodily intelligence. The emergence of contentful thinking in human cognitive architecture is a late evolutionary phenomenon dependent upon our abilities to use language. Realizing the need to find a balanced approach between continuity, on the one hand, and an indubitable qualitative shift between non-human (non-symbolic) and human (symbolic) cognition, on the other, the authors of this volume explore the manifold manners in which pragmatist thinkers contribute to a both non-reductionist and naturalistic picture of human nature.

## 3 Open questions: identifying the core issues

It would be rather pretentious and futile merely to contend that the pragmatists had gotten things right already a century before 4EA did. Underlining, as most contributors to this issue do, that pragmatism is the appropriate philosophical foundation for embodied cognition, is therefore endangered by two fatal misunderstandings. (1) Pragmatists are highly skeptical about giving timeless answers in an ever-changing world and specifically life-world. Ongoing dynamic interactions may over time deeply alter the structure of the coupling between an organism and its environment. It is therefore very likely that, since we live in a world that differs from the decades around the turn to the 20$^{th}$ century in many crucial ways, some important aspects of pragmatism need to be rectified. Talking about "foundation" should not mean excluding mutuality. To give just one example, pragmatists give high priority to bodily, face-to-face interactions, whereas the interactive behavior and informationally sophisticated lifestyle of many human beings in the 21th century is shaped increasingly by technological props and gadgets. It is an open and empirical question if, and if yes to what extent, such technological (and the accompanying social) developments alter the importance of elementary embodied cognition. (2) One main lesson to be learned from classical pragmatism is that human experience comes in many forms ("legetai pollachos", as Aristotle said about Being) and that full acknowledgement of all these forms is essential for understanding human cognition. The fact that John Dewey, in his unfinished drafts to a new introduction to *Experience and Nature* from 1949, contemplated renaming his book *Culture and Nature* is evidence for his awareness that cognitive processes are embedded in the multifarious spheres of meaning human cultures produce. Cognitive science may profit from pragmatism not only for its conceptual foundations, but also for getting the larger picture right, that is for not losing sight of the fact that the mind-body-situation coupling is always part of the cultural process of sense-making which reaches from the everyday-challenges of ordinary life to those interpretations of the *conditio humana* known as comprehensive world views and/or religions.

Against the backdrop of these considerations, for the editors the following considerations identify some of the core issues the future dialogue between pragmatism and embodied cognitive science ought to tackle.

## 3.1 Beyond skilful coping—the importance of higher-order ideals, reflexive normativity and culture in general

For pragmatists, the cognition of human beings is action-oriented all the way down—and all the way up, too. This insight should caution us against prematurely identifying mindedness with skillful coping, the piecemeal solving of everyday problems coming up in the course of life. As already underlined above, human cognition is capable of long-term planning, of developing normative and evaluative attitudes and generally of engaging in practices that are rather remote from the immediate needs of coping with the environment. Not only (cognitive) science itself, but art, religion and many other reflective cultural practices offer convincing examples. The important and deeply pragmatist insight into the ubiquitous influence of bodily interactions, as well as in the common root of both human and non-human cognition in these sensory-motor feedback loops, should not lead us to neglect the pervasive influence of culture and its second-order evaluations on social and individual behavior/cognition.

Since the demise of German idealism after Hegel's death, many cultural theorists have shown that the symbolic forms of culture, albeit deeply imbedded in and crucially dependent upon the interacting body, develop certain semi-independence from the organic life process. Georg Simmel calls it the "axial turn" (*Achsendrehung*), Marburg Neokantians (foremost Cassirer), Phenomenologists, Wilhelm Dilthey with his philosophy of life and last but not least both American transcendentalism and pragmatism all agree that culture and the way it shapes cognition cannot be captured in terms of skillful coping alone. More recently, Michael Tomasello has shown in numerous books (Tomasello 1999; 2008; 2014) that human beings are naturally adopted for growing up in normatively structured cultures. It is by a somewhat ironic twist of history, that Hubert Dreyfus developed his coping model of human cognition with strong reference to Martin Heidegger, thereby implicitly ratifying the dubious "victory" of the latter in the "Davos Disputation" with Cassirer. Without convincing arguments, but supported by an existential stance Cassirer unfortunately seemed to lack in the eyes of the audience, Heidegger brushed aside cultural normativity and opted for existential "being" (*Dasein*) in its nakedness. He thereby not only paved the way to his later fatal engagement with National Socialism but unnecessarily severed embodiment and culture.

The pragmatists, being (with the exception of Peirce) ardent democrats, would never commit such an unfortunate mistake. They saw human cognition as always action-oriented, and situated coping as the "original situation" of cognition, but they were far from playing down the importance of cultural values. John Dewey's "theory of valuation" is an impressive example of the manner in which the pragmatists succeeded in reconciling situatedness and the quest for reflective, second-order ideals and values (as distinct from mere preferences). To put it differently: Dewey's pragmatism is both anti-reductionist and naturalistic. Embodied cognitive science can thus draw on pragmatist resources in order to avoid succumbing to an oversimplified picture of cognition. Mind is for action, yes, but action, as distinguished from mere non-reflective behavior, exhibits value-orientation (to use Max Weber' term), long-term-planning and even the formation of comprehensive worldviews. It is, in a word, a *cultural* activity. And embodied cognition research should not relegate culture to a higher sphere and focus entirely on basic cognition. Understanding genuinely human cognition presupposes avoiding such a new dualism and concentrating, as Dewey did, on the interactions between situated coping and higher-order ideals, values, norm and institutions. We should not be willing to pay the price for overcoming the Cartesian divide by creating new dualisms between situated cognition on the one hand and the cultural spheres of meaning on the other. Mark Johnson, in his recent publications, has already taken significant steps towards including the more subtle and advanced forms of human experience. In this volume, Tibor Solymosi, also argues for recovering philosophy from cognitive science by learning the lesson from it, and then leaving it behind and proceeding in a creative, imaginative Deweyan project towards a more democratic society. The possibility, however, that living in the space of culture modifies the structure of the interaction between human organisms and their environment in a top-down manner also creates new ventures for interdisciplinary research. Pragmatist cognitive science may thus enter into dialogues hitherto unthought of, for example with the philosophy of culture (one of its most prominent exponents, Ernst Cassirer, even took a scarcely explored pragmatist turn towards the end of his life). It is only after having integrated the Gibsonian "space of affordances" with the normatively structured "space of culture" (Luft 2015) that the full impact of pragmatist insights will have reached embodied cognition research.

## 4 The contributions to this volume

The first part of the book, which presents pragmatism as an important philosophical foundation of the situated cognitive science, is opened by *Shaun Gal-*

*lagher*'s essay on the pragmatist interventions into enactive and extended conceptions of cognition. In his paper, Gallagher analyses important philosophical lessons which contemporary situated cognition theories can draw from the pragmatist philosophy. More particularly, he leans toward a position which could be called an integrative conception of extended cognition. The following paper written by *Joshua A. Skorburg* assesses Gallagher's strategy and offers a critical perspective on the use of Deweyan pragmatism to resolve these tensions in the 4EA. Skorburg argues that although pragmatism and the so-called 4EA program may form a united front against methodological individualism, classical cognitivism, traditional internalism, and the like, the 4EA approach is not without its own internal tensions. Some tensions between extended and enactive views on cognition are also addressed in *Anthony Chemero*'s paper on phenomenology, pragmatism and extended cognition. More precisely, Chemero endeavors to set out a defensible version of the hypothesis of extended cognition. To do so, he draws together William James' radical empiricism, on the one hand, and Maurice Merleau-Ponty's phenomenology of the lived body, on the other. The first part of this volume is concluded by *Pierre Steiner*'s paper in which he addresses the notoriously problematic notion of mental representation. Some 40 years ago, pragmatism made it back on the philosophical main stage especially due to Richard Rorty's criticism of representationalism. In this sense, Steiner's essay not only continues this good old pragmatist tradition but also develops it by analyzing James' and Dewey's views on the topic with regard to the sciences of the mind.

The second part of this book aims at providing a larger picture on the debate on 4EA cognition. *Mark Johnson*'s opening essay "Pragmatism and Embodied Cognitive Science" points to the limitations as well as implications of the pragmatist turn in cognitive sciences. Johnson's long-standing work on the topic of a pragmatist cognitive science has now given him the opportunity to sum up his thoughts. Exactly this is the goal of his paper. After briefly identifying the shortcomings of what he calls the "first generation cognitive science", Johnson enumerates his strongest arguments about how the second (pragmatistically informed) cognitive science overcomes the difficulties of the classical approach and permits us to recognize the aesthetic, qualitative character of cognition. Also *Erkki Kilpinen*, in his contribution, tackles the question of the position of pragmatism in the embodied cognitive science debate. According to him, certain difficulties in assessing the relation between pragmatist philosophy and cognitive science, particularly embodied cognitive science, are due to the fact that pragmatism is a minority position in contemporary philosophy. Kilpinen identifies the primary point of contact between pragmatism and embodied cognitive science in their almost identical perspective on action.—Can neuroscience one day replace philosophy? Is there something with which a pragmatistically informed philosophy can alleviate the tensions between neuroscience and philosophy in the social context? These questions are tackled by *Tibor Solymosi* in his essay. Solymosi outlines a reconstruction of philosophy via cognitive science in order to restore philosophy's traditional aspiration toward justice. To this end, he seeks new possibilities of application for Dewey's transactional conception of experience.—The issue of a situated mind and the question of agency is addressed in *Teed Rockwell*'s paper. There, he explores and questions the philosophical assumptions of game theory—the branch of computer science that assumes that society can only be understood as the interaction of isolated rational autonomous agents. Rockwell argues that just as extended cognition theories recognize that the line between the mind and the world is a flexible one, dynamic social theories would recognize that the line between the mind and the mind is equally flexible. The contribution of *Guido Baggio* goes in a similar direction, although he tackles the question of behavioral analysis from the point of view of George H. Mead. As Baggio argues, Mead's anti-individualist conception of the self and his notion of social action could shed new light on some important methodological questions that contemporary so-called neuroeconomics is concerned with.

The third part of the volume is dedicated to specific analyses of the notion of embodiment in the work of particular pragmatist thinkers. The philosophy of all major American pragmatists is investigated from multiple philosophical perspectives with embodiment and cognitive situatedness as their common denominator. The first essay is devoted to one of the most neglected but at the same time arguably one of the most original philosophers of embodiment—George Herbert Mead. In the essay "Mind, Symbol, and Action-Prediction: George H. Mead and the 'Embodied' Roots of Language" *Roman Madzia* presents Mead's radically embodied and anti-intellectualist view of the development of language and symbolic behavior. The relation between Dewey's enactivism and Greek thought is discussed in the essay of *Matthew Crippen*. In his paper, Crippen investigates some possible Greek influences on Dewey's understanding of the notion of 'skill' and 'habit' and, subsequently, finds common philosophical grounds of some Greek thinkers, Dewey and contemporary enactivism. Charles S. Peirce as a thinker of embodiment is presented in the essay by *Tullio Viola*. His paper explores Peirce's relevance to embodied cognitive science by discussing his notion of *abduction*, or hypothetical inference. Viola also tries to pin down some important divergences between Peirce's account of cognition and that of some phenomenologists. For *Rebekka Hufendiek*, in turn, it is William James' embodied theory of emotions which deserves the attention of the theorists of embodiment. In her paper, she presents William James' theory of emotions and analyzes the central role it ascribes to the bodily reactions in emotions. William James' philosophy is the central theme also of the final text of the volume. In her essay, *Ling Zhu* concentrates on James' view of the role played by feelings in his famous notion of the 'stream of thought'. This analysis is made on the background of the distinction between 'substansive' and 'transitive' parts of the stream of thought.

## References


Clark, A. (1997). *Being There: Putting Brain, Body, and World Together Again.* Cambridge, MA: MIT Press.

Clark, A. (2008). "Pressing the Flesh: A Tension in the Study of the Embodied, Embedded Mind." In: *Philosophy and Phenomenological Research*, 76(1), 37–59.

Clark, A. (2015). *Surfing Uncertainty: Prediction, Action, and the Embodied Mind.* New York: Oxford University Press.

Dretske, F. (1995). *Naturalizing the Mind.* Cambridge, MA: MIT Press.

Dreyfus, H. L. (1979). *What Computers Can't Do: The Limits of Artificial Intelligence.* New York, NY: Harper & Row.

Dreyfus, H. L. (1991). *Being-in-the-world. A Commentary on Heidegger's Being and Time, Division I.* Cambridge, MA—London, UK: MIT Press.

Dreyfus, H. L. (1993/2002). "Heidegger's Critique of Husserl/Searle Account of Intentionality." In: H. L. Dreyfus & M. Wrathall (Eds.). *Heidegger Reexamined. Volume 1. Dasein, Authenticity, and Death.* New York, NY: Routledge, 135–156.

Engel, A. K. et al. (2013). "Where's the Action? The Pragmatic Turn in Cognitive Science." In: *Trends in Cognitive Science* 17(5), 202–209.

Franks, D. (2010). *Neurosociology: The Nexus Between Neuroscience and Social Psychology.* New York: Springer.

Gallagher, S. (2005). *How the Body Shapes the Mind.* New York, NY: Oxford University Press.

Gallagher, S. (2012). "Social Cognition, the Chinese room, and the Robot Replies." In: Z. Radman (Ed.). *Knowing Without Thinking: Mind, Action, Cognition, and the Phenomenon of the Background.* Basingstoke & New York: Palgrave Macmillan, 83–97.

Gibson, J. J. (1979). *The Ecological Approach to Visual Perception.* Boston, MA: Houghton-Mifflin.

Hutto, D. D., & Myin, E. (2013). *Radicalizing Enactivism. Basic Minds Without Content.* Cambridge, MA: MIT Press.

Johnson, M. (2007). *The Meaning of the Body: Aesthetics of Human Understanding.* Chicago, IL: University of Chicago Press.

Kurthen, M. (1994). *Hermeneutische Kognitionswissenschaft. Die Krise der Orthodoxie.* Bonn: Djre Verlag.

Lakoff, G., & Johnson, M. (1999). *Philosophy in the Flesh: The Embodied Mind and Its Challenge to Western Thought.* New York, NY: Basic Books.

Luft, Sebastian (2015). *The Space of Culture. Towards a Neo-Kantian Philosophy of Culture (Cohen, Natorp & Cassirer).* Oxford: Oxford University Press.

Noë, A. (2004). *Action in Perception.* Cambridge, MA: MIT Press.

Noë, A. (2009). *Out of Our Heads: Why You Are Not Your Brain, and Other Lessons from the Biology of Consciousness.* New York, NY: Hill and Wang.

Noë, A. (2012). *Varieties of Presence.* Cambridge, MA: Harvard University Press.
Rockwell, W. T. (2005). *Neither Brain nor Ghost. A Non-dualist Alternative to the Mind-brain Identity Theory.* Cambridge, MA: MIT Press.
Rowlands, M. (2003). *Externalism. Putting Mind and World Back Together Again.* Montreal: McGill-Queen's University Press.
Solymosi, T. (2011). "Neuropragmatism, old and new." In: *Phenomenology and Cognitive Sciences* 10(2), 347–368.
Tomasello, M. (1999). *The Cultural Origins of Human Cognition.* Cambridge MA/London: Harvard University Press.
Tomasello, M. (2008). *Origins of Human Communication.* Cambridge MA/London: MIT Press.
Tomasello, M. (2014). *A Natural History of Human Thinking.* Cambridge MA/London: Harvard University Press.
Varela, F., Thompson, E., & Rosch, E. (1991). *The Embodied Mind. Cognitive Science and Human Experience.* Cambridge, MA: MIT Press.
Wilson, M. (2002). "Six Views of Embodied Cognition." In: *Psychonomic Bulletin & Review* 9 (4), 625–636.

# Part I: Pragmatism as a Philosophical Foundation of Cognitive Science

Shaun Gallagher

# Pragmatic Interventions into Enactive and Extended Conceptions of Cognition

**Abstract:** Clear statements of both extended and enactive conceptions of cognition can be found in John Dewey and other pragmatists. In this paper I'll argue that we can find resources in the pragmatists to address two ongoing debates: (1) in contrast to recent disagreements between proponents of extended vs enactive cognition, pragmatism supports a more integrative view—an enactive conception of extended cognition, and (2) pragmatist views suggest ways to answer the main objections raised against extended and enactive conceptions—specifically objections focused on constitution versus causal factors, and the mark of the mental.

Recently developed enactive and extended approaches to cognition have different roots. Enactive approaches typically point to phenomenology and theoretical biology; extended mind approaches are more influenced by analytic philosophy of mind, computational models and cognitive science more generally. Rarely noted, however, or sometimes only noted in passing, pragmatism is something of a forerunner of both of these approaches. Clear statements of both extended and enactive conceptions of cognition can be found in Charles Sanders Pierce, John Dewey, Georg Herbert Mead, and other pragmatists. Although today enactive and extended approaches disagree with each other on a number of issues, the fact that pragmatists could embrace both views suggests that there may be resources in pragmatism that can help to adjudicate some of the current debate and to develop a more integrated perspective.

I will also suggest that one can draw on the same resources to address some of the objections that have been leveled at enactive and extended models of the mind. My aim is not to provide a full historical account of how pragmatism anticipates the more recent developments; rather, my aim is to put pragmatism to work on clarifying and integrating these approaches, and thereby resolving some objections that have been raised against them. I'll begin, however, with some brief indication of the pragmatic anticipation of enactive and extended approaches.

---

This article was originally published in: Philosophical Issues, 24 (1), 2014, 110–126.
Research on this paper was supported by a Humboldt Foundation Anneliese Maier Research Award (2012–17) and by a research stay at the Ruhr Universität at Bochum.

# 1 Pragmatism and enactive perception

The origin of the concept of enactive perception, and more generally, enactive cognition is often indicated by pointing to the work of Francisco Varela, (e.g., Varela, Thompson and Rosch 1991), and then back to its phenomenological roots in Husserl, Heidegger, and Merleau-Ponty. Husserl (1989) had developed the idea of the "I can" as part of the structure of embodied perception. On his view, I perceive things in my environment in terms of what *I can* do with them. Heidegger (1962) went further toward a pragmatic view in suggesting that our primary way of being-in-the-world—our way of relating to various events and objects—is in the mode of the ready-to-hand (*Zuhanden*) attitude. In almost all everyday engagements, I understand the world in terms of pre-reflective pragmatic, action-oriented use, rather than in reflective terms of an intellectual or overly cognitive attitude of conceptual contemplation or scientific observation. Merleau-Ponty (1962), drawing on both Husserl and Heidegger, works this out in some detail in his analysis of perception, and focuses on the important role played by the embodied motor system. A number of theorists who help to develop an enactivist view of cognition were influenced by Merleau-Ponty's account—not only Varela et al. (1991), but also J. J. Gibson (1977), who developed the notion of affordances,[1] Hubert Dreyfus (1992), both in his critique of classical cognitivism and his emphasis on embodied practice, and Marc Jeannerod, who attended Merleau-Ponty's lectures and later developed a distinction between semantic and pragmatic perception (Jeannerod 1994; 1997; also see Noë 2004; Gallagher 2005a; Thompson 2007).[2]

That the roots of enactivism are to be found in phenomenology is an uncontroversial claim; the enactivists explicitly acknowledge this. Moreover, anyone familiar with the phenomenologists will clearly see that certain pragmatic aspects of phenomenological philosophy are the ones that lead forward to the development of the enactivist approach. For that reason it is surprising that the philosophical tradition of pragmatism is hardly ever mentioned by the enactivists. With the exception of William James (and usually his *Principles of Psychology*)

---

1 Anthony Chemero (private correspondence) notes that Gibson distributed copies of Merleau-Ponty's *Phenomenology of Perception* to his students. For a sustained discussion of Gibson from a pragmatist perspective see F. T. Burke. (2013a), Chemero (2009); Heft (2001) and Rockwell (2005).

2 I'm leaving out Jerome Bruner (1966), who offers a concept of early developing enactive (action-based) representation involving motor memory as in some way basic to development. Bruner draws directly from Piaget, and was possibly influenced by Dewey (see his 1961, "After Dewey, What?), but it's not clear that the phenomenologists formed part of his thinking.

the pragmatists are rarely cited. There is no mention of Peirce, Dewey, or Mead, for example, in Noë (2004), Thompson (2007) or Hutto and Myin (2013). Varela et al. make one general and undeveloped reference to pragmatism (Varela et al. 1991, 30–31), but without direct relevance to the enactive aspects of this work.[3]

Notwithstanding this lack of acknowledgment for the pragmatists, however, it's very clear that one can find the central concepts of enactivism already discussed by Peirce, Dewey and Mead. Peirce, for example, foreshadows the externalist turn that is so important for enactivism: "just as we say that a body is in motion, and not that motion is in a body, we ought to say that we are in thought, and not that thoughts are in us" (Peirce, CP 5.289, n. 1). Dewey, in turn, argues for an understanding of perception starting "not with a sensory stimulus, but with a sensori-motor coordination [...] it is the movement which is primary, and the sensation which is secondary, the movement of body, head and eye muscles determining the quality of what is experienced" (Dewey 1896, 358–359). Here, sensory-motor contingencies, which play a most obvious role in Noë's (Noë 2004; O'Regan and Noë 2001) version of enactivism, are given a primacy in perception.

Re-echoing this idea, and building on the insight found in his famous essay, "The reflex arc concept in Psychology," that one's active response defines the nature of what the organism takes as a stimulus, Dewey offers a characterization of the role of the brain in cognition that comes close to embodied-enactivist views today.

> The advance of physiology and the psychology associated with it have shown the connection of mental activity with that of the nervous system. Too often recognition of connection has stopped short at this point; the older dualism of soul and body has been replaced by that of the brain and the rest of the body. But in fact the nervous system is only a specialized mechanism for keeping all bodily activities working together. Instead of being isolated from them, as an organ of knowing from organs of motor response, it is the organ by which

3 My intent is not to do an exhaustive survey in this regard, but I will note the following. In the recent volume of essays, *Enaction: Toward a New Paradigm for Cognitive Science*, Andreas Engel (2010; also see Engel 2013) does highlight the relevance of the pragmatists, but only in the most general way, and with an indication that this relevance deserves "further exploitation" (Engel 2010, 222). Di Paulo, Rohde and De Jaegher (2010), in the same volume, make a passing reference to Dewey. Richard Menary (2011) points out the affinity between Dewey's account of experience and recent developments in externalist accounts of cognition including extended and enactive accounts of cognition. Mark Johnson in his *The Meaning of the Body: Aesthetics of Human Understanding* celebrates Dewey and at one point equates Dewey's notion of transaction with enaction (Johnson 2008, 274). Again, however, Chemero (2009) gives the most insightful account that connects the radical embodied approach, with which he associates enactivism, with the pragmatists.

> they interact responsively with one another. The brain is essentially an organ for effecting the reciprocal adjustment to each other of the stimuli received from the environment and responses directed upon it. Note that the adjusting is reciprocal; the brain not only enables organic activity to be brought to bear upon any object of the environment in response to a sensory stimulation, but this response also determines what the next stimulus will be. (Dewey 1916, 336–337)

This is fully consistent, for example, with enactivist arguments that cognition is not just a matter of brain processes—the cognitive agent is more than a brain in a vat; rather, the brain is one part of a body in which dynamical regulation goes both ways, with the brain both biologically and functionally dependent on the rest of the body which is in dynamical interaction with the environment (see, e.g., Cosmelli and Thompson 2010; Gallagher 2005b).

Part of what it means for the body to be dynamically interacting, or as Dewey would say, transacting with the environment is captured by George Herbert Mead in his characterization of the reachable peripersonal space around the body, or what he called the "manipulatory area." He suggested, very much in the spirit of enactivism, that what is present in perception is not a copy (or representation) of the perceived, but "the readiness to grasp" what is seen (Mead 1938, 103). Accordingly, the perception of objects outside of the manipulatory area is always relative to

> the readiness of the organism to act toward them as they will be if they come within the manipulatory area [...].We see the objects as we will handle them [...]. We are only 'conscious of' that in the perceptual world which suggests confirmation, direct or indirect, in fulfilled manipulation (Mead 1938, 104–105).

Just as concepts like the manipulatory area, response dependent stimulation, the contingent nature of sensory-motor coordination, and Peircean externalism prefigure related concepts in recent enactivism, other pragmatist concepts anticipate recent work on the extended mind.

## 2 Pragmatism and the extended mind

If extended mind theorists are somewhat better at acknowledging their prefigurement in the pragmatists, this only goes so far. Although the classic paper by Clark and Chalmers (1998) makes no mention of the pragmatists, Clark's book begins with an invocation from Dewey.

> Hands and feet, apparatus and appliances of all kinds are as much a part of it [thinking] as changes in the brain. Since these physical operations (including the cerebral events) and equipments are a part of thinking, thinking is mental, not because of a peculiar stuff which enters into it or of peculiar non-natural activities which constitute it, but because of what physical acts and appliances do: the distinctive purpose for which they are employed and the distinctive results which they accomplish. (Dewey 1916, 8–9)

But this beginning is the end of it—there is no further discussion of Dewey or other pragmatists in *Supersizing the Mind.*[4] Dewey's quote foreshadows much of what Clark wants to say, however. Richard Menary (2007) is perhaps the one figure closely associated with extended mind approaches who understands the pragmatic connections. Indeed, in his book, *Cognitive Integration,* he appeals to Dewey's notion of organism-environment transactions to work out a characterization of how embodied cognition incorporates the environment. I'll return to his analysis below. Menary also points to Peirce's 'continuity principle', which requires that there be no deep metaphysical discontinuity between the mind and the world, as an important source for understanding a neutral ground between internalist and externalist conceptions of representation (Menary 2007, 129). Indeed, I think it's quite clear that Peirce anticipates what was to become the extended mind.

> A psychologist cuts out a lobe of my brain (*nihil animale a me alienum puto)* and then, when I find I cannot express myself, he says, 'You see, your faculty of language was localized in that lobe.' No doubt it was; and so, if he had filched my inkstand, I should not have been able to continue my discussion until I had got another. Yea, the very thoughts would not come to me. So my faculty of discussion is equally localized in my inkstand. (Peirce CP, 7.366)

It is a short distance from the inkstand to Clark and Chalmers' example of Otto's notebook. Peirce anticipates something similar. "In my opinion it is much more true that the thoughts of a living writer are in any printed copy of his book than they are in his brain" (Peirce, CP 7.364).

Indeed, Peirce's notion that artifacts can play a cognitive role suggests an extension of thought into social practices and institutions, such as the institution of science (see Gallagher 2013a and the special issue of *Cognitive Systems Research* Vols. 25–26 for more on the socially extended mind). For Peirce, "it is no figure of speech to say that the alembics and cucurbits of the chemist are instruments of thought, or logical machines" (Peirce 1887, 168). The manip-

4 One has the impression that this invocation, which precedes the forward by Chalmers, is an afterthought motivated by Pierre Steiner calling Clark's attention to Dewey (Clark 2008, xviii).

ulation of artefacts such as chemical flasks and tubing evoke a very specific way of thinking in the chemist. In this regard, "thought is not just expressed in work, it is executed in work."[5] For Peirce, as Aydin (Aydin 2013, 16) explains, "the mind has an artifactual character."

Although Karl Popper is not usually considered a pragmatist in the classical sense, he was very much influenced by just these Peircean ideas, as Skagestad (1993; 1999) and Aydin (2013) suggest, and this is clearly reflected in a number of his statements, which may be considered an extension of Peirce. For example, Popper suggests that some cognition is exo- (but not necessarily extra-) somatic.

> Yet the kind of exosomatic evolution which interests me here is this: instead of growing better memories and brains, we grow paper, pens, pencils, typewriters, dictaphones, the printing press, and libraries. [...] The latest development (used mainly in support of our argumentative abilities) is the growth of computers. (Popper 1972, 238–239; also see 225, n. 39)

Dewey's contention that the mind involves more than "armchair" cognition follows directly from Peirce's artifactual externalism since not only the body but "apparatus and appliances of all kinds" are involved, as much as the brain is. F. Thomas Burke (2013b) points out that both Dewey and Mead, in the pre-computational age, promoted a more Darwinian conception of functionalism. Agents are inclined to use whatever is available to solve survival problems, and this extends to "apparatus and appliances of all kinds." The point is not to compute solutions on such instruments, but to incorporate them into the extended phenotype (Dawkins 1982), an idea that may be more consistent with Kim Sterelny's niche construction model of evolution, which emphasizes active agency in constructing an adaptive fit between agent and world (Sterelny 2010), than with Clark's functionalism.

5 Peirce, cited in David Kirsh (Kirsh 2005, 157). Kirsh explains: "C. S. Peirce, [Peirce] in his prescient way, was fond of saying that a chemist as much thinks with test tube and beaker as with his brain. His insight was that the activity of manipulating tools—in Peirce's case, manipulating representation rich tools and structures such as measuring devices, controllable flames, the lines in diagrams, written words— this activity is part of the overall process of thought. There is not the inner component, the true locus of thought, and its outer expression. The outer activity is a constituent of the thought process, though for Peirce it had to be continually re-interpreted to be meaningful."

# 3 Dewey's notion of situation

Turning from these historical connections, my focus in this section is on Dewey's notion of situation. This concept will provide a productive way to effect a rapprochement between enactive and extended theories in the following section.

For Dewey's understanding of cognition, the unit of explanation is not the biological individual, the body by itself, or the brain, but the organism-environment. Organism and environment are not two self-sufficient or easily distinguishable items. Rather, they are always found together in a dynamical transactional relation.[6] They are, in effect, coupled in a way such that to pull them apart is to destroy them, or to treat them as theoretical abstractions. An organism never exists (and can never exist) apart from some environment; an environment is what it is only in conjunction with a particular organism that defines it. "In actual experience, there is never any such isolated singular object or event; an object or event is always a special part, phase, or aspect, of an environing experienced world—a situation" (Dewey 1938a, 67). Although 'organism' seems a very biological term, by characterizing it in relation to the environment in this way Dewey's concept is very much akin to the notion of the lived body (*Leib*) as found in phenomenology and as distinguished from the objective body (*Körper*). Neither the organism nor the environment should be taken in strictly objective terms precisely because they are co-relational and are defined relative to each other.

> The statement that individuals live in a world means, in the concrete, that they live in a series of situations. The meaning of the word "in" is different from its meaning when it is said that pennies are "in" a pocket or paint is "in" a can. It means [...] that interaction is going on between individuals and objects and other persons. The conceptions of situation and of interaction are inseparable from each other. (Dewey 1938b, 43)

The specific ways in which the coordinated coupling of organism-environment gets established are tied to how they can successfully work for purposes of survival. Dewey's concept of situation arises when the coupling of the organism-environment becomes problematic or starts to break down. To the extent that the organism-environment is considered a self-organizing system, what some enactivists, following Varela, would call an 'autopoietic' system, even when functioning well it is in a precarious situation (see e.g., De Jaegher and Di Paulo 2007). When it starts to go wrong, this is what Dewey calls a problematic situation and it calls for a kind of re-pairing, a reestablishment of a workable coupling. These are ongoing processes that give rise to experience. Cognition, in such cases, is a

6 For Dewey's anticipation of dynamical systems approaches, see Rockwell (2005).

form of inquiry, understood as a hands-on practical activity through which we transform the problematic situation into one that is less confused and more comprehensible, and where ideas for successful action start to emerge. An idea is not primarily an intellectual entity in the head, but "an organic anticipation of what will happen when certain operations are executed under and with respect to observed conditions" (Dewey 1938a, 109). In this regard, Dewey was influenced by Peirce's view that in coping with a problematic situation, we use physical tools as well as ideas to physically reshape the environment. This includes linguistic tools, which in communicative contexts may be used to reshape the dynamics of the situation. For Dewey, ideas, as well as gestures and speech acts, are themselves tools for this kind of interaction.

The important thing to note in Dewey's concept of situation is that the situation is not equivalent to the environment. That is, it is not the organism that is placed in a situation. Rather the situation is constituted by organism-environment, which means that the situation already includes the agent or experiencing subject. In this regard, for example, if I am in a problematic situation, I cannot strictly point to the situation because my pointing is part of the situation. I cannot speak of it as some kind of objective set of factors because my speaking is part of it. My movement is a movement of the situation. Accordingly, the trick to solving a problematic situation is not simply to rearrange objects in the environment, but to rearrange oneself as well—to make adjustments to one's own behaviors. Indeed, any adjustment one makes to objects, artifacts, tools, practices, social relations or institutions, is equally an adjustment of oneself.

Another important thing to note about this concept of situation is that, for Dewey, it is almost always social. That is, for the human being the environment is never just the physical environment; it is physical and social. Even if there are no others present in an environment, in a literal or objective sense, it still involves social dimensions because it is coupled to a social organism. Likewise, as we attempt to resolve a problematic situation, our action involves a social dimension even if we are going at it alone. This idea is taken over and made clear by Mead.

> [I]t is not necessary that we should talk to another to have these ideas. We can talk to ourselves, and this we do in the inner forum of what we call thought. We are in possession of selves just insofar as we can and do take the attitudes of others toward ourselves and respond to those attitudes [...]. We assume the generalized attitude of the group, in the censor that stands at the door of our imagery and inner conversations, and in the affirmation of the laws and axioms of the universe of discourse [...]. Our thinking is an inner conversation in which we may be taking the roles of specific acquaintances over against ourselves, but usually it is with what I have termed the "generalized other" that we converse, and so attain to

> the levels of abstract thinking, and that impersonality, that so-called objectivity that we cherish. (Mead 1964, 287–288)

Thus, for Mead, one's self and one's mind arise in human behavior in a way that is shot through with sociality. As Burke (2013b) points out, the pragmatists are espousing a form of social externalism. This means what counts as important in any situation will depend on how the living organisms are dynamically coupled to each other and to their environments. What counts as an object, for example, depends on what it *affords*, to introduce Gibson's term. An affordance, however, is defined, not objectively, but always relative to the animal—to the kind of body one has. A chair affords sitting (or solving a problematic situation) only if the organism has bendable joints. Various affordances are also modified in some way by who else may be around. A heavy object may afford lifting, and that may resolve the problematic situation, but only if someone else is around to help lift it. My action, or our conjoint action, or a collective action can define what a particular object means in relation to the situation. "Objects exist in nature as patterns of our actions" (Mead 1964, 289).

This is clearly consistent with an enactive viewpoint; we enact the meaning of the world through our individual and joint actions (see the notion of participatory sense making in De Jaegher and Di Paulo 2007). Nor is this view inconsistent with an extended mind concept. Otto's notebook, in Clark and Chalmers' (1998) famous example, affords him the ability to resolve a problematic situation. Other people can and often do serve the same function, and in larger ways, Mead's notion of the generalized other can act as an extended archive of knowledge. This indeed leads to the notion of a socially extended mind where we can dynamically engage with complex social institutions and social practices to accomplish certain cognitive tasks or to solve problems (Gallagher 2013a).

# 4 Rapprochement

As represented in the common refrain that takes embodied, embedded, enactive and extended cognition as a unitary "4-E" alternative to classical cognitivism, enactivist approaches are often associated with the extended mind hypothesis. Yet there are important differences between enactivist and extended models, and in some regards, specific differences about how essential the notion of embodiment is for understanding cognition. Enactivists argue against the functionalism of extended mind theorists who discount any special role of the living body in cognition, given that one may be able to substitute prosthetic parts to take over or augment bodily functions (see, e.g., Thompson and Stapleton 2009). A related disagreement concerns the role of representation in action-oriented perception and basic cognition (see Gallagher 2007; Hutto and Myin 2013; Rowlands 2013), or the role of higher-order representations in transforming processes that are specific to bodily action (e.g., Clark 2008). Whether such differences should be the pivots of an uneasy detente, or whether one could work out an enactivist version of extended mind might find some clarification by turning to the pragmatists, who not only express enactivist leanings, but also embrace the notion of an extended mind.

In many regards much depends on how we understand the coupling relation between organism-environment. The fact that most extended mind proponents adopt a functionalist framework, which includes representational mechanisms that discount the role of the body, does not mean that this is the only way to understand the extended mind. If we follow Clark then we would have to accept that the coupling is accomplished by higher-order representations that "float free of the full spectrum of fine sensorimotor detail" (Clark 2008, 179). Internal representations sort out differences in input signals that may be due to bodily differences. So animals (or robots) with different bodies may, in principle, experience the same world as long as a certain set of representations (apparently running different algorithms in each case) filter or properly adjust, for example, visual input that may be different because of different eye structure. Differences are ironed out at the representational level, in which case the same shape, structure or design of the body is not so important. In contrast, if we follow Thompson and Varela's (2001) notion of radical enactive embodiment, the body makes all the difference and it reaches into the experience of the organism. That is, as the theoretical biologists (from von Uexküll to Maturana and Varela) tell us, each species of animal enacts a different world (see Chemero 2009, 185 ff. for discussion). What the frog perceives is quite different from what the human perceives, primarily because the frog has a different body (which includes different eye structure and a very different brain), which defines different affordances.

Can a pragmatist like Dewey adjudicate on this issue? Notions like organism-environment and situation, as defined by Dewey, suggest that one can have a theory of extended mind that is based on enactive principles rather than functionalist principles. The life-worlds of frogs, robots and humans will differ insofar as they are constituted by organism-environment transactions. Pragmatically speaking it would make no evolutionary sense to think that frogs, robots and humans would need to inhabit the same meaningful world rather than an organism-relative world. Organism and environment must be thought together in a dynamical relation, so that if differences or adjustments in organism are made, one finds correlative adjustments and differences in the experienced environment.

Accordingly the animal enacts an ecological and cognitive niche appropriate to its embodied needs; the organism defines the manipulatory area of its reach and therefore the practical valence of things in its surrounding world. For the animal, in any case, a change in its neural representations will not change the very physical fact of how far its limb or (in the case of the frog) its tongue reaches into the environment. Changes in neural representations (if there be such things) will normally reflect the shape and structure of the body, and plastic changes in the brain will occur only in correlation to bodily changes. This, at least, is a matter of pragmatic (and evolutionary) principle. This may be different for non-enactive (non-Brooksian) robots where representations are the rule. In the case of cyborg existence it may depend on whether the principles of operation follow biological plasticity or computational adjustments. But if executive control is left to computational adjustments, biological plasticity will accommodate the difference, or the system will likely stop working or will work in a pathological way. One way or another, however, the organism will be doing important work.[7]

In no way, however, does a substitution of enactive principles for functionalist principles rule out the concept of extended mind. Rather, it simply reinforces one of the main points behind both approaches, namely, that we need to think of the mind differently from the traditional internalist models. As extended and enactive, the mind is *situated* in the way that Dewey defines this notion. The situation includes not just our notebooks, computers, and other cognitive technologies, and not just the social and cultural practices and institutions that help us solve a variety of cognitive problems, it also includes us. We are *in-the-world* in a way that is not reducible to occupying an objective position in the geography of surrounding space, and at the same time the world is irreducible to an abstraction of itself represented in one's brain. We, as minded beings, are definitively "out there", dynamically coupled to artifacts, tools, technologies, social practices and institutions that extend our cognitive processes. Enactive and extended conceptions are, or at least should be of one mind in this regard.

## 5 Responding to Objections

Can these considerations about the pragmatist precursing of enactive-extended cognition provide a response to the current cursing criticism of such approaches? Pragmatism reinforces in many ways the enactive-extended model, and champions an integrative approach that follows along the lines of proposals made

7 See Menary (2013) for more on neural and phenotypic plasticity.

by theorists like Menary (2007). Menary, who is much influenced by pragmatism, proposes his own integrationist responses to objections raised against the extended (and by implication, the enactive) approach to cognition.

Consider the coupling-constitution objection (Adams and Aizawa 2008). The objection is that extended mind proposals confuse causal relations for constitutive relations. Adams and Aizawa characterize the position they criticize as follows: "If a cognitive agent causally interacts with some object in the external world in some 'important' way—if that agent is coupled to an object—then that agent's cognitive processing is constituted by processes extending into that object" (Adams and Aizawa 2010). Menary (2010c) rightly replies that this characterization assumes a pre-formed cognitive agent, fitted out with internal mental representations, causally interacting with an external world that is supposedly independent of them. Then the problem becomes one of explaining how that agent can incorporate elements of the environment into its own cognitive apparatus. The pragmatist would ask why one should assume that agents are cognitively independent of, or merely in a causal relation to their environment. For the pragmatist, the organism and environment are not two things that are merely causally related to each other, but are mutually constituted in this relation—organism-environment (as explained in the previous section). The organism is not a cognitive agent before coupling to an environment; the environment is an essential, constitutive, element in making the organism what it is. Specifically, the organism's cognitive capacities are transformed by the way it is coupled with (transacts with, or manipulates) structures available in the environment.

Adams and Aizawa's idea that the cognitive agent is pre-formed independently of the environment is tied to their view that all cognition is in the brain. Unless one accepts a reductionistic identity theory, however, neural processing is itself a causal factor. This does not rule out the idea that it could also be constitutive, since causal factors may be constitutive and not *merely* causal. The question then is what makes neural processes (or any factor) constitutive? Menary (2010c) suggests that Adams and Aizawa fail to provide a way to distinguish merely causal from constitutional relations. I think there are three things to say in this regard. First, Adams and Aizawa do have a theory of what distinguishes mental from non-mental phenomena, i.e., non-derived intentionality. In that sense they point to an answer about what makes something constitutive of mind, even if they don't provide a general principle about what distinguishes the merely causal from the constitutive. Second, however, whether one should think of non-derived intentionality as something that is reducible to brain processes is a separate question. Brains, as noted above, do what they do only by being integrated in the system of brain-body-environment, and only in their dynamical, non-linear transactions with bodily and environmental processes.[8] In this regard there is no in-principle reason to think of cognition as belonging to anything less than the system as a dynamic whole. Third, even if non-derived intentionality were a mark of the mental, and even if we could agree on what non-derived intentionality is, it's not clear that there is only one mark of the mental rather than a pattern of elements that constitute the mind.

What is non-derived intentionality? And is it *the* mark of the mental, or only one, non-necessary aspect of the mental? One can raise the second question because the notion of intentionality is not completely settled. Some philosophers, e.g., Husserl (Husserl 1982, §36) and Searle (1992), contend that some mental phenomena are non-intentional. Tim Crane (1998), for example, claims that the experience of pain is a mental experience, but is not necessarily intentional. As Shapiro (Shapiro 2009, 268) notes, "there is today no received theory of how original content comes to be in the first place." Accordingly, there are debates about what is derived versus non-derived or original intentionality. Adams and Aizawa would argue that something like beliefs and other propositional attitudes are characterized by non-derived intentionality; but pragmatists and neo-pragmatists would treat the intentionality of propositional attitudes as derived from a more original form of embodied intentionality, what phenomenologists like Husserl and Merleau-Ponty call 'motor intentionality'. Robert Brandom makes this point very clearly with reference to classical pragmatism.

> A founding idea of pragmatism is that the most fundamental kind of intentionality (in the sense of directedness towards objects) is the *practical* involvement with objects exhibited by a sentient creature dealing skillfully with its world (Brandom 2008, 178).

Brandom, as a neo-pragmatist, pictures this intentionality as more basic than propositional-based "*semantic* intentionality." It involves feedback-governed processes that extend across body and environment, and which exhibit

> a complexity [that] cannot in principle be specified without reference to the changes in the world that are both produced by the system's responses and responded to [...]. [Such practices] are 'thick', in the sense of essentially involving objects, events, and worldly states of affairs. Bits of the world are *incorporated* in such practices (Brandom 2008, 178).

**8** Chemero (Chemero 2009) argues that the presence of just such non-linear relations, or, as Orestis Palermos (Orestis Palermos 2013) argues, the continuous mutual interactions which entail such relations, may count as an objective criterion of constitution.

On this view, non-derived intentionality may not be what Adams and Aizawa think it is.

One can also ask whether intentionality, whether derived or non-derived, is necessarily the only mark of the mental. If one carefully considers such phenomena as emotions and the experience of self, one can argue that these phenomena, and more generally the mind itself, are really constituted by a pattern of factors, no one of which is necessary, but some number of which are sufficient for that constitution (Gallagher 2013b; Newen et al. 2015). Emotion, for example, can be characterized as involving the following elements: autonomic processes (e. g., James 1884), action and action-tendencies (Frijda 1986); overt expressions; phenomenal feeling; cognitive aspects such as attitudes, shifts of attention and changes to perception; and intentionality (Goldie 2000). Is any one of these the unique mark of emotion? Newen et al. (2015) argue that not every emotion has a distinct autonomic pattern, and different emotions need not have different autonomic patterns (Prinz 2004); that some emotions, e. g., happiness, may not include action tendencies; that some may not include expressive features; that in some, albeit, rare cases, typical physiological, expressive and cognitive aspects may be present without the phenomenal aspect (e. g., in those disposed to repress fear [Sparks et al. 1999]). Indeed, any particular feature, including the cognitive and intentional features, may or may not be present, and yet the presence of a variable number of these features would be sufficient to constitute a particular emotion. In that case, a pattern of marks rather than one unique mark may delineate mental from non-mental phenomena.

Similar considerations apply to the self where we may find a plurality of aspects organized in typical patterns that include embodied, minimal experiential, affective, intersubjective, psychological/cognitive, and/or narrative aspects (Gallagher 2013b). James (1890) added extended aspects suggesting that what we call self may include physical pieces of property, such as clothes, homes, and various things that we own, since we identify ourselves with our stuff and perhaps with the cognitive technologies we use, or the cognitive institutions we work in.

One could argue that more generally this same pattern theory applies to the mind itself. The mind (or the mental) is constituted by a pattern of aspects or elements not all of which are necessary in every case, but some of which are constitutionally sufficient. This puts pressure on the idea that there is one unique mark of the mental. To complete the pragmatist picture in this regard, let me add that emotion, self, and more generally the mind itself constitutionally involve situational aspects. For example, Dewey, in his early essays on a theory of emotion (1894; 1895), not only emphasizes the action-oriented aspect of emotion, but also points out that emotions are not reducible to a set of bodily states since the body is always coupled to an environment, and always includes situational aspects. In regard to the self, these are aspects that play some (major or minor) role in shaping who we are. They may include the kind of family structure and environment where we grew up; cultural and normative practices that define our way of thinking and living, and so on. Even in defining something like belief, we should not think that it is reducible to functional states of the brain, since belief may be constituted in terms of dispositions or action-tendencies that vary across different situations.

## Conclusion

In this paper I have shown that many of the insights of enactive and extended approaches to cognition were foreshadowed in pragmatists such as Peirce, Dewey and Mead. This is more than an interesting historical fact, however. The work of the pragmatists suggests ways to flesh out the various embodied and situated accounts of cognition so that enactive and extended versions are not necessarily in conflict. I've argued that Dewey's notion of situation and organism-environment may be a way to develop a theory of the extended mind that is based on enactivist rather than functionalist principles. Furthermore, this pragmatist-enactivist version of extended cognition can better respond to objections raised by critics who actually share some of the functionalist views of the original extended mind hypothesis. Specifically, a pragmatist view would argue that non-derived intentionality is not reducible to either physical or functional processes in the brain, but that it is better understood as a more embodied form of transaction with the environment. Non-derived intentionality is the action-oriented coupling of organism-environment. Further, even on this notion of intentionality, neither non-derived nor derived intentionality is the sole mark of the mental. As pragmatists like Dewey demonstrate, intentionality is only one possible element of a pattern that constitutes the mind.

## References

Adams, F., & Aizawa, K. (2008). *The Bounds of Cognition.* Malden, MA: Blackwell.
Adams, F., & Aizawa, K. (2010). "The value of cognitivism in thinking about extended cognition." In: *Phenomenology and the Cognitive Sciences* 9 (4), 579–603.
Aydin, C. (2013). "The artifactual mind: overcoming the 'inside–outside' dualism in the extended mind thesis and recognizing the technological dimension of cognition." In: *Phenomenology and the Cognitive Sciences.* doi 10.1007/s11097-013-9319-x.
Bruner, J. S. (1961). "After John Dewey, What?" Bank Street College of Education.
Bruner, J. S. (1966). *Toward a Theory of Instruction.* Cambridge, MA: Harvard University Press.

Brandom, R. B. (2008). *Between Saying and Doing: Towards an Analytic Pragmatism.* Oxford: Oxford University Press.
Burke, F. T. (2013a). *What Pragmatism Was.* Bloomington: Indiana University Press.
Burke, F. T. (2013b). *Extended mind and representation.* Draft manuscript. Retrieved October 26, 2013. http://www.academia.edu/3670335/Extended_Mind_and_Representation.
Burke, F. T., & Everett, S. (2013). "Social-psychological externalism and the coupling-constitution fallacy." In: F. T. Burke & K. P. Skowronski (Eds.). *George Herbert Mead in the Twenty-First Century.* Lanham, MD: Lexington Books, 107 – 135.
Chemero, A. (2009). *Radical Embodied Cognitive Science.* Cambridge, MA: The MIT Press.
Clark, A. (2008). *Supersizing the Mind: Embodiment, Action, and Cognitive Extension.* Oxford: Oxford University Press.
Clark, A., & Chalmers, D. (1998). "The extended mind." In: *Analysis* 58 (1), 7 – 19.
Cosmelli, D., & Thompson, E. (2010). "Embodiment or envatment? Reflections on the bodily basis of consciousness." In: J. Stewart, O. Gapenne, & E. A. Di Paolo (Eds.). *Enaction: Toward a New Paradigm for Cognitive Science.* Cambridge, MA: The MIT Press, 361 – 385.
Crane, T. (1998). "Intentionality as the mark of the mental." In: A. O'Hear (Ed.). *Contemporary Issues in the Philosophy of Mind.* Royal Institute of Philosophy Supplement (No. 43). Cambridge: Cambridge University Press, 229 – 251.
De Jaegher H., Di Paolo E. A. (2007). "Participatory sense-making: an enactive approach to social cognition." In: *Phenomenology and the Cognitive Sciences* 6, 485 – 507.
Dawkins, R. (1982). *The Extended Phenotype.* Oxford: Oxford University Press.
Dewey, J. (1894). "The theory of emotion: I: Emotional attitudes." In: *Psychological Review* 1 (6), 553.
Dewey, J. (1895). "The theory of emotion: II: The significance of emotions." In: *Psychological Review* 2 (1), 13.
Dewey, J. (1896). "The reflex arc concept in psychology." In: *Psychological Review* 3 (4), 357 – 370.
Dewey, J. (1916a/1985). *Democracy and Education.* ed. by J. A. Boydston, & P. Baysinger. Carbondale: Southern Illinois University Press.
Dewey, J. (1916b). *Essays in Experimental Logic.* Chicago: University of Chicago Press.
Dewey, J. (1938a). *Logic: The Theory of Inquiry.* New York: Holt, Rinehart & Winston.
Dewey, J. (1938b). *Experience and Education.* New York: Macmillan.
Di Paolo, E. A., Rohde, M., & De Jaegher, H. (2010). "Horizons for the enactive mind: Values, social interaction, and play." In: J. Stewart, O. Gapenne & E. A. Di Paolo (Eds.). *Enaction: Toward a New Paradigm for Cognitive Science.* Cambridge, MA: The MIT Press, 33 – 87.
Dreyfus, H. L. (1992). What Computers Still Can't Do: A Critique of Artificial Reason. Cam-bridge, MA: The MIT Press.
Engel, A. K., Maye, A., Kurthen, M., & König, P. (2013). "Where's the action? The pragmatic turn in cognitive science." In: *Trends in Cognitive Sciences* 17 (5), 202 – 209.
Engel, A. K. (2010). "Directive minds: how dynamics shapes cognition." In: J. Stewart, O. Gapenne and E. A. Di Paolo (Eds.). *Enaction: Toward a New Paradigm for Cognitive Science.* Cambridge, MA: The MIT Press, 219 – 243.
Gallagher, S. (2013a). "The socially extended mind." In: *Cognitive Systems Research* 25 – 26, 4 – 12.

Gallagher, S. (2013b). “A pattern theory of self.” In: *Frontiers in Human Neuroscience* 7 (443), 1–7. doi: 10.3389/fnhum.2013.00443.
Gallagher, S. (2008). “Are minimal representations still representations?” In: *International Journal of Philosophical Studies* 16 (3), 351–369.
Gallagher, S. (2005a). *How the Body Shapes the Mind.* Oxford: Oxford University Press.
Gallagher, S. (2005b). “Metzinger’s matrix: living the virtual life with a real body.” In: *Psyche* 11 (5), 1–9.
Gibson, J. J. (1977). “The concept of affordances.” In: R. E. Shaw & J. Bransford (Eds.). *Perceiving, Acting, and Knowing.* New Jersey: Lawrence Erlbaum Associates, 67–82.
Heft, H. (2001). *Ecological psychology in context: James Gibson, Roger Barker, and the legacy of William James’s radical empiricism.* Mahwah, NJ: L. Erlbaum.
Heidegger, M. (1962). *Being and Time.* Trans. J. Macquarrie, & E. Robinson. New York: Harper and Row.
Husserl, E. (1989). *Ideas Pertaining to a Pure Phenomenology and to a Phenomenological Philosophy—Second Book: Studies in the Phenomenology of Constitution.* Trans. R. Rojcewicz, & A. Schuwer. Dordrecht: Kluwer Academic.
Husserl, E. (1982). *Ideas Pertaining to a Pure Phenomenology and to a Phenomenological Philosophy.* Trans. F. Kersten. The Hague: Nijhoff.
Hutto, D. D., & Myin, E. (2012). *Radicalising Enactivism.* Cambridge, MA: The MIT Press.
James, W. (1890). *The Principles of Psychology.* New York: Basic Books, 1950.
Jeannerod, M. (1994). “The representing brain: Neural correlates of motor intention and imagery.” In: *Behavioral and Brain Sciences* 17 (2), 187–201.
Jeannerod, M. (1997). *The Cognitive Neuroscience of Action.* Oxford: Blackwell Publishing.
Johnson, M. (2008). *The Meaning of the Body: Aesthetics of Human Understanding.* University of Chicago Press.
Kirsh, D. (2005). “Metacognition, distributed cognition and visual design.” In: P. Gardenfors & P. Johansson (Eds.). *Cognition, Education, and Communication Technology.* London: Routledge, 147–180.
Mead, G. H. (1938). *The Philosophy of the Act.* Chicago: University of Chicago Press.
Mead, G. H. (1964). *Selected Writings.* Ed. A. J. Reck. Chicago: University of Chicago Press.
Menary, R. (2007). *Cognitive Integration: Mind and Cognition Unbounded.* Basingstoke: Palgrave Macmillan.
Menary, R. (Ed.). (2010a). *The Extended Mind.* Cambridge, MA: The MIT Press.
Menary, R. (2010b). “The extended mind and cognitive integration.” In: R. Menary (Ed.). *The Extended Mind.* Cambridge, MA: The MIT Press, 227–244.
Menary, R. (2010c). “The holy grail of cognitivism: A response to Adams and Aizawa.” In: *Phenomenology and the Cognitive Sciences* 9 (4), 605–618.
Menary, R. (2011). “Our glassy essence: the fallible self in pragmatist thought.” In: S. Gallagher (Ed.). *The Oxford Handbook of the Self.* Oxford: Oxford University Press, 609–632.
Menary, R. (2013). “The enculturated hand.” In: Z. Radman. *The Hand: An Organ of the Mind.* Cambridge, MA: The MIT Press, 349–368.
Merleau-Ponty, M. (1962). *Phenomenology of Perception.* Trans. C. Smith. London: Routledge and Kegan Paul.
Newen, A., Welpinghus, A., & Jukel, G. (2015). “Emotion recognition as pattern recognition: the relevance of perception.” In: *Mind and Language* 30(2), 187–208.

Noë, A. (2004). *Action in Perception.* Cambridge, MA: The MIT Press.
O'Regan, J. K., & Noë, A. (2001). "A sensorimotor account of vision and visual consciousness." In: *Behavioral and Brain Sciences* 24 (5), 939–972.
Orestis Palermos, S. (2013). "Loops, constitution, and cognitive extension." In: *Cognitive Systems Research.* http://dx.doi.org/10.1016/j.cogsys.2013.04.002.
Peirce, C. S. (1887). "Logical machines." In: *American Journal of Psychology* 1 (1), 165–170.
Peirce, C. S. (1931–1935, 1958). *Collected Papers of C. S. Peirce.* ed. by C. Hartshorne, P. Weiss & A. Burks. Cambridge: Harvard University Press. (Abbreviated: 'CP' followed by the conventional '[volume].[page]'-notation.)
Popper, K. R. (1972). *Objective Knowledge: An Evolutionary Approach.* Oxford: Oxford Univer-sity Press.
Prinz, J. (2004): *Gut Reactions. A Perceptual Theory of Emotion.* New York, NY: Oxford University Press.
Rockwell, W. T. (2005). *Neither Brain nor Ghost: A Nondualist Alternative to the Mind-Brain Identity Theory.* Cambridge, MA: The MIT Press.
Rowlands, M. (2013). *The New Science of the Mind.* Cambridge, MA: The MIT Press.
Searle, J. R. (1992). *The Rediscovery of the Mind.* Cambridge, MA: The MIT Press.
Shapiro, L. A. (2009). "A review of Frederick Adams and Kenneth Aizawa. The Bounds of Cognition." In: *Phenomenology and the Cognitive Sciences* 8 (2), 267–273.
Skagestad, P. (1993). "Thinking with machines: Intelligence augmentation, evolutionary epistemology, and semiotic." In: *The Journal of Social and Evolutionary Systems* 16 (2), 157–180.
Skagestad, P. (1999). "Peirce's inkstand as an external embodiment of mind." In: *Transactions of the Charles S. Peirce Society* 35 (3), 551–561.
Sparks, G. G., Pellechia, M., & Irvine, C. (1999). "The repressive coping style and fright reactions to mass media." In: *Communication Research* 26, 176–192.
Sterelny, K. (2010). "Minds—extended or scaffolded." In: *Phenomenology and the Cognitive Sciences* 9 (4), 465–481.
Stewart, J., Gapenne, O., & Di Paolo, E. A. (Eds.). (2010). "Introduction." In: *Enaction: Toward a New Paradigm for Cognitive Science.* Cambridge, MA: The MIT Press, vii-xiv.
Thompson, E. (2007). Mind in Life: Biology, Phenomenology, and the Sciences of Mind. Cambridge, MA: Harvard University Press.
Thompson, E., & Stapleton, M. (2009). "Making sense of sense-making: Reflections on enactive and extended mind theories." In: *Topoi* 28 (1), 23–30.
Thompson, E., & Varela, F. J. (2001). "Radical embodiment: neural dynamics and consciousness." *Trends in Cognitive Sciences* 5 (10), 418–425.
Varela, F. J., Thompson, E. T., & Rosch, E. (1991). *The Embodied Mind: Cognitive Science and Human Experience.* Cambridge, MA: The MIT Press.

Joshua A. Skorburg

# Pragmatism, Embodiment, and Extension

**Abstract:** While pragmatism and the so-called 4E program may form a united front against methodological individualism, classical cognitivism, traditional internalism, and the like, the 4E approach is not without its own internal tensions. One such tension, between Embodied and Extended, is brought to light by Clark (2008), who argues in favor of the latter. Dempsey and Shani (2013) reply that Clark's functionalism undercuts what should be a more fundamental commitment to Embodied. With respect to this tension, Gallagher (2014) claims that "there may be resources in pragmatism that can help to adjudicate some of the current debate and to develop a more integrated perspective" (Gallagher 2014, 110). In this paper, I assess Gallagher's strategy and offer a critical perspective on the use of Deweyan pragmatism to resolve these tensions in the 4E program.

## Introduction: Jesuits, Nazis, and the sciences of the mind

Consider the following commonality between some 16th Century Spanish Jesuits and some 19th Century Germans who turned out to be influential members of the Nazi party: in establishing their respective settlements in Paraguay, both groups initially struggled to cultivate *Ilex paraguariensis*—better known today as yerba mate. *Ilex paraguariensis* is a small tree native to the humid woodlands of eastern Paraguay. Like coffee and tea, it contains caffeine as well as other psychoactive alkaloids such as theophylline and theobromine (Folch 2010, 9). Macintyre remarks that yerba "is drunk all the year round, all day, by everybody. It is also the main topic of conversation, the social glue that holds the country together" (Macintyre 1993, 42).

Macintyre describes the German settlement in Paraguay, Nueva Germania, as "good land for yerba [...] which all the German colonists had become addicted," and adds that the cultivation of yerba was "back-breaking work." He continues, "the Jesuits had known how to cultivate the tea, indeed they had established large plantations, but when they had been thrown out of the country [...] the secret had gone with them [...]. For some reason just planting yerba seeds didn't work—they steadfastly refused to germinate" (Macintyre 1993, 161). It took "six or eight years of tireless experimentation" before the secret was uncovered. Fortuitously, one of the German colonists noticed that the areas where the birds roosted were the same areas that produced the largest quantities of yerba. “He deduced that the birds’ digestive system was acting on the seeds they had eaten and accelerating germination. He concocted a mixture of acid and charcoal, and, by seeping the yerba seeds in it, achieved the same result” (Macintyre 1993, 166). Indeed, Folch confirms that “according to common colonial wisdom seeds are encased in a shell so hard that germination is impossible unless the seeds have passed through the intestines of birds, where the acidity wears down the lining to allow the seeds to burst through the case” (Folch 2010, 11).

This is a nice story because (among other things) there is an important and instructive similarity between the cultivation of yerba and the scientific study of the mind: looking solely at internal structures and processes in the brain will not tell the whole story. There is nothing *inside* the yerba seed that would indicate it needed to be passed through the digestive tract of a bird in order to facilitate germination. To understand how to germinate the seed, the German colonists had to look to the wider context of the seed’s interactions with various parts of the environment—to structures and processes *external* to the seed itself.[1]

Arguably, this is an insight shared by classical pragmatist theories of mind and the so-called “4E” conception of the mind: the mind as Embodied, Embedded, Extended, and Enactive.[2]

In the Introduction to the special 4E edition of *Phenomenology and the Cognitive Sciences*, Menary notes that “one reason that the four E’s are grouped together is that they are all held to reject or at least radically reconfigure traditional cognitivism, coupled with a methodological individualism” (Menary 2010, 459). Put another way, the four E’s share, in some form or another, an assumption that a comprehensive understanding of the mind will require looking beyond the brain, and beyond the bounds of skin and skull.[3]

It has been well documented by scholars from various traditions that classical pragmatism was an important historical antecedent of this 4E view.[4] I take it

1 Thanks to John Sarnecki for helping develop this metaphor.

2 For the sake of convenience, I will refer each “E” in this way, rather than using clunky phrases like the “embodied conception of the mind” or the “hypothesis of extended mind,” etc.

3 I hasten to add that as is the case with most catchy and convenient labels for philosophical positions, “4E” can tend to obscure just as much as it clarifies (this is of course also true of the “traditional cognitivism” and “methodological individualism” labels). Nonetheless, Menary’s claim does provide a prima facia reason why the four E’s are grouped together in the first place.

4 See, for example, Burke (2013), Chemero (2011), Johnson (2008), Menary (2007), Madzia (2013), Rockwell (2005) Skorburg (2013), and Vaesen (2014).

these claims are pretty well established by now, and so this paper will not present another argument of the "James/Peirce/Dewey-said-it-first" variety.

While pragmatism and the 4E program may form a united front against methodological individualism, classical cognitivism, traditional internalism, and the like, the 4E approach is not without its own internal tensions. One such tension, between Embodied and Extended, was brought to light by Clark (2008), who argues that the distinction between the two cuts in favor of the latter. More recently, other theorists have pushed back against the functionalist underpinnings of Clark's view. Dempsey and Shani (2013), for instance, argue that Clark's functionalism undercuts what should be a more fundamental commitment to Embodied. Similarly, Thompson and Stapleton (2009), and Gallagher (2014) take pains to distance various brands of Enactive from what they take to be problematic strands of overly-functionalist Extended.

These tensions are the focus of the present paper. Of particular interest is Gallagher's claim that "there may be resources in pragmatism that can help to adjudicate some of the current debate and to develop a more integrated perspective." These resources, he continues, "can address some of the objections that have been leveled at enactive and extended models of the mind [...]. My aim is to put pragmatism to work on clarifying and integrating these approaches and thereby resolving some objections that have been raised against them" (Gallagher 2014, 110 – 111).

Gallagher marshalls the resources of pragmatism to deal with the tension between Extended and Enactive, but his strategy, I will argue, may also be applicable to the tension between Extended and Embodied, as iterated by Clark (2008) and Dempsey and Shani (2013), respectively. The purpose of this paper, then, is to offer a critical perspective on the use of pragmatism to resolve these tensions in the 4E program. More specifically, I will raise two concerns: (1) in light of Rowlands's (2010) distinction between theories of dependence and theories of constitution, I suggest that textual evidence may only support reading Dewey as a proponent of the former. Put another way, it might be the case that Dewey advocates for something like Embedded, rather than Extended. (2) I suggest that Gallagher's appropriation of Dewey's notion of the situation to smooth out tensions between Enactive and Extended might actually serve to exacerbate the well-known problem of "cognitive bloat".

Here is the plan for the paper. First, I will briefly summarize Clark's distinction between what he calls body-centric or Special Contribution Stories (SCS) and extended-functionalist or Larger Mechanism Stories (LMS). This distinction more or less maps onto Embodied and Extended, respectively. The lines along which this distinction is drawn, on Clark's view, favor LMS. In response to Clark, Dempsey and Shani (2013) argue that (a more nuanced version of) SCS, or what they term "strong embodiment" (SE) ought to be favored. Having thus elaborated the tension between SE and LMS, and hence, Embodied and Extended, following Menary (2010), I argue that a shared commitment to challenging strictly internalist models of the mind provides a *pro tanto* reason to attempt to reconcile LMS and SE. I then consider whether Gallagher's (2014) proposed rapprochement is up to the task. I conclude with a few reservations about Gallagher's strategy, and the resources of pragmatism for 4E, more generally.

# 1 The tension

Clark describes the tension between the extended and embodied conceptions of the mind in terms of Larger Mechanism Stories and Special Contribution Stories. The former is characterized as follows:

> Aspects of the body and world can, at times, be proper parts of larger mechanisms whose states and overall operating profile determine (or minimally, help determine) our mental states and properties. Call this the *Larger Mechanism Story (LMS)* [...] larger systemic wholes, incorporating brains, bodies, the motion of sense organs, and (under some conditions) the information-bearing states of non-biological props and aids, may sometimes constitute the *mechanistic supervenience base* for mental states and processes. (Clark 2008, 39–40)

The latter is characterized as follows:

> Specific details of human embodiment make a special and ineliminable contribution to our mental states and properties. Call this the *Special Contribution Story (SCS)* [...] specific features of the body (and perhaps the world) make a *persistent, non-trivial, and in some sense special* contribution to our mental states and processes. (Clark 2008, 39–40)

I assume at the outset that this is neither a zero sum game (why would the four E's have been grouped together in the first place?) nor an entirely false characterization on Clark's part (we can either say it is the physical properties, or the functional role that is the difference that makes a difference, but not both). In other words, there is a genuine tension here, and what is at stake is the role of the body: is the body, "just one element in a kind of equal-partners dance between brain, body, and world, with the nature of the mind fixed by the overall balance achieved," (Clark 2008, 56–57) as Clark suggests? Or must the body "lead this dance because it is what realizes the autonomous organization necessary for individual agency and sense-making," as Thompson and Stapleton (Thompson & Stapleton 2009, 28) suggest?

In the next two sections, I briefly summarize how the tension thus described can be taken to favor LMS or SE.

## 2 Clark's argument for LMS

In the first place, it should be noted that Clark's functionalism is ostensibly not meant to supplant the arguments for Embodied. His purpose is largely reconciliatory in that he attempts to lay out "a way of understanding the embodied, embedded approach that sees it as extending, rather than undermining, a broadly functionalist story" (Clark 2008, 44). To support this broadly functionalist story, Clark first takes aim at a common (mis)characterization of the debate between embodied mind theories, on the one hand, and multiple realizability theories, on the other.

In brief, the former holds that human "mindware" requires, and is reflected in, human "bodyware". The latter theories are "platform neutral" in that they hold that "mindware" could be run on, or realized by, various kinds of "hardware." Much evidence has been brought to bear in favor of embodied mind theories thus described—O'Regan and Noë (2001) and Lakoff and Johnson (1999) being the most prominent examples mentioned by Clark. The aim here is not to discredit these insights, but rather, to caution against making unwarranted inferences from them. As Clark says:

> From the fact that (as seems highly likely) our human experience really does depend in part on many idiosyncratic aspects of our embodiment, it does not follow that *only* a creature thus embodied could have those very experiences. The very most that follows is that, for a creature like us, all other things being equal, *we* would not have that experience were it not for such-and-such an idiosyncratic fact. Other creatures, for all that we have so far had cause to accept, may still have the very same experience courtesy of quite different forms of embodiment and sensory set-up. (Clark 2008, 42)

In other words, one cannot infer the falsity of *functionalism* on the basis of the idiosyncrasies of human embodiment (i.e. the location of our eyes producing a certain kind of visual experience, or the location of our ears producing a certain kind of auditory experience) which seem to discredit strong or global theories of *multiple realizability.* As Clark points out, his brand of extended functionalism is "logically independent, but thematically related" (Clark 2008, 43) to multiple realizability, which is to say that one can be an extended functionalist without buying strong or global multiple realizability.

This is important because, as will become clear in the next section, when considering Clark's positive arguments in favor of his brand of functionalism, it will not suffice to provide arguments against multiple realizability in an attempt to undermine his position.

According to Clark, "arguments in favour of LMS appeal mainly, if not exclusively, to the *computational role* played by certain kinds of non-neural events and processes in online problem solving" (Clark 2008, 44). Clark and Chalmers's thought experiment with Otto, his notebook, and Inga serves as a keystone here.[5] Their (in)famous proposal is that:

> If, as we confront some task, a part of the world functions as a process which, *were it done in the head*, we would have no hesitation in recognizing as part of the cognitive process, then that part of the world *is* (so we claim) part of the cognitive process. Cognitive processes ain't (all) in the head! (Clark & Chalmers 1998, 8)

There is no shortage of claims one may object to here, perhaps the most pressing being that there is no argument here, just an intuition pump. Suffice it to say that, if nothing else, Otto's notebook *does* shine a light on deeply held assumptions about the boundaries between mind, body, and world, even if this thought experiment does not do the work of actually reconstructing these boundaries. Luckily, we can set aside many of the admittedly controversial issues surrounding Otto's notebook, as Clark also presents empirical evidence in support of his argument for LMS, to which I now turn.

Ballard et al. (1997) conducted a series of experiments wherein a computer screen is divided into three spaces: target space, work space, and reserve space. Subjects are given a pattern of colored blocks in the target space and asked to reproduce this pattern as quickly as possible in the work space by selecting individual blocks from the reserve space. The task is performed using mouse clicks, and the subject's eye movements are constantly tracked. According to Ballard and colleagues:

---

5 We are asked to imagine two characters: Otto, who suffers from Alzheimer's and uses a notebook to write down everything he would have normally committed to memory and Inga, whose biological memory is wholly intact. Both Otto and Inga hear about an exhibit at the Museum of Modern Art they wish to see. Otto consults his notebook to "remember" that the MoMA is on 53rd St. while Inga "consults" her biological memory to remember that the MoMA is on 53rd St. According to Clark and Chalmers: "in relevant respects the cases are entirely analogous: the notebook plays for Otto the same role that memory plays for Inga. The information in the notebook functions just like the information constituting an ordinary non-occurrent belief; it just happens that this information lies beyond the skin" (Clark & Chalmers 1998, 13). What is important here is that Otto's notebook, a non-neural structure, performs the same computational role that Inga's biological memory does in recalling the address of the MoMA.

> A striking feature of task performance is that subjects behaved in a very similar, stereotypical way, characterized by frequent eye movements to the model pattern. Observation of individual eye movements suggest that the information is acquired incrementally during the task and even modest demands on visual memory are avoided. For example, if the subject memorized and copied four subpatterns of two blocks, which is well within memory limitations, one would expect a total of four looks in to the model area. Instead, subjects sometimes made as many as 18 fixations in the model area in the course of copying the pattern, and did not appear to memorize more than the immediately relevant information from the model [...] these fixations allows subjects to postpone the gathering of task-relevant information until just before it is required. (Ballard et al. 1997, 731)

This led Ballard and colleagues to conclude that “eye movements, head movements, and memory load trade off against each other in a flexible way” (Ballard et al. 1997, 732). Here is how Clark marshalls this evidence in support of LMS:

> Two morals matter for the story at hand. The first is that visual fixation here is playing an identifiable computational role [...]. The second is that repeated saccades to the physical model thus allows the subject to deploy what Ballard et al. dub ‘minimal memory strategies’ to solve the problem. The idea is that the brain creates its programs so as to minimize the amount of working memory that is required, and that the eye motions are recruited to place a new piece of information into memory [...]. The Ballard et al. model is an example of what might be called an extended functionalist approach. It analyses a cognitive task as a sequence of less intelligent sub-tasks, using recognizable computational and information-processing concepts, but applies those concepts within a larger organizational whole. (Clark 2008, 45)

Notably, Ballard et al. and Clark recognize the importance of bodily actions in structuring and completing the block copying task, but what makes this task what it is has less to do with some unique contribution of the body, and more to do with the functional profile of the component parts involved in the task. That is, in order to complete the task, information has to be stored (i.e. the color of the block, or the location of the block with relation to other blocks) and then processed. The best way to understand the data is that it is the *functional role* rather than the *physical properties* of the information storing and processing mechanisms that explains how subjects complete the task. What matters is the ability of the mechanism to store information—a role fulfilled by the reliable presence of the block pattern in the target space—not the physical makeup of the information storing mechanism. The information storage mechanism is realized through the combination of eye saccades and the reliable presence of the block pattern in the target space on the computer screen. The fact that the computer component of this information processing mechanism is silicon-based rather than carbon-based makes no difference in our understanding of its operation. Note also, that this kind of functional analysis is in no way committed to claims of phenomenal duplication—the claim that a computer program could "realize" the qualitative human experience of memory and task completion.

The fact that, as Ballard et al. pointed out above, the demands of the block completion task are well within the limitations of working memory capacity, and yet, subjects still used repeated saccades rather than memorizing the pattern, points to a crucial fact about human mental makeup—what Clark elsewhere calls the "007 Principle":

> In general, evolved creatures will neither store nor process information in costly ways when they can use the structure of the environment and their operations upon it as a convenient stand-in for the information-processing operations concerned. That is, know only as much as you need to know to get the job done. (Ballard et al. 1997, 46)

This, I take it, is the "hard core" of LMS (and arguably, Darwinian-inspired pragmatist theories of mind as well). If Clark is on the right track with the 007 Principle, then our theories of mind must be able to account for the ways in which, as organisms embedded in an environment, we *opportunistically* recruit and utilize features of this environment to decrease internal processing loads, despite that the fact that the makeup of these environmental features are quite different than our own biological makeup. Indeed, Clark contends that the special contribution of the body to the mind is precisely this ability to opportunistically recruit and exploit features of the non-bodily environment to optimize internal functioning.

In this section, I have tried to show how Clark argues that the tension between LMS and SE favors LMS. I have also attempted to show that the "hard core" of LMS lies not in some notion of multiple realizability (though LMS is certainly amenable to at least some accounts of multiple realizability, it does not logically entail them, nor is it logically entailed by them), but rather in something like the 007 Principle. Thus, I take it that the upshot of Clark's view is that any theory of mind must take account of the ways in which organisms opportunistically recruit features of the environment to decrease internal processing loads, much like the Ballard et al. experiments demonstrate. In the next section, I turn to Dempsey and Shani's argument for why the tension between LMS and SE favors SE.

## 3 Dempsey and Shani's argument for SE

Dempsey and Shani take up the debate between LMS and SCS on (mostly) the same terms that Clark lays out, with two notable exceptions—the first being terminological:

> We propose the term *strong embodied cognition thesis* ('strong embodiment' for short) [...] as close in spirit to Clark's body-centric SCS. Strong embodiment is 'strong' in that, by embracing SCS, it assigns embodiment a degree of significance in the shaping of the character of mind which is unmatched by extended functionalism. (Dempsey & Shani 2013, 597)

In what follows, then, I will refer to Dempsey and Shani's notion of "strong embodiment" (SE) on their own terms, rather than Clark's more general characterization (SCS), so as to be as charitable as possible to their position. Here is a succinct formulation of their view: "the kernel of strong embodiment is the claim that experience and embodiment are lawfully interdependent," and "this raises the question, what is the proper manner of interpreting, and of evaluating, this interdependence?" The answer, they argue, is that "the claim that experience and embodiment are lawfully interdependent ought to be interpreted as applying to coherent streams of consciousness occurring under normal ecological circumstances in which agents are free to use their bodies spontaneously" (Dempsey & Shani 2013, 600).

The second difference between Clark's and Dempsey and Shani's characterization is that the latter couch the debate in terms of phenomenal duplication. In their own words:

> Clark maintains that the *quality of experience* is a composite function combining the details of physical embodiment, downstream internal processing, and the relevant external props in the agent's natural or artificially constructed environment. Clark's appeal to this composite function opens up the possibility that sameness of experience may not correspond to sameness of embodiment since the specific details of bodily structure and functioning are merely one contributing factor in the making of experience. (Dempsey & Shani 2013, 598, emphasis mine)

While I do not think Clark would object this characterization of his position, I do want to flag the fact that while Dempsey and Shani's focus on phenomenal or qualitative experience could very well undermine strong or global multiple realizability-style arguments (indeed, as they correctly point out, these theories of mind are "notorious for their apparent inability to accommodate conscious experience" [Dempsey & Shani 2013, 614]), they do not immediately or obviously address what I take to be the "hard core" of LMS, namely, something like the 007 Principle.

At any rate, as proponents of SE characterize the debate, LMS would be undermined and SE subsequently favored if *either* it could be shown that sameness of phenomenal or qualitative experience *does* require sameness of embodiment, *or* that sameness of phenomenal or qualitative experience *could not* be compensated for by adjustments in downstream internal processing or relevant external props in the environment. Dempsey and Shani opt for the latter, and the key to this strategy is to link LMS to a notion of multiple realizability. They write: "extended functionalism presupposes GMR [Global Multiple Realizability, which they define as: "the view that complete phenomenal lives are, in principle, realizable across radically different physical embodiments" (Dempsey & Shani 2013, 602)], or something near enough, whereas strong embodiment is incompatible with any notion of MR [Multiple Realizability]" (Dempsey & Shani 2013, 603). Thus, their argument in favor of SE relies heavily on discrediting the plausibility of LMS via a refutation of multiple realizability:

> Clark is arguing that the contributions of the body are always (in principle) negotiable courtesy of the compensatory effects of inner processing or of environmental scaffolding. The central question, then, is whether or not such differences in embodiment can be compensated for. (Dempsey & Shani 2013, 605)

Dempsey and Shani's argument turns on the claim that while it is, in principle, possible to compensate for differences in embodiment, these compensations are so rare and contrived that they do not present a real threat to the "hard core" of SE. That is, SE "ought to be endorsed for the simple reason that it is better attuned to the realistic conditions under which experiences unfold in normally functioning embodied creatures," and it is only under "carefully designed artificial conditions" (Dempsey & Shani 2013, 600–601) that anything like the phenomenal duplication required for multiple realizability, and hence, LMS, could take place.

Further, they conclude that the contrived conditions under which limited cases of phenomenal duplication are possible actually bolsters the argument for SE: "compensating for the special contribution of the body involves either preventing it from making a contribution in the first place, or transforming it into another type of body, and in both cases this serves only to show just how privileged this contribution really is" (Dempsey & Shani 2013, 611).

One might worry here that Dempsey and Shani's case for the primacy of SE relies too heavily on a negative argument against LMS (and one which arguably does not address its most important insight), and is too light in terms of a positive argument for SE. As I mentioned at the outset, though, I do not take this debate to be a zero-sum game and so the shortcomings of these arguments for SE by no means settle the debate once and for all in favor in LMS. What I do take to be the upshot of Dempsey and Shani's arguments against LMS is their pragmatic appeal:

> SE is primarily a thesis concerning the manner in which embodiment constrains cognition and experience in normal psychological agents operating under normal psychological circumstances [...]. Hence, the possibility that there may be rare exceptions to the norm involving brief isolated episodes of experience occurring accidentally, or under carefully designed experimental control, is, we think, no good reason to deny that experience and embodiment are intimately linked. (Dempsey & Shani 2013, 601)

Indeed, much of literature on philosophical functionalism in the last 50 years would have done well to heed this appeal to actual agents acting in actual situations as setting the standard (how many more papers do we need on Martian pain, anyway?). As relates to the present paper, this appeal serves to ground the debate, first and foremost, in terms of organism environment transactions, to borrow a phrase from Dewey. While the degree to which the actual, bodily, fleshy components of this engagement *determine* character of mental life remains an open question, we can appreciate the constraints that this view places on some of the (admittedly far-fetched) thought experiments brought to bear in favor of functionalism, not the least of which is Clark's own snake-computer example.[6] Here, I fully concede that Dempsey and Shani's criticism of LMS hits its mark.

In this section, we have seen how Dempsey and Shani argue that the tension between LMS and SE favors SE. Their argument attempts to show the implausibility of Clark's claims that downstream processing and non-bodily props can compensate for changes in embodiment so as to yield sameness of phenomenal experience. Their argument that the idiosyncrasies of our embodiment make ineliminable contributions to our mental life sets another standard for a theory of mind: we must respect the *actual* engagements of organisms with their environment. That is, thought experiments can only go so far in showing that minds are (or are not) constituted by bodily structures and processes beyond the brain.

---

6 "Imagine now a case in which we have [...] a snake-like creature lying on top of an advanced touch-screen like environment. In this flat-screen setting every little wriggle of the snake can cause specific external symbolic tokens to appear elsewhere on the screen [...]". (Clark 2008, 50)

## 4 Why bother?

Having now entertained arguments from each side of the debate, it is fair to ask, "why bother?" That is, why not just claim that strong Embodiment and strong Extended are incompatible? Or, at the other end of the spectrum, that the entire debate is based on a false dichotomy? Or that the debate as I have laid it out is really just a debate over the proper grain of analysis?

A (partial) answer to these questions, I suggest, is that both the Embodied and Extended share a fundamental insight: the mind is made up of, quite literally *constituted* by, structures and processes beyond the brain. Put another way, both the Embodied and Extended theorists could take the yerba example from the introduction as motivating their respective position. Indeed, as Dempsey and Shani state, "both perspectives are, on the face of it, complimentary since each seeks to extend the mind beyond the brain proper and to consider the contributions that the brain's proximate and distal environments make on both cognition and phenomenology" (Dempsey & Shani 2013, 592). Additionally, insofar as Menary's claim that both approaches "reject or at least radically reconfigure traditional cognitivism, coupled with a methodological individualism" hits its mark, then both approaches are (more or less) committed to what Rowlands (2010) has identified as theories of *constitution* as opposed to *dependence*. The latter kinds of theories make the relatively weak claim that minds *depend* on bodies and environments—an insight easily incorporated by the traditional cognitivist or methodological individualist. The former kinds of theories make the relatively stronger claim that the mind is *constituted by* or *made up of* bodily and/or worldly structures and processes. Broadly, it is this sort of view which the traditional cognitivist or methodological individualist would want to reject, and that the 4E-inclined theorist would be compelled to defend.

If this is on the right track, then there should be good reason to attempt to integrate (or refrain from disintegrating) Embodied and Extended. In a similar project to my own, Gallagher (2014) argues that, with respect to the tension between Enactive and Extended, the pragmatist tradition contains valuable resources for just this kind of integration, and so *mutatis mutandis*, perhaps this same strategy can alleviate the tension just described between Extended and Embodied.

# 5 Gallagher's rapprochement

In line with Dempsey and Shani (2013), Gallagher writes, "enactivists argue against the functionalism of extended mind theorists who discount any special role of the living body in cognition" (Gallagher 2014, 117), but "the fact that most extended mind proponents adopt a functionalist framework, which includes representational mechanisms that discount the role of the body, does not mean that this is the only way to understand the extended mind" (Gallagher 2014, 118). So Gallagher, like Clark and Dempsey and Shani, is most interested in the reconciliation of the views in tension. At the heart of this reconciliatory effort for Gallagher are Dewey's notions of organism-environment and the situation.

In the first place, Gallagher notes, "for Dewey's understanding of cognition, the unit of explanation is not the biological individual, the body itself, or the brain, but the organism-environment" (Gallagher 2014, 115). This sentiment is nicely elaborated in a little passage from Dewey's *1900–1901 Lectures on Ethics:*

> So it is not simply that we happen to have *an* organism drop down into *an* environment and then these two react upon each other. It is quite the opposite. Organism and environment are the two things which converge in the life process. We do not begin with the two things and have them react and produce the life process. (Dewey 1900/1991, 364)[7]

Gallagher is exactly right to identify this level of what Dewey often calls the larger organic whole or unity as the proper unit of analysis for talking about cognition. When organism and environment are thus understood as co-extensive and mutually-defining, their interactions lead to continual changes, adjustments, and adaptations in one another. Cognition, on this view, is a pattern of organized response in the organism that arises from, and also helps to direct these adjustments.

---

7 Cf. Dewey (1898/1976): "No matter how much one school may insist on the organism, no matter how much another school may insist on the environment, it is certainly true that both are parts of a common world...From the larger standpoint we must have one thing, one reality, the world at large, *and the distinction between the organism and the environment and their adjustments to each other must be capable of definition and interpretation from the standpoint of this larger whole.*" (Dewey 1898/1976, 272, emphasis mine). Also cf. Alexander: "The organism and its environment are mutually implicated at each moment; they are aspects of one situation fundamentally related through the act. The organism is just this ability to draw on a range of material in the world and transform the energy in that material into an organized pattern of activity. An environment is in turn that range of energy which is available to the organism and necessary for its survival" (Alexander 1987, 135).

This view of organism-environment naturally leads Gallagher to Dewey's notion of a situation, which

> arises when the coupling of the organism-environment becomes problematic or starts to break down […]. When it starts to go wrong, this is what Dewey calls a problematic situation and it calls for a kind of re-pairing, a reestablishment of workable coupling […]. Cognition in such cases, is a form of inquiry, understood as a hands-on practical activity through which we transform the problematic situation into one that is less confused and more comprehensible. (Gallagher 2014, 115–116)

When we combine the ideas that organism-environment is the proper level of analysis, and the situation is constituted by the ongoing adjustments and interactions of organism-environment, then according to Gallagher, we have a way of understanding Extended that is grounded in Enactive, rather than Clark-style functionalism:

> organism and environment must be thought together in a dynamical relation, so that if differences or adjustments in organism are made one finds correlative adjustments and differences in the experienced environment. Accordingly the animal enacts an ecological and cognitive niche appropriate to its embodied needs (Gallagher 2014, 118).

If Gallagher's argument goes through here, then perhaps what I have identified as the hard core of LMS can be preserved, while still giving pride of place to the body and allowing it to lead the dance. On this view, different creatures will *enact* their worlds in different ways. The kinds of problematic situations faced by a field mouse (i.e. avoiding aerial predators) are obviously quite different than the kinds of problematic situations faced by humans (who by and large don't have to worry about aerial predators). Given these different embodied needs, different parts of the environment will afford different resources for resolving problematic situations. Quite consistent with Dewey's view is the idea that *any number of things can afford such resources:* bodily components, environmental features, ideas, concepts, tools, artifacts, other people, social institutions, etc.[8] Indeed, this line of thought in Dewey squares very nicely with Clark's version of LMS, where different parts of the body, brain, and world can interact to form "larger systemic wholes, incorporating brains, bodies, the motion of

**8** This, I take it, is the upshot of Hickman's (1990) reading of Dewey as a philosopher of technology. Future work on Dewey should assess the degree to which Dewey's characterization of the social environment can incorporate the role of technology (in the narrow sense of computers, smartphones, social media, etc.).

sense organs, and (under some conditions) the information-bearing states of non-biological props and aids" (Clark 2008, 40).

Now perhaps one could object that giving the body pride of place in this way does not actually square as neatly with LMS as is suggested here. Or one might object that invoking enactivist principles somehow undermines the case for SE. Or that a Dewey-inspired enactivism is orthogonal to the tension between LMS and SE. These (or other) objections may very well undermine the case for a Deweyan integration of the 4E program. But for the sake of argument, let's assume that this same strategy of appealing to Dewey's notions of organism-environment and the situation can help alleviate the tension between LMS and SE in the same way that these concepts "provide a productive way to effect a rapprochement between enactive and extended theories" (Gallagher 2014, 115). However one would deal with the objections, we can assume that if Gallagher is on the right track, we will have made headway toward keeping the E's intact, which, as noted above, we should be motivated to do. But at what cost?

## 6 Prospects for Pragmatism and 4E

In the previous section I argued that Gallagher's proposed rapprochement between Enactive and Extended, via Dewey's notions of organism-environment and the situation, might well provide a model for effecting a rapprochement between LMS and SE. I suspect that 4E theorists with pragmatist sympathies will find this strategy rather attractive. But there might be reasons to think that what's best for integrating Enactive and Extended would be different than what's best for integrating Extended and Embodied. In this section I want to suggest that even if we can dispel this worry, there might still be additional cause for concern.

More specifically, I want to raise two possible concerns about the appropriation of Dewey's notions of organism-environment and his corollary concept of body-mind. First is a worry about what exactly Dewey anticipated in the 4E program. Part of what ties Embodied and Extended together is a commitment to theories of wide constitution, as opposed to theories of mere dependence. That is, that minds are quite literally made up of, or constituted by structures beyond the bounds of skull and skin, not just dependent on them. The worry I want to raise is that textual evidence in Dewey might not warrant constitution claims, and might only offer evidence for dependence. A related worry is that appeal to Dewey's notions of organism-environment, the corollary concept of body-mind, and the situation, while perhaps providing a valuable resource for integrating perspectives within the 4E program, might come at the cost of further opening the program to persistent and worrisome criticisms.

First, the worry about constitution vs. dependence. The very same logic and metaphysics which allows Dewey to construct the concept of organism-environment allows Dewey to develop an all-important corollary concept: the body-mind.[9] This is particularly important in the context of the LMS/SE tension. In a passage often quoted to demonstrate Dewey's anticipation of various versions of the embodied mind thesis, Dewey (1925/2008) writes:

> The world is subject matter for knowledge, because mind has developed in that world; a body-mind, whose structures have developed according to the structures of the world in which it exists, will naturally find some of its structures to be concordant and congenial with nature, and some phases of nature with itself [...]. In ultimate analysis, the mystery that mind should use a body, or that a body should have a mind, is like the mystery that a man cultivating plants should use the soil; or that the soil which grows plants at all should grow those adapted to its own physico-chemical properties and relations [...]. Every 'mind' that we are empirically acquainted with is found in connection with some organized body. Every such body exists in a natural medium to which it sustains some adaptive connection: plants to air, water, sun, and animals to these things and also to plants. Without such connections, animals die; the 'purest' mind would not continue without them (Dewey 1925/2008, 211–212).

The language of a mind *using* a body and a body *having* a mind are what I find potentially worrisome. Even the most staunch functionalist, committed to strong global multiple realizability could admit that every mind we are so far acquainted with is found *in connection* with an organized body. Nothing here forecloses on the possibility that such a mind *could in principle* be realized in a different kind of organized body. This gets back to Clark's point that yes, "our human experience really does depend in part on many idiosyncratic aspects of our embodiment," but importantly "it does not follow that only a creature thus embodied could have those very experiences" (Clark 2008, 42). But on the other hand, Dempsey and Shani's admonition that our theories of mind must be based in *ac-*

**9** In *Experience and Nature*, Dewey (1925/2008) defines the concept as follows: "Body-mind designates an affair with its own properties. A large part of the difficulty in its discussion—perhaps the whole of the difficulty in general apart from detailed questions—is due to vocabulary. Our language is so permeated with consequences of theories which have divided the body and from each other, making separate existential realms out of them, that we lack words to designate the actual existential fact. The circumlocutions we are compelled to resort to [...] thus induce us to think that analogous separations exist in nature [...] but body-mind simply designates what actually takes place when a living body is implicated in situations of discourse, communication and participation" (Dewey 1925/2008, 217).

*tual* engagements of *actual* organisms in *actual* environments may be able to guard against this move. Neither side, at any rate, would deny that minds depend on some kind of body, and that one couldn't *have* a mind without some kind of body. But SE is committed to a much stronger claim than minds *using* a body, or a body *having* a mind. Indeed these kinds of claims imply, embodiment theories of *dependence*, rather than *constitution*. And as should be clear by now, SE falls squarely in the latter camp. To be sure, this is not necessarily a slight on Dewey. Rather, if my worry here has some substance, it only means that we might not be able to marshal the resources of Dewey's conception of body-mind in service of the 4E program in the ways Gallagher and others might like.

But perhaps this worry is irrelevant. Indeed, this criticism may amount to faulting Dewey for, in 1925, not making use of a technical distinction coined in 2010. If this is true, then all the better for our prospects of using Dewey to help resolve tensions in the 4E program. But even if the dependence/constitution criticism misses its mark, however, I don't think we're out of the woods yet. There is another worry related to appropriating Dewey's thought in the context of Extended. Namely, that one might think the textual evidence only supports reading Dewey as a forerunner of Embedded. The most well-known formulation of the latter view is espoused by Rupert: "cognitive processes depend very heavily, in hitherto unexpected ways, on organismically external props and devices and on the structure of the environment in which cognition takes place" (Rupert 2004, 393). On Rupert's view, these props, devices, and structures are not themselves *constitutive* of the mental. Rather, the mental merely *depends* on them, albeit in important, significant, and surprising ways.[10] It seems both plausible and safe to read Dewey in this vein. To read Dewey as making stronger claims about the mind being *constituted* by bodily structures or environmental props would require a liberal interpretation which may be hard to ground in the text. Again, this is not a slight against Dewey. Rather, if my worries are grounded, it simply suggests that Dewey's conceptual resources might be better suited for talking about the mind as Embedded, rather than Extended.[11]

---

**10** Rupert's view develops out of concern for the strength of arguments that are brought to bear in favor of constitution theories. It is much trickier, he thinks, to argue that organismically external props and devices are *constitutive* of the mind, than it is to argue that these props and devices significantly influence the mind. So if the framework of Embedded can cover the same range of cases as Extended, then by "the methodological principle of conservatism" (Rupert 2004, 395), he argues, we should endorse the former over the latter.

**11** Closely related to Embedded is Sterelny's (2010) notion of cognitive scaffolding. It is outside the scope of the present paper to pursue this line any further, but it seems likely there is a profit-

The second set of concerns has to do with Gallagher's appeal to the situation as the proper unit of analysis for talking about cognition. While we may be convinced by Gallagher that this is a promising solution for ironing out wrinkles in the 4E program, the adoption of this solution might give rise to a family of other problems: those of cognitive bloat variety. Rowlands defines this worry as follows:

> the admission of extended cognitive processes places us on a slippery slope. Once we permit such processes, where do we stop? Our conception of the cognitive will become too permissive, and we will be forced to admit into the category of the cognitive all sort sorts of structures and processes that clearly are not cognitive. (Rowlands 2010, 86)[12]

I take this to be a particularly pressing worry because the tractability of the 4E program hinges on the ability to provide operationalizable definitions of cognitive processes, such that they are subject to empirical verification.[13] The lack of a principled criteria for what counts as cognitive and what doesn't would severely hinder the prospects of the 4E program as a viable scientific approach to studying the mind.

As regards the present project, one might worry that an appeal to organism-environment or the situation may lack principled criteria of individuation, or a way to distinguish cognitive from non-cognitive processes at level of organism-

---

able comparison to be made between Dewey and Sterelny in this regard. Relatedly, Skorburg (2013) argues that Menary's (2007) notion of cognitive integration, rather than Clark and Chalmers's (1998) notion of cognitive extension, is the proper framework for understanding Dewey's theory of mind.

**12** Rowlands's solution is an ownership condition: "an adequate criterion of the cognitive will contain an ownership condition: any process that is to count as cognitive must be owned by a cognizing organism or subject" (Rowlands 2010, 97). Palermos's (2014) feedback loops criterion, where the bodily and/or worldly components continually affecting and being affected by each other are taken to be necessary and sufficient for wide constitution, is, in my view, the most convincing solution to the problem of cognitive bloat.

**13** Wagman and Chemero make this point strikingly clear: "although there is still open dispute over whether cognitive systems are extended, even the foes of extended cognition agree that it is an empirical matter whether there are extended cognitive systems [...]. This state of play, with everyone agreeing that the claim that whether cognition is extended is an empirical matter, would seem to leave open the possibility of a scientific resolution to the debate. That is, if experiments designed to detect the presence of extended cognition do in fact detect extended cognition, this would seem to constitute empirical confirmation of the existence of extended cognition. Should this happen, it would seem that the debate over extended cognition would simply be over" (Wagman & Chemero 2014, 106).

environment transactions. Gallagher appeals to the following passage from Dewey's *Experience and Education* in his efforts to reconcile Extended and Enactive:

> The statement that individuals live in a world means, in the concrete, that they live in a series of situations. The meaning of the word 'in' is different from its meaning when it is said that pennies are 'in' a pocket or paint is 'in' a can. It means [...] that interaction is going on between individuals and objects and other persons. The conceptions of situation and of interaction are inseparable from each other (quoted in Gallagher 2014, 115).

We can gather that interaction is central to what Dewey means by situation, but surely not all elements that might be considered part of a situation interact with one another. To be more concrete, we can say that an advocate of LMS would not argue that the clouds in the sky form a proper component of an information processing mechanism and similarly, an SE advocate would not argue that the length of one's toenails profoundly shapes our rational capacities. But an advocate of LMS might argue that a laptop could be considered a proper component of an information processing mechanism and similarly, an SE advocate might argue that the position of one's eyes on the face does profoundly shape the kinds of concepts we reason with. So where do we draw the lines in these two cases? The worry here is that Dewey's notion of the situation does not provide any resources to answer this question. Or worse, Dewey's notion of the situation is intentionally vague and blurs such lines by its very definition. Indeed, I think this is how Dewey is often read, and if this much is right, we may find ourselves slipping down the slope to cognitive bloat, where we have no criteria to rule out aspects of the situation which ought not to be considered constitutive of the mental.

The same rebuttal might be made here again, however: that this worry amounts to unfairly imposing anachronistic demands on Dewey's thought. But again, even if this is so, I still think a general worry still persists that the grain of analysis of organism-environment or the situation may just be too coarse for the practical, logistical demands of the cognitive sciences.[14] While appealing to the level of organism-environment may indeed be the best strategy to reconcile Extended and Enactive, and perhaps even Embodied and Extended, it may come at the cost of being even more vulnerable to charges of cognitive bloat, and

14 This is not to say, however, that such a perspective cannot be extremely useful in other contexts. For example, the notions of organism-environment and situation may provide just the right grain of analysis for examining the ways in which cognitive processes are shaped not just by bodily engagements and various computational technologies, but by broader social and cultural forces.

other related worries.[15] I do not want to suggest that such charges are necessarily conclusive, but rather, that an appeal to Dewey might not be the most effective recourse in this context.

## Conclusion

In this paper I have argued that the 4E program is unified in its stance against traditional cognitivism, methodological individualism, and classical internalism, and in this sense is indebted to the tradition of classical pragmatism, and especially the work of John Dewey. Despite these commonalities, however, the 4E program has its own internal fissures. Recently, many disagreements have centered on the functionalist commitments of Clark's formulations of Extended. Two such disagreements are taken up by Dempsey and Shani (2013) with respect to Embodied, and Gallagher (2014) with respect to Enactive. I have argued that perhaps Gallagher's strategy of appealing to Dewey's concepts of organism-environment, body-mind, and the situation to effect a rapprochement between Extended and Enactive might also work to alleviate the tension between Extended and Embodied, as represented by Clark's notion of LMS, and Dempsey and Shani's notion of SE, respectively.

Lastly, I have argued that even if we grant that Gallagher's strategy is applicable to the LMS/SE debate (which it may or may not be), this may still come at a cost. Namely, that appealing to these aspects of Dewey's thought might only provide support for claims of dependence, and not constitution, in which case his insights might be better suited for Embedded than Extended, and hence, might not have much bearing on the tensions between LMS/SE or Extended/Enactive.

Perhaps more worrisome is that even if we can dispense with all of these worries, appealing to Dewey's notion of the situation in order to reconcile internal tension in the 4E program may come at the cost of being left open to the persistent and formidable charges of cognitive bloat. Dewey's framework may actually magnify this worry, by not establishing boundaries and criteria, but instead purposefully blurring and obfuscating the boundaries.

Given my own pragmatist sympathies, I hope these worries can be quelled and objections answered. The upshot of this paper should not be that we can't or shouldn't appeal to the tradition of pragmatism (of the "classical" or "neo" varieties), but rather, that we need to think carefully about where and how prag-

15 E.g. Adams and Aizawa's (2001) well-known charge of the coupling-constitution fallacy.

matism can actually make meaningful contributions to the cognitive sciences. I don't think this contribution is useful in the form of "Peirce/James/Dewey-said-it-first" arguments, nor is it useful to think that pragmatism always has the cure to what ails the cognitive sciences. And at the other end of the spectrum, it is equally mistaken to treat pragmatism as a mere historical relic irrelevant to today's concerns. In the end, I hope to have brought a critical perspective to bear in order to bolster the contribution that the tradition of pragmatism can ultimately make to advance the scientific study of the mind.

## References

Adams, F. & Aizawa, K. (2001). "Defending the Bounds of Cognition." In: Richard Menary (Ed.). *The Extended Mind.* Cambridge: MIT Press, 67–80.

Alexander, T. (1987). *John Dewey's Theory of Art, Experience, and Nature: The Horizons of Feeling.* Albany: State University of New York Press.

Ballard, D. et al. (1997). "Deictic Codes for the Embodiment of Cognition." In: *Behavioral and Brain Sciences* 20, 723–767.

Burke, F.T. (2013) *What Pragmatism Was.* Bloomington: Indiana University Press.

Chemero, A. (2011). *Radical Embodied Cognitive Science* (Reprint edition). Cambridge, MA: A Bradford Book.

Clark, A. (1997). *Being There: Putting Brain, Body, and World Together Again.* MA: Massachusetts Institute of Technology Press.

Clark, A. & Chalmers, D. (1998). "The Extended Mind." In: *Analysis* 58, 7–19.

Clark, A. (2008). "Pressing the Flesh: A Tension in the Study of the Embodied, Embedded Mind." In: *Philosophy and Phenomenological Research* 76(1), 37–59.

Dempsey, L. & Shani, I. (2013). "Stressing the Flesh: In Defense of Strong Embodied Cognition." In: *Philosophy and Phenomenological Research* 86, 590–617.

Dewey, J. (1925/2008). *The Later Works of John Dewey,* vol. 1. ed. by J. A. Boydston. Carbondale: Southern Illinois University Press.

Dewey, J. (1898/1976). *Lectures on Psychological and Political Ethics: 1898.* ed. by D. F. Koch. New York: Hafner Press.

Dewey, J. (1900/1991). *Lectures on Ethics: 1900–1901.* ed. by D. F. Koch. Carbondale: Southern Illinois University Press.

Folch, C. (2010). "Stimulating Consumption: Yerba Mate Myths, Markets, and Meanings from Conquest to Present." In: *Comparative Studies in Society and History* 52, 6–36.

Gallagher, S. (2014). "Pragmatic Interventions into Enactive and Extended Conceptions of Cognition." In: *Philosophical Issues* 24, 110–126.

Hickman, L. (1990). *John Dewey's Pragmatic Technology.* Bloomington: Indiana University Press.

Johnson, M. (2008). *The Meaning of the Body: Aesthetics of Human Understanding.* University of Chicago Press.

Lakoff, G. & Johnson, M. (1999). *Philosophy in the Flesh: The Embodied Mind and its Challenge to Western Thought.* New York: Basic Books.

Macintyre, B. (1993). *Forgotten Fatherland: The Search for Elisabeth Nietzsche.* New York: Harper Perennial.
Madzia, R. (2013). "Chicago Pragmatism and the Extended Mind Theory: Mead and Dewey on the Nature of Cognition." In: *European Journal of Pragmatism and American Philosophy* 5(1), 279 – 297.
Menary, R. (2010). "Introduction to the Special Issue on 4E Cognition." In: *Phenomenology and the Cognitive Sciences* 9, 459 – 463.
Menary, R. (2007). *Cognitive Integration: Mind and Cognition Unbounded.* Baskingstoke: Palgrave Macmillan.
O'Regan, J.K. & Noë, A. (2001). "A Sensorimotor Approach to Vision and Visual Consciousness." In: *Behavioral and Brain Sciences* 24(5), 939 – 973.
Palmeros, S.O. (2014). "Loops, Constitution, and Cognitive Extension." In: *Cognitive Systems Research* 27, 25 – 41.
Rockwell, W.T. (2005) *Neither Brain nor Ghost: A Non-Dualist Alternative to the Mind-Brain Identity Theory.* Cambridge: MIT Press.
Rowlands, M. (2010). *The New Science of the Mind: From Extended Mind to Embodied Phenomenology.* MA: Massachusetts Institute of Technology Press.
Rupert. R. (2004), "Challenges to the Hypothesis of Extended Cognition." In: *Journal of Philosophy* 101, 389 – 428.
Sterelny, K. (2010). "Minds: Extended or Scaffolded?" In: *Phenomenology and the Cognitive Sciences* 9:4, 465 – 481.
Thompson, E. & Stapleton, M. (2009). "Making Sense of Sense-Making: Reflections on Enactive and Extended Mind Theories." In: *Topoi* 28, 23 – 30.
Vaesen, K. (2014). "Dewey on Extended Cognition and Epistemology." In: *Philosophical Issues* 24, 426 – 438.
Wagman, J. & Chemero, A. (2014). "The End of the Debate over Extended Cognition" In: T. Solymosi & J. Shook (Eds.). *Neuroscience, Neurophilosophy, and Pragmatism: Brains at work with the world.* New York: Palgrave, 105 – 124.

Anthony Chemero and Stephan Käufer

# Pragmatism, Phenomenology, and Extended Cognition

**Abstract:** In this paper, we will set out a defensible version of the hypothesis of extended cognition. To do so, we will draw together two separate streams of philosophical and psychological work from the 20th century: William James's radical empiricism and Maurice Merleau-Ponty's phenomenology of the lived body. The confluence of these two streams can be seen in J.J. Gibson's ecological approach to the perception.

## Introduction

The hypothesis of extended cognition, in its currently debated form, was introduced by Andy Clark and David Chalmers (1998; see also Clark 1997; Haugeland 1993). The story Clark and Chalmers tell involves a contrast between two imaginary people, Otto, with a severely impaired memory, and Inga, with an intact memory. Because Otto's memory is faulty, he carries a notebook with him in which he records important facts and appointments. Imagine that Otto is very proficient at using his notebook, and that he relies on it to go about his daily tasks successfully. Imagine, in fact, that Otto is so reliably and confidently uses his notebook that it plays the same role in his abilities that Inga's memory-enabling brain areas do. If this is the case, Clark and Chalmers suggest, then we should be equally willing to call Otto's notebook and Inga's brain areas components of their abilities. Since it seems obvious that Inga's brain areas are part of her as a rememberer, so too is Otto's notebook part of him as a rememberer. This is the hypothesis of extended cognition: sometimes things outside the biological body are literally parts of the cognitive system; i.e., the thing that does the thinking sometimes includes nonbiological stuff like notebooks and smartphones.

Many philosophers find this thought experiment immediately compelling. First, we understand humans and human behavior as extremely malleable and flexible. In fact, this flexibility is arguably the key difference between humans and other animals. The extended cognition hypothesis is compelling because it is a straightforward continuation of this flexibility. The second reason many philosophers find the extended cognition hypothesis compelling is more theoretically loaded. To many philosophers of mind and cognitive science, it seems obvious that there is nothing magic about the material brains are made of, and given this, there is no reason that some other kind of stuff could play the same role. This is the case, we suggest, because the default position in the philosophy of mind and cognitive science is some kind of philosophical functionalism.[1] Notebooks and smartphones can literally be parts of cognitive systems whenever they play the same functional role that brain areas play in cognitive systems. For better or worse, the hypothesis of extended cognition is a natural corollary of philosophical functionalism (Sprevak 2009; Wheeler 2010). Unfortunately, though, this makes it easy for those who don't find the thought experiment compelling to argue against it. First, one might simply draw on the decades of arguments against philosophical functionalism (see Polger 2012). Take away the functionalism, and the thought experiment seems less compelling. Secondly, one might accept the functionalism and argue that it is not the notebook or smartphone that plays the same functional role as Inga's brain areas. This argumentative strategy works because Otto uses his notebook by perceiving it. One might then argue that it is the perception of the notebook, not the notebook itself, that plays the same functional role as the brain areas. The perception of the notebook is a safely biological phenomenon, presumably some event in Otto's brain, so the inference that the notebook itself is part of the cognitive system is blocked (Chemero 2009; Rupert 2009; Steiner 2015). In this paper, we will set out a more defensible version of the hypothesis of extended cognition. To do so, we will draw together multiple separate streams of philosophical and psychological work from the 20$^{th}$ century: one whose source is William James's radical empiricism and another whose source is Gestalt psychology. The confluence of these two streams can be seen in Maurice Merleau-Ponty's phenomenology of the lived body and J.J. Gibson's ecological approach to the perception, and especially their discussions of tools.[2] First, though, it is important to see to see the scientific backdrop against which both these streams emerged.

## The foundation of scientific psychology

Wilhelm Wundt founded the world's first experimental psychology lab in 1879 at Leipzig and widely known as the "founding father of psychology". The specific

1 Note that what we call 'philosophical functionalism' is not the same thing as the kind of psychological approach called 'functionalism', and defended by pragmatists in the late 19th and early 20th centuries.

2 These historical figures are discussed in more detail in Käufer and Chemero 2015.

assumptions of Wundt's psychology were controversial, and later psychology (especially Jamesian psychology and Gestalt psychology) were explicit responses to Wundt's methods and theoretical foundations. Wundt's psychology is built upon advances in psychophysics brought about by Ernst Weber and Gustav Fechner in the middle of the 19th century. Weber showed that there are experimentally determinable *just noticeable differences* between stimuli, the smallest difference in the intensity of the stimuli that leads to a change in experience. What Fechner (1860/1912) called "Weber's Law" states that for any given stimulus *s* (say, a tone with a particular volume) the difference between *s* and the just noticeably different stimulus *s'* (the tone that is just sufficiently louder than *s* that the difference between *s* and *s'* is detectable) is proportional to the intensity of *s*. Fechner generalized Weber's law from just noticeable differences to all sensory differences, demonstrating a clear mathematical relationship between stimulus and sensation.

This psychophysics forms the basis for Wundt's experimental psychology. First, Wundt makes a distinction between inner and outer experience, acknowledging that inner experience is indeed inaccessible to experimental methods. Indeed, what is generally known as Wundtian psychology, the kind that was undertaken in his laboratory at Leipzig and in which he trained nearly every member of the first generation of scientific psychologists, is his science of outer experience. Wundt's psychology of inner experience, called *Völkerpsychologie*, had virtually no influence on later psychology (Leahey 2000). *Völkerpsychologie* (1912) applies historical and anthropological methods. Inner experience is inaccessible through psychophysics, but Wundt argues that it is possible to experimentally access outer experience using the methods of psychophysics. Psychophysics uses trained introspectors, in highly controlled circumstances, to experiment upon the basic elements of conscious experience. Subjects, for example, are presented with a series of tones that are increasingly high in pitch and are asked to report when then experience a pitch difference; or they are asked to press a key when a light appears on a screen; or they are asked to press when a blue light (but not a red light) appears on a screen. Experiments such as this are used to examine the variety of sensations humans are capable of having, and the temporal duration of sensory processes.

Using psychophysics to study outer experience places strict limits on the scope of experimental psychology. According to Wundt, psychology is the study of the elements of immediate conscious experience.

> Psychological analysis leaves us with two such elements [of conscious experience], of specifically different character: with sensations, which as the ultimate and irreducible elements of ideas we may term the objective elements of the mental life, and with feelings,

> which accompany these objective elements as their subjective complements, and are referred not to external things but to the state of consciousness itself. In this sense, therefore, we call blue, yellow, warm, cold, etc., sensations; pleasantness, unpleasantness, excitement, depression, etc., feelings. (Wundt 1874, 12)

Sensations and feelings are the atoms of conscious experience. They are combined to form the molecules: sensations combine to form ideas; feelings combine to form complex feelings. Wundt's experimental psychology was, thus, reductionist in that it focused on the smallest possible bits of conscious experience. It was also atomistic in that it thought that the elements maintained their identities in whatever context they appeared. ("As conscious contents, blue is and remains blue, and the idea of an object is always a thing ideated in the outside world, whether the external stimulus or the thing outside of us be really present or not." Wundt 1874, 13–14)

Wundt's pioneering scientific psychology was, thus, a reductionistic, atomistic science of outer experience. It was a successful science, and Wundt was very influential: he published more than 60,000 pages during his lifetime, and educated thousands of students in his methods (Fancher 1995). One of these students was William James, the "founding father "of American psychology, who became an opponent of Wundt's methods which he mocked as "brass instrument psychology". Another student was Oswald Külpe, who founded the Psychology Laboratory at Würzburg, where he taught two other opponents of Wundt, the founders of Gestalt psychology, Max Wertheimer and Kurt Koffka.

## Stream 1: James to Holt to Gibson

At the same time that Wundt developed his scientific psychology, William James was teaching a very different version of the science, one whose primary influence was Darwin. For Wundt, scientific psychology was nothing other than the use of the methods of psychophysics to study simple sensations. As he wrote in his 1890 textbook *The Principles of Psychology*, James doubted that there even were simple sensations, and thought that Wundt's focus on them was harmful to the science.

> Most books start with sensations, as the simplest mental facts, and proceed synthetically, constructing each higher stage from those below it. But this is abandoning the empirical method of investigation. No one ever had a simple sensation by itself. Consciousness, from our natal day, is of a teeming multiplicity of objects and relations, and what we call simple sensations are results of discriminative attention, pushed often to a very high degree. It is astonishing what havoc is wrought in psychology by admitting at the outset

> apparently innocent suppositions, that nevertheless contain a flaw. The bad consequences develop themselves later on, and are irremediable, being woven through the whole texture of the work. The notion that sensations, being the simplest things, are the first things to take up in psychology is one of these suppositions. (James 1890, 225)

Focusing exclusively on simple sensations, James argued, missed most of what a science of the mind should be about.

> Psychology is the Science of Mental Life, both of its phenomena and of their conditions. The phenomena are such things as we call feelings, desires, cognitions, reasonings, decisions, and the like; and, superficially considered, their variety and complexity is such as to leave a chaotic impression on the observer. (James 1890, 1)

The science imagined by James was thus much broader in scope than the studies of sensations being undertaken at Wundt's Institute in Leipzig. Not mentioned in this quotation, but getting explicit discussion in the text are habits, emotions, consciousness, instinct, perception, imagination, and memory, among many others, along with the mental lives of infants and non-human animals. Of course, James thought we should study sensations—he devotes a chapter to them—but not as the sole or foundational subject matter as Wundt did.

Because James was explicitly Darwinist in approach, he argued for understanding each of these aspects of the mind as *adaptations* to the environment. Just as your lungs are a life-long adaptation to our oxygen-rich atmosphere, your belief that you are holding a book is a temporary adaptation to your current situation. Aspects of our mental life, in other words, were to be understood in terms of what they are for.[3] Understanding the mind as a collection of adaptations makes the contrast with Wundtian psychology even starker. Particular adaptations must be understood in the context of the environment, and in the context of the whole animal. Lungs have their functions only in oxygen-rich environments, and only when paired with hearts and circulatory systems. Moreover, adaptations can only be understood historically, over developmental and evolutionary time. Understanding how lungs came to have the functions that they do requires understanding how our lungs develop and how they functioned in our evolutionary ancestors. While Wundt and his followers focused very narrowly on instantaneous simple sensations, Jamesian psychologists studied as-

3 Because of this, Titchener, a Wundtian psychologist in the United States, mockingly called the Jamesian approach "functionalism," a name that stuck. Note that psychological functionalism is not the same as philosophical functionalism discussed above. To avoid confusion, we call functionalist psychology 'Jamesian psychology' in what follows.

pects of the mind in the context of the other aspects of the mind, in the context of the environment, over developmental and evolutionary timescales.

James published his textbook at the time when psychology began to become a discipline separate from philosophy. James stayed with philosophy, and his own work became increasingly philosophical. He called his later philosophical views "radical empiricism." He characterizes radical empiricism as consisting "first of a postulate, next of a fact, and finally of a generalized conclusion" (James 1904, XII). His postulate is that all and only things that are experienced exist. His fact is that relations are experienced. It follows that relations exist. His generalized conclusion is that the world is experienced directly. The world on this view is nothing more or less than what we experience, and our experiences of the world are nothing more or less than the world itself. Although James himself didn't use this term, Jamesian radical empiricism amounts to a sort of neutral monism—there is only one world, and it is neither material nor mental. The world experienced and the experiences themselves are the very same thing.

> As 'subjective' we say that the experience represents; as 'objective' it is represented. What represents and what is represented is here numerically the same; but we must remember that no dualism of being represented and representing resides in the experience *per se*. In its pure state, or when isolated, there is no self-splitting of it into consciousness and what the consciousness is 'of'. Its subjectivity and objectivity are functional attributes solely, realized only when the experience is 'taken,' *i.e.*, talked-of, twice, considered along with its two differing contexts respectively, by a new retrospective experience, of which that whole past complication now forms the fresh content. The instant field of the present is at all times what I call the 'pure' experience. It is only virtually or potentially either object or subject as yet. For the time being, it is plain, unqualified actuality, or existence, a simple *that*. (James 1904, 484)

This passage is from "Does Consciousness Exist?" (1904). James's answer to the question is that consciousness does not exist, at least if what you mean by consciousness is what Wundt meant by consciousness, something inside your mind composed of simple sensations and feelings and representing the external world.

E.B. Holt was a student of William James who attempted to push radical empiricism to its limits and to make it into a psychological theory. Holt's work was very interesting in itself, though it has been until recently neglected by historians of psychology (Heft 2001; Charles 2012). For current purposes, though, Holt's most important role was as James Gibson's teacher at Princeton. In the early 1920s, Gibson studied philosophy at Princeton as an undergraduate, and became a member of the first cohort of graduate students in psychology there. Holt was one of his professors. Another was Herbert Langfeld, who had studied with Brentano. Gibson's thesis was an experimental refutation of work on form perception by the Gestalt psychologist Wulf, who had been a student of Koffka. Gibson wrote of this in his autobiography (Gibson 1967): "Form perception was learned. Otherwise one fell into the arms of Immanuel Kant. I was a radical empiricist, like Holt... Little did I know that I would be facing Koffka himself weekly across a seminar table." As we will see below, the Jamesian Holt and Koffka are the two main influences on Gibson's ecological theory.

We can introduce the ecological approach as consisting in three principles.

*Principle 1: Perception is direct.* To claim that perception is direct is to claim that perception is not the result of inferences performed on sensory representations. When an animal perceives something directly, the animal is in contact with that thing. This implies, of course, that the perceiving isn't inside the animal, but rather is part of a system that includes both the animal and the perceived object. This just is Jamesian radical empiricism.

*Principle 2: Perception is for action.* The purpose of perception is for the control of action. We perceive the world in order to act in it. In fact, a good deal of action is also for perception or cognition. We lean to the left in order to get a better view; we re-arrange the Scrabble tiles to help us figure out what word to play. This intimate, two-way connection between perception and action has an immediate ring of evolutionary plausibility.

*Principle 3: Perception is of affordances.* The primary objects of perception are affordances. Affordances are environmental opportunities for action, relationships between an animal's abilities and features of the world. Given the height, strength, balance, and flexibility of a typical adult human, the staircases in most buildings afford climbing for typical adult humans. Notice, though, that the same set of stairs might not afford climbing to a toddler or an elderly person, because their abilities are different. The walls in the building afford climbing to ants, but not dogs or humans. What a creature can do determines which affordances are available to it (or him or her). Thus affordances are both features of the environment and dependent in some sense on animals. This makes it difficult to say exactly what affordances are. Gibson himself describes affordances in an easy to understand way and a difficult to understand way. First, he says, "the affordances of the environment are what it offers the animal, what it provides or furnishes, either for good or ill" (Gibson 1979, 127).

This makes perfect sense. As animals move around, they see what they can do. Walking into a party, we see places to sit and stand, stuff to eat and drink, people to talk to; these are affordances.

However, a few pages on, when Gibson tries to say what affordances are in detail, he becomes much harder to understand.

> [A]n affordance is neither an objective property nor a subjective property; or it is both if you like. An affordance cuts across the dichotomy of subjective-objective and helps us to under-

> stand its inadequacy. It is equally a fact of the environment and a fact of behavior. It is both physical and psychical, yet neither. An affordance points both ways, to the environment and to the observer. (Gibson 1979, 129)

The latter is puzzling, to put it mildly. However it becomes much easier to understand when one remembers Gibson's intellectual lineage, and puts him in the context of Jamesian neutral monism. Affordances are pure experience.

## Stream 2: Gestalt to Merleau-Ponty

Like Jamesian functional psychology, Gestalt psychology is an explicit rejection of the views and methods of Wundt. Recall that Wundt's psychology is reductionist in that it focuses solely on the smallest elements of conscious experience, the atomistic sensations from which experiences are built. The Gestaltists, like William James, oppose this focus on atomistic sensations and feelings.

One of the specific points of Wundt's atomistic psychology that the Gestalt psychologists identify as faulty is called the "constancy hypothesis" (Köhler 1913). This hypothesis is described clearly in Koffka (1923), the first article in Gestalt psychology to have been published in English. In Wundt's psychology, outer experiences are molecular elements, composed of simple sensations. Each experience is a bundle of simple sensations.

> In accordance with the method by which sensations have been investigated, it has been necessary to refer to the stimulus-side in defining the principle which underlies this concept. More explicitly, this relation of the sensation to its stimulus is expressed by a generally accepted rule, termed by Köhler the 'constancy-hypothesis'; that the sensation is a direct and definite function of the stimulus. Given a certain stimulus and a normal sense-organ, we know what sensation the subject must have, or rather, we know its intensity and quality [...] (Koffka 1923, 534).

According to the constancy hypothesis, the stimulus determines what molecule of simple sensations our experiences will be, and it does so in a point-by-point way. That is, to each atomic sensation will correspond some feature of the world. Imagine looking at a monitor displaying nothing but a red circle on a white background. The experience you have will be the collection of simple sensations corresponding to, and only to, the pixels of the image on the monitor.

There was already good reason to doubt the constancy hypothesis in the late 19th century, even before Wundt opened his institute in 1878. In 1870, the physiologist Ludimar Hermann published a paper describing what has come to be known as the Hermann grid. The grid is a light-colored or white lattice against a black background. Looking at the grid, most people see gray blobs at the intersections of the white lattice. Yet these blobs are not due to gray ink blobs on the page. The perceptual experience of the blogs is therefore inconsistent with the constancy hypothesis. This inconsistency was discussed by Ernst Mach in his *Analysis of Sensations* (1886), using the example of what are now known as Mach bands (Mach 1865). Mach bands are visible in an image of an elongated rectangle, where the left and right ends are black and white, respectively. Starting near the middle of the rectangle, the black and white are separated by, from the left, successively brighter shades of gray, so that the black gradually shades into the white. Looking at the rectangle, most people see a band of extra dark black just to the left of the gradual shading and a band of extra bright white just to its right. These are Mach bands. There is no change to the ink on the page at either location. Mach suggested that these bands were instances of "spatial forms" (*Gestalten*), extra qualities that come along with the more mundane sensations arising from the ink on the page. Spatial forms are inconsistent with the constancy hypothesis insofar as they are a simple sensation that does not correspond to any individual element in the world. Rather, this sensation derives from an arrangement of those elements. (Mach also discusses "tone forms" and makes analogous points).

Mach's discussion of spatial and tone forms is the launch point for "On Gestalt Qualities," the 1890 essay by Christian von Ehrenfels. Ehrenfels points out that by discussing spatial and tone forms as *sensations*, Mach's analysis implies that the forms are not the result of mental analysis or the application of concepts but are instead simply sensed. Given this, Ehrenfels asks about the nature of the sensation: "Is a melody (i) a mere sum of elements, or (ii) something novel in relation to this sum, something that certainly goes hand in hand with but is distinguishable from the sum of elements?" (trans. Smith 1988, 83).

To answer this question, Ehrenfels considers transposing a melody from one key to another. Imagine, for example, "Happy Birthday" sung by a five year old and by an opera baritone. These two different performances will share none of the same notes, but will still be the same melody. Indeed, it is the same melody if played on a guitar, a piano, or by an orchestra. However, suppose we took the notes that make up "Happy Birthday" in the key of C and play them in a different order. That would not be the same melody. So you can change all the notes and maintain the same melody; and you can keep all the same notes and produce a different melody. So the melody must be something other than the notes that make it up. This something else is the Gestalt or form quality.

As noted above, the Gestalt psychologusts were influential on Gibson. They were also profoundly influential on Merleau-Ponty. Gestalt psychology is foundational in Merleau-Ponty's first book, *The Structure of Behavior* (1942), in which Merleau-Ponty claims that behavior is itself a Gestalt.

> [...] behavior is detached from the in-itself (en soi, [physical]) and becomes the projection outside the organizm of the *possibility* which is internal to it. The world, inasmuch as it harbors living beings, ceases to be a material plenum consisting of juxtaposed parts; it opens up as the place where behavior appears. (Merleau-Ponty 1942, 47)

and

> [B]ehavior is not a thing, but neither is it an idea. It is not the envelope of pure consciousness and, as the witness of behavior, I am not a pure consciosuness. It is precisely this which we wanted to say in stating that behavior is a form. (Merleau-Ponty 1942, 49)

In *The Structure of Behavior*, Merleau-Ponty grounds his phenomenology in the form of behavior. English "form" is from French "*forme*" which is from German "*Gestalt*". Behavior for Merleau-Ponty is a Gestalt. Behavior is neither material, nor mental. It is more than a series in time (i.e., more than the sum of its temporal parts). The world is the place where behavior occurs. Note the similarity to what James says about "pure experience" and what Gibson says about affordances.

In his later work, *The Phenomenology of Perception* (1962), Merleau-Ponty writes that there is "an entire philosophy emerging from the critique of the 'constancy hypothesis'" (PP, 509n). The philosophy that emerges from this criticism, he claims, is his phenomenology of the *lived body.* Merleau-Ponty discusses the lived body via a distinction made by Edmund Husserl, that between *Körper* and *Leib.* The *Körper* is the body as a physical object, taking up space in the world; the *Leib* is the lived body, which is the set of abilities to engage with the world that makes experience possible. The lived body opens up the possibilities that make up the world. We do not entertain these possibilities mentally, by thinking about them or imagining them. Instead, we are open to them through the skills and habits of the lived body. Our constant readiness to deploy those skills is the "body schema." It makes the world intelligible and hence constitutes the possibility of relating to objects and regions within it. "The 'body schema' is, in the end, a manner of expressing that my body is in and towards the world" (PP, 103). Most important for important purposes, as we will see below, is that Merleau-Ponty argues that the lived body is *malleable.*

## A point of confluence

Though they came from different traditions, both Gibson and Merleau-Ponty were influenced by Gestalt psychologists, especially early in their careers. In fact, Gibson's views and Merleau-Ponty's views are in some respects very similar. That said, for the most part, Gibson did not get his ideas from Merleau-Ponty or other phenomenologists, but from decades of applying the Jamesian tradition in psychology to our perceptual experience. However, Gibson was aware of Merleau-Ponty's work. Indeed, according to a regularly repeated anecdote, Gibson once tried to scare off prospective psychology graduate students, by telling them they needed to understand *The Phenomenology of Perception* to work in his lab. Gibson also taught a seminar on *The Phenomenology of Perception.*

Recently, William Mace (2014) has uncovered direct evidence of Merleau-Ponty's influence on Gibson, finding Gibson's notes on *The Phenomenology of Perception* from the early 1970s. These include at least one specific case of an idea that appears in print for the first time in Gibson's *The Ecological Approach to Visual Perception* (1979). Gibson's notes on Merleau-Ponty's pages 255 and following say "a brilliant description of *depth-perception*, showing the mistakes made by both Berkeley and Kant." Gibson's notes continue that "The essence of the problem of depth is *occlusion* (not, as I thought, the *recession* of the ground and the gradients corresponding)." In Gibson's earlier work, he had argued that we perceive by being sensitive to texture gradients. For example, on a patterned carpet, nearby squares take up more space in the visual field than more distant squares. You can perceive the distance of an object on the carpet by in virtue of the apparent size of the squares it is sitting on, so that if it is sitting on apparently smaller squares it is farther away. In the notes, Gibson describes his realization that it is occlusion that gives information about distance: things nearer by occlude more distant things from any particular point of view.

Gibson's notes continue "A man 'transcends' occlusion only by *changing* his point of observation—by *touring*, by being a *tourist*". That is, by moving around, a person or animal can see distance by changing which objects occlude which other objects. For example, you lean to the left to see the rest of the object that is partly occluded by your laptop monitor. Or you can walk around the fence to see what is behind it.

This is interesting for three reasons. First, as mentioned above, Gibson's 1979 book, includes extended discussions of the relation between occlusion and perceived depth and of the effects of moving on occlusion. Second, as these notes make plain, Gibson came to realize the importance of occlusion in depth perception while reading Merleau-Ponty. Third, and surprisingly, given the second point, the pages of *The Phenomenology of Perecption* that Gibson's notes concern have no discussion of occlusion. What the notes make clear, then, is that Gibson read Merleau-Ponty carefully and productively. It is also clear that, despite the similarities of their views, Gibson was not repeating Merleau-Ponty's ideas. As we will argue below, it is these similarities that lead us to a better, more defensible version of the hypothesis of extended cognition.

## Merleau-Ponty and Gibson on tools

As noted above, Merleau-Ponty's phenomenology of the lived body portrays the lived body as changing over time. Merleau-Ponty discusses several cases that make this point. These examples are illustrations of the specific bodily or motor "understanding" that opens up meaningful situations, as well as the experience of our bodies fusing or merging with tools.

> A woman may, without any calculation, keep a safe distance between the feather in her hat and things which might break it off. She feels where the feather is just as we feel where our hand is. If I am in the habit of driving a car, I enter a narrow opening and see that I can 'get through' without comparing the width of the opening with that of the wings, just as I go through a doorway without checking the width of the doorway against that of my body. (PP, 165)

There is also the famous example of the blind person with a cane:

> The blind man's stick has ceased to be an object for him, and is no longer perceived for itself; its point has become an area of sensitivity, extending the scope and active radius of touch, and providing a parallel to sight...If I want to get used to a stick, I try it by touching a few things with it, and eventually I have it 'well in hand', I can see what things are 'within reach' or out of reach of my stick. (PP, 144)

These examples show that the mastery and immediate understanding we have of our own bodies extends to clothes, tools, and machines that we use and that envelope us. Merleau-Ponty says that "habit expresses our power of dilating our being-in-the-world, or changing our existence by appropriating fresh instruments" (PP, 143). This "dilating" or extension of our bodily understanding into the world also happens whenever we use tools.

Gibson's discussion of tools reaches similar conclusions. He considers tools in terms of his theory of affordances. An affordance is a perceivable fit between an animal's abilities and the situation it is in. A tool is both something that has affordances, and alters the affordances available. You can pick up a corkscrew or a wine bottle; you can wield either; you can strike the table with either. But once picked up, the tool changes the affordances available to the user: once you pick up the corkscrew, you can use it to open the bottle of wine. You-plus-corkscrew have abilities, and hence there are affordances available to you-plus-corkscrew, that were unavailable to you alone.

> When in use, a tool is a sort of extension of the hand, almost an attachment to or a part of the user's own body, and thus is no longer a part of the environment of the user. But when not in use, the tool is simply a detached object of the environment, graspable and portable, to be sure, but nevertheless external to the observer. This *capacity to attach something to the body* suggests that the boundary between the animal and environment is not fixed at the surface of the skin but can shift. More generally it suggests that the absolute duality of "objective" and "subjective" is false. When we consider the affordances of things, we escape this philosophical dichotomy. (Gibson 1979, 41)

In the passages quoted in this section, Merleau-Ponty and Gibson describe what is now known as the hypothesis of extended cognition. "[T]he boundary between the animal and the environment is not fixed at the surface of the skin but can shift." In the final section, we will argue that their understanding of the hypothesis of extended cognition is superior to that of Clark and Chalmers.

## An alternative version of extended cognition

The key difference between the version of extended cognition from Gibson and Merleau-Ponty and the more familiar Clark and Chalmers version is over whether the nonbiological parts of the environment that extend cognition are perceived, or whether we *perceive with them* (Chemero 2015). As noted above, when notebooks and smartphones seem to extend cognition by serving as external, nonbiological memory stores, they must be perceived to do their job. Thus the opponent of extended cognition is open to arguing that it is the perceived smartphone, not the smartphone itself, that is part of the cognitive system. So it might be argued that nothing outside the biological body is part of the cognition, so the hypothesis of extended cognition is false.

Replacing Otto the notebook or smartphone user with the cane navigator as the central exemplar of the extended cognition avoids this objection. A cane navigator uses a cane to determine whether she can fit through the doorway. To do so, she explores the doorway by tapping along its interior edges with the cane. Doing so, she perceives the doorway's width with the cane, but she does not perceive the cane itself. As Merleau-Ponty suggests, she perceives the world at the end of the cane. Later on, when she is done with the doorway, she can stop using the cane to perceive the world, and can make the cane into her object of perception. If she is blind, she can explore it with her fingers. If she is a sighted experimental subject, she can also remove her opaque glasses and just look at the cane.

There are other advantages to adopting this version of the hypothesis of extended cognition. First, it preserves the key psychological idea that guides the extended cognition thesis. Humans are so cognitively flexible that they can even alter what they experiences as themselves, and they can do so quickly and more or less effortlessly. The blind person picks up the cane, and feels her boundaries expand. So the intuition that extended cognition is just a continuation of human behavioral and cognitive flexibility is preserved.

Another advantage of this version of extended cognition is that it is independent of philosophical functionalism. A blind person who is adept at using a cane to navigate does not perceive the cane, but the world at the end of the cane. When he or she does so, the cane becomes part of an extended cognitive system. Note that this has nothing to do with the functional role of the cane. Nothing in the brain of a person not navigating with a cane plays the same functional role as the cane does for a person navigating with a cane. Thus the Gibson and Merleau-Ponty version of the hypothesis of extended cognition does not depend upon the problematic assumptions of philosophical functionalism.

A third advantage of this alternative version of the hypothesis of extended cognition is that it is empirically tractable. There have been decades of research on sensory substitution. It has been shown repeatedly that people can adapt quickly to devices that allow them to perceive the world with nonbiological devices. There is no magic here. The blind person and her cane is a well-worn example, but there are many others. Amputees can perceive with prosthetics, and surgeons can perceive organs with laparoscopic tools. We can also perceive the ground through shoes and the road through a car or bicycle. We never say "The tire of the bicycle I was riding hit a bump;" we say "I hit a bump." It is continuous with the ability of both human and non-human animals to perceive the world by means of non-innervated appendages such as fingernails, claws, whiskers, antennae and quills. The only difference between whiskers and canes is that I can put the cane down. That hardly seems like a reason to disqualify one from being part of an embodied cognitive system, but not the other.

## References

Charles, E. (2012). *A New Look at New Realism: The Psychology and Philosophy of E. B. Holt.* Edison, N.J.: Transaction Publishers.

Chemero, A. (2009). *Radical Embodied Cognitive Science.* Cambridge: MIT Press.

Chemero, A. (2015). "The world at the end of the cane." In: *The Philosophers' Magazine* 68, 1, 84–87.
Clark, A. (1997). *Being There.* Cambridge: MIT Press.
Clark, A. and D. Chalmers (1998). "The Extended Mind." In: *Analysis* 58 (1), 7–19
Ehrenfels, M. (1890/1988). "On Gestalt Qualities." In: B. Smith (Ed. and Trans.). *Foundations of Gestalt Theory.* Munich and Vienna: Philosophia Verlag. (Original work published as "Über Gestaltqualitäten." In: *Vierteljahrsschrift für wissenschaftliche Philosophie* 14, 1890, 249–292)
Fancher, R. (1995). *Pioneers of Psychology*, 3rd edn. New York: Norton.
Fechner, G. T. (1860). *Elements of Psychophysics.* Sections VII ("Measurement of Sensation") and XVI ("The Fundamental Formula and the Measurement Formula") (trans. Herbert S. Langfeld, first appearing in B. Rand (Ed.) (1912). The Classical Psychologists).
Gibson, J.J. (1979). *The Ecological Approach to Visual Perception.* Boston: Houghton-Mifflin.
Haugeland, J. (1993). "Mind embodied and embedded." In: Yu-Houng H. Houng & J. Ho (Eds.). *Mind and Cognition: 1993 International Symposium.* Academica Sinica. 233–267.
Heft, H. (2001). *Ecological Psychology in Context: James Gibson, Roger Barker, and the Legacy of William James's Radical Empiricism.* Mahwah, N. J.: Erlbaum.
James, W. (1890). *The Principles of Psychology.* Boston: Henry Holt.
James, W. (1904a). "Does consciousness exist?" In: *Journal of Philosophy, Psychology, and Scientific Methods* 1, 477–491.
James, W. (1904b). "A world of pure experience." In: *Journal of Philosophy, Psychology, and Scientific Methods* 1, 533–543, 561–570.
Käufer, S. & Chemero, A. (2015). *Phenomenology.* London: Polity Press.
Koffka, K. (1923). "Perception: An Introduction to the Gestalt-Theorie." In: *Psychological Bulletin* 19, 531–585.
Köhler, W. (1913). "Über unbemerkte Empfindungen und Urteilstäuschungen." In: *Zeitschrift für Psychologie* 66, 51–80.
Leahey, T. H. (2000). *A History of Modern Psychology*, 3rd edn. New York: Pearson.
Mace, W. (2014). "Gibson in context." Presentation at the North American Meeting of the International Society for Ecological Psychology, Oxford, Ohio.
Mach, E. (1865). "On the effect of the spatial distribution of the light stimulus on the retina." In: F. Ratliff (1965). *Mach Bands—Quantitative Studies on Neural Networks in the Retina.* San Francisco: Holden-Day, 253–271.
Mach, E. (1886). *The Analysis of Sensations.* Trans. by C. M. Williams. La Salle: Open Court, 1984.
Merleau-Ponty, M. (1942/1963). *The Structure of Behavior.* Trans. Alden Fisher. Pittsburgh: Duquesne University Press, 1963.
Merleau-Ponty, M. (1945/2012, cited as PP). *Phenomenology of Perception.* Trans. Donald Landes. London/New York: Routledge.
Polger, T. W. (2012). "Functionalism as a philosophical theory of the cognitive sciences." In: *Wiley Interdisciplinary Reviews: Cognitive Science* 3 (3), 337–348.
Rupert, R. D. (2009). *Cognitive systems and the extended mind.* New York: Oxford University Press.
Sprevak, M. (2009). "Extended cognition and functionalism." In: *The Journal of Philosophy* 106 (9), 503–527.

Steiner, P. (2015). “A problem for representationalist versions of extended cognition.” In: *Philosophical Psychology* 28(2), 184 – 202.
Wheeler, M. (2010). “In defense of extended functionalism.” In: R. Menary (Ed.). *The Extended Mind.* Cambridge: MIT Press. http://hdl.handle.net/1893/11399.
Wundt, W. M. (1874/1902/1904). *Principles of Physiological Psychology* (Edward Bradford Titchener, trans.) (from the 5th German edn., published 1902; 1st German edn. published 1874) http://psychclassics.yorku.ca/Wundt/Physio/.
Wundt, W. M. (1912/1916). *Völkerpsychologie.* Trans. as: *Elements of Folk-Psychology.* by E. L. Schaub. London: Allen.

Pierre Steiner

# Embodied Cognitive Science, Pragmatism, and the Fate of Mental Representation

**Abstract:** How much was classical pragmatism compatible with anti-representationalism as it is for instance endorsed today by radical embodied cognitive science? And should contemporary pragmatism be anti-representationalist? The debate between representationalism and anti-representationalism is crucial when one wants to clearly define the originality and the scope of pragmatism within contemporary 4E (embodied, embedded, extended, enacted) cognitive science. After having distinguished some varieties of representations and, accordingly, some varieties of representationalism and anti-representationalism, this chapter reconstructs and critically considers original arguments *pro* and *contra* cognitive representationalism that can be found in three pragmatist authors: Rorty, Dewey and James.

## Introduction

In his 2009 book *Radical Embodied Cognitive Science*, Anthony Chemero argued that the time may have come for a new kind of cognitive science (whose name corresponds to the title of his book). Along with the idea that cognition should be described in terms of agent-environment dynamics, radical embodied cognitive science endorses a specific version of anti-representationalism. This is methodological anti-representationalism: according to this view, it is both possible and preferable to *explain* cognition, perception and action without appealing to mental representations. Chemero sees his anti-representationalism as a descendent of American naturalism, including James and Dewey's psychologies (Chemero 2009, 18–19).

Subsequently, in a 2013 paper, Teed Rockwell has explicitly criticized Chemero's anti-representationalism, by advocating the necessity of representationalism in order to account for Deweyan inquiry, and defined his concept of "mental representations" with reference to analog representations, in alleged continuity with James' radical empiricism. It is important to note that, in other contexts, Rockwell has been—and still is—an ardent defender of dynamical systems theory in the definition of cognition (Rockwell 2005). We thus have two philosophers who agree on the necessity of rejecting a classical (i. e. internalist and symbolic) model of cognition, who share a commitment to dynamical systems theory, and who each see their own work as continuous with some trends of classical pragmatism. However, they differ on the issue of representationalism: should the appeal to "mental representations" play any role in our best cognitive theories?

My goal here is not to settle or to refine the dispute between Chemero and Rockwell.[1] I rather take this recent dispute as an opportunity to reflect on the possible relations there might be between pragmatism and embodied cognitive science on the very issue of representationalism.

Extending the realizers of cognition out of the skulls of cognizers—in bodily engagement with the environment, in the manipulation of artifacts, in social interactions—has been a defining mark of the "4E" (embodied—embedded—enactive—extended) cognition research program, putting it in contrast with the main pillars of traditional cognitivist and connectionist frameworks. The rediscovery of the relevance of the pragmatist tradition for 4E cognitive science is now flourishing. Pragmatists such as Dewey are now often depicted as avant-garde or precursory critiques of classical cognitive science (Johnson 2006), as precursors of enactive and extended conceptions of cognition (Gallagher 2014), but also of situated cognition (Gallagher 2007), neurophilosophy (Solymosi 2011) and dynamical systems theory (Rockwell 2005, 177). And if that was not enough, their insights on the constitutive relations between perception, cognition and action would now be confirmed by empirical findings in computational neuroscience (Rockwell 2014). Still, the recent 4E wave in cognitive science is by no means monolithic. The issue of computation (is 4E cognition a turn or a reform within the computational tradition, or should it break with the very idea of computation?) is a decisive issue, and so is the issue of representation. Along with authors like Dan Hutto and Erik Myin (2013), Chemero (2009, 30) is one of the few that lucidly invite us not to lose track of the following situation: the main debates and tensions are perhaps not between "the" 4E wave and the basic tenets of cognitivism; they rather lie within the 4E wave itself, between for instance representationalist models that are in continuity with the computational theory, and more radical non-representationalist models (such as the ones defended by Chemero and Hutto & Myin's radical enactivism).[2] Can the resources of classical pragmatism make a difference in this debate?

Section I recapitulates what I take as basic clarifications if one wants to understand what the controversies between representationalism and anti-representationalism in cognitive science amount to. In section II, I investigate the scope

1 Chemero wrote a reply to Rockwell (in the same issue of the journal in which Rockwell [2013] was published), and Rockwell later replied to this reply (same journal, issue 61, 2014/1).
2 On this issue, see Steiner (2010) and Steiner (2015).

and the limits of Rorty's anti-representationalism if one wants to import it into the field of cognitive science. I then turn to the anti-representationalist commitments that one may find in Dewey's philosophy (section III), before finally turning to the relations between James' denunciation of the "psychologist's fallacy" and cognitive anti-representationalism. For reasons of space and clarity, other pragmatist variations around representation such as Peirce's semiotics,[3] Mead's theory of mind,[4] Putnam's realism or Brandom's inferentialism[5] will not be discussed here.

## 1 Mental representations and representationalism: some clarifications

Mental representations are classically defined as physical structures (vehicles) playing a role in cognitive processing in virtue of some content whose existence does not depend on the existence of social and linguistic practices and on the ability of the organism to take part in those practices. Typically, these mental representations consist in subpersonal and intracranial processes that are naturally or intrinsically contentful.[6] Note that according to this definition, external (public and/or shareable) representations such as models, images or natural language sentences are not mental representations: one can therefore be a critic of representationalism without denying the existence and the cognitive importance of external representations. From Peirce's semiotic model of the mind to Dewey's theory of inquiry and Mead, underlining the constitutive role for cognition of external representations such as symbols, pictures, charts, or spoken words was moreover a landmark of the classical pragmatist tradition.

A distinction should be made between two kinds of stances in the defence or in the rejection of the appeal to mental representations in cognitive science: an ontological stance, and a methodological stance. Ontological representationalism considers that mental representations exist, whereas ontological anti-repre-

3 Before being sometimes rediscovered or praised in the 4E paradigm, we should not forget that Peirce has been read and used in classical cognitive science, sometimes for refining representationalism (see Von Eckhardt 1995).

4 On this issue, see Madzia (2013).

5 But see Steiner (2014b) for the radical relevance of Brandom's inferentialism for 4E cognitive science.

6 A synonym for "made out of natural content" is "intrinsically having content": the possession of content by physical vehicles does not depend on the existence of linguistic, representational or symbolic human practices. "Intrinsically" does *not* mean here "non-relational".

sentationalism denies their existence. A methodological version is agnostic on the issue of the reality of mental representations: they are not taken as literally existing (we may not be warranted in positively asserting their reality). Still, methodological representationalism argues that mental representations (and their manipulation, production, or retrieval) must necessarily be posited for the explanation or prediction of the performances of cognitive systems; whereas methodological anti-representationalism argues that they do not need to be invoked (it is possible to posit them, but there are more helpful theoretical posits). Controversies on the very existence of mental representations are thus orthogonal to controversies about their explanatory need and relevance. Methodological anti-representationalism is not against the *existence* of mental representations: it may consider that the ontological debate between representationalism and anti-representationalism is vain, for instance because of a lack of clear definition of what a representational property is. In both methodological stances, nothing ontological is inferred from the presence or the absence of the concept "mental representation" in successful explanatory and predictive practices. Methodological versions of representationalism and anti-representationalism may be said to be anti-realist in the following sense: they deny that theories involving the positing (or the non-positing) of mental representations are truth-conditioned descriptions of their intended domain (observable and unobservable), and that their predictive or explanatory successes entail that the entities they posit have "real" counterparts. It is well known that there are different kinds of anti-realism in philosophy of science: instrumentalism, but also pragmatism, phenomenalism, interpretationism, constructive empiricism, fictionalism and idealism, to name but a few. Depending on the kind of anti-realist commitments one assumes when methodological representationalism is endorsed, mental representations can have different status: they can be conceived as models, fictions, useful falsehoods, explanatory tools, instruments of calculation, descriptive labels, etc. The fact that one posits mental representations—and not nails, flies or cucumbers—for achieving predicting and explanatory purposes can be explained (but not justified) by mentioning how it is often comforting to rely on representational systems such as language for modeling and defining thought and its intentionality (Sellars 1956, § 50–52, and § 57–58 is a central reference on this topic).

Be it ontological or just methodological, the debate between representationalism and anti-representationalism is not about the *role* of representation in the definition of cognitive phenomena, or about the *formats* and the *properties* of mental representations. A self-proclaimed defender of representationalism (like Andy Clark) may very well agree that

(a) cognition and cognitive phenomena such as perception, language-understanding or problem-solving are not representational, in the sense of being functionally isolated from action or from the active and embodied engagement of cognitive creatures in the world; and that

(b) cognition and cognitive phenomena such as perception, language-understanding or problem-solving are not representational, in the sense of consisting in the production, the manipulation or the retrieval of symbolic, abstract, action-neutral and detailed mental representations of the environment; but he may simultaneously claim that

(c) cognition and cognitive phenomena such as perception, language-understanding or problem-solving are representational, in the sense of involving the production, the manipulation or the retrieval of mental representations as contentful physical structures, whatever their formats and roles in cognitive processing.

(c) is representationalism (as I mean it here), and can receive an ontological reading, or a methodological reading.

These latter distinctions about claims concerning the role of representation in the definition of cognition, the formats of representations, and their (ontological or methodological) inclusion in cognitive theories are not always respected or clarified in the literature. For instance, in his 2009 book mentioned in the introduction, Chemero (Chemero 2009, 19–20) attributes anti-representationalist commitments to John Dewey's "The reflex arc concept in psychology", from the correct observation that Dewey, in this paper, argues that perception should not be defined or study as composed out of elements or entities such as stimuli or ideas that are independent of the activity of the organism, or as processes whose main business is to construct and use representations of the world. Those are thesis (a) and (b) defined above, but these theses do not entail a commitment to (c).

Analogously, a part/whole fallacy is committed when one believes that the criticism and the rejection of internalist and symbolic accounts of mental representations are criticism and rejection of representationalism *tout court*. Vehicle-internalist and cognitivist theories of cognition are not the exclusive owners of representationalism. On the contrary: many versions of extended, distributed or situated cognition may endorse representationalism. For most proponents of extended cognition, even if there may be cases or aspects of extended cognitive processing that do not (just) involve mental representations as made out of natural content and thus that do not require representationalist explanations, the existence of mental representations (realized in intracranial and subpersonal processes) and the necessity of referring to them when one tries to explain a great variety of cognitive phenomena are not at issue.[7] Unlike classical AI representations, these representations do not need to be complete, inert, propositional, denotational, action- and perception-neutral, stable, complex, detailed, digital, discrete, amodal, syntactically structured, or symbolic. Mental representations can be built and used on the fly; they can be modal (even when they are categorical), minimal (content-sparse), superposed, partial, action-oriented, context-dependent, embodied, analogue, distributed, or sub-symbolic.

It is not difficult to agree with John Shook and Tibor Solymosi that classical pragmatists "denied that cognition consists entirely of internal representations about static matters rationalistically manipulated within a Cartesian theatre" (Shook & Solymosi 2014, 2). But it remains to be seen whether this denial equates with a denial of representationalism *tout court*. Representationalism is a polymorphous and plastic thesis, sometimes looking like a Lernaean Hydra: shallow and maximalist versions of representationalism are easy to see and to dislodge, but only at the benefit of other versions which are much more tenacious, refined and pervasive—and that can be found in the works of pragmatism-inspired authors (see for instance Schulkin 2014).

The main claim of ontological anti-representationalism is that mental representations do not exist so that, by implication, cognitive processing does not involve the production or the use of mental representations. In their recent book *Radicalizing Enactivism. Basic Minds without Content*, Dan Hutto and Erik Myin have defended such a claim, as applied to basic cognition (typically exemplified in perceptual experience, sensori-motor coordination, reaching and grasping, or keeping track of another's gaze). Radical enactivism admits that enculturated or linguistically-scaffolded minds may involve or be informed by contents or representations, but not as made out of natural content (Hutto & Myin 2013, ix, xviii, 82). The authors do not only claim that it is possible to explain basic cognition without endorsing representationalism (a methodological claim): they put forward an ontological argument against representationalism, based on the current failures of the various philosophical attempts (Dretske, Millikan, Fodor, ...) to naturalize representational content. Since representational content has no place in a naturalistic ontology, there are good reasons to think it does not exist as an entity conveyed or produced by natural processes, including subpersonal and intracranial ones.

---

7 See for instance Clark (Clark 2008, 19, 26, 153); Rowlands (2010, chapter 5); Wheeler (2005, chapter 8); and Wilson (2004, chapter 8).

Methodological anti-representationalists will rather propose an explanatory strategy against representationalism, saying that it is not necessary to posit the existence of mental representations and to use the concept "mental representation" when one wants to explain the operations that make cognition possible. We have other explanatory resources which do not involve or presuppose an appeal to representational properties (Calvo Garzon 2008). In order to demonstrate its viability on a global or on a local scale, this explanatory strategy must provide a sufficient range of examples and applications. Its general (or local) scope will often be achieved by accumulating local explanatory or predictive successes. In any case, it is by its practical fruits that anti-representationalism must be considered and defended, not by the ontological criticism of representationalism.

Now that we have made these clarifications,[8] we can turn to the pragmatist tradition (or at least to some of its proponents). Pragmatism does not come with scientific realism, at least if the latter is defined as the claim that the primary function of scientific theories and models is to describe some reality that is what it is apart from human needs and interests, and that the unobservable entities that are posited in our best scientific theories really exist outside of those theories and of our scientific practices. It would therefore be quite dubious to attribute a defence of ontological representationalism to pragmatists. But the rejection of scientific realism does not entail the rejection of representationalism as a *methodological* claim. Still, are there good reasons to believe that pragmatism is compatible with methodological representationalism? And, more fundamentally, does the distinction between ontology and methodology make sense in a pragmatist framework?

## 2 Rorty and anti-representationalism: from persons to cogs and levers?

Perhaps more than any other pragmatists, Richard Rorty has developed a systematic critique of representationalism in epistemology and semantics. With anti-essentialism, anti-dualism and anti-foundationalism, anti-representationalism is a pillar of Rorty's pragmatist criticism of the philosophical tradition. Rorty sometimes defined representationalism as "the doctrine that our knowledge of the world passes through a representing medium which may or may not distort what the world is really like" (Rorty 2000b, 216). This understanding of represen-

8 See Steiner (2014) for more developed distinctions in the representationalism vs. anti-representationalism debate.

tationalism is continuous with its characterization by another contemporary pragmatist, namely Joseph Margolis. Margolis defines the representationalism criticized by pragmatism as follows: “the theory that whatever we may claim as knowledge of the world is invariably mediated by some direct, or more direct, awareness of the contents of the would-be cognizing mind” (Margolis 2007, 196).

Mental representations would be here epistemic intermediaries or mediations between the cognizing agent and the world. They would raise the veil-of-ideas problem: since we only know *X* by representations of *X*, do we ever access *X* epistemically? Direct or naïve realisms would provide an alternative to this representationalism (Putnam 2002), but they are not the positions that are favoured by Rorty. Indeed, Rorty prefers to attack and to reject the broader picture in which representationalism and the veil-of-ideas problem defined above are framed, and is led to define a new and broader sense of representationalism, encompassing the previous definition.

In this broader sense, representationalism is defined by two claims:

(1) The claim that there is an intrinsic nature of non-human reality, and thus a philosophical distinction between appearance and reality (Rorty 2007, essay 9; Rorty 1990a, introduction);

(2) The claim that the primary function of knowledge, belief and language is to represent such reality, so that the cognitive value of every belief, theory and assertion should be assessed with reference to its matching or correspondence with this reality.

Rorty sees Dewey and Davidson as the main sworn opponents to these two claims: Dewey because of his rejection of spectator accounts of knowledge, and Davidson because of his dismissal of correspondence theories of truth (and, more broadly, of his rejection of the representational functions of language and belief).[9] According to Rorty, anti-representationalism—as a rejection of this broader sense of “representationalism”—enables us to give up the reality-appearance distinction, and to replace it with a distinction between useful descriptions and useless descriptions of the world (Rorty 1991a, 4–5; Rorty 1991b; Rorty 1990b). In an anti-representationalist framework, a distinction between “things as they are” and “things as we describe them” is non-sense. The same verdict applies to questions like “are your representations exact?” or “do you reach the world as it is in itself?” (Rorty 1990a). Anti-representationalism does not re-

9 Today, a new pragmatist as Huw Price also defines pragmatism in terms of anti-representationalism as applied to language, and more precisely as opposed to “the idea that the function of statements is to ‘represent’ worldly states of affairs and that true statements succeed in doing so” (Price 2013, 24).

ject the existence or the relevance of representations as things we do and make (sentences, assertions, models, pictures, ...); it criticizes the idea that these representations ought to be defined and assessed from their ability to picture or to correspond with phenomena that would be epistemically independent from our representational practices, and yet epistemically accessible. Models, sentences, vocabularies or pictures answer to things in the world that are not representations, but this contact is not epistemic or semantic, it is just causal. Representation is not as an *epistemic* relation to something in the world (Rorty 2000a, 18). Similarly, knowledge is not a matter of getting reality right, but of acquiring habits of action for coping with reality (Rorty 1990a, I).

This broader Rortyan characterization of both representationalism and anti-representationalism reveals that Rorty's primary target, when it comes to mental representations, are not mental representations as theoretical entities posited in the context of scientific explanations, but mental representations as *epistemic* objects or relations defining the very possibility or purpose of personal-level phenomena such as perceiving, believing or knowing. Before and after Rorty, many philosophers, from John Dewey, Martin Heidegger, Ludwig Wittgenstein, Maurice Merleau-Ponty and Emmanuel Lévinas to Hubert Dreyfus, Charles Taylor, Hilary Putnam, Robert Brandom, John McDowell or Charles Travis have developed non-representationalist approaches of conceptual, perceptual, doxastic or epistemic intentionality. Their strategy consists in arguing that "mental representation" is the wrong conceptual unit for defining the relations which take place between epistemic (or perceiving, ...) agents and their environments. This strategy is based on considerations on the (alleged) nature of perception (if one endorses anti-representationalism for perception), knowledge or belief, and sometimes on the (alleged) nature of their intentionality. But these philosophical considerations on perception, knowledge or reasoning as *personal-level* phenomena do not seem to entail a rejection of representationalism as a definition of what makes these phenomena causally possible on a subpersonal level. On the contrary: representationalism, as a hypothesis concerning the existence of subpersonal (and possibly intracranial) mental representations endowed with natural content, may always be invoked for explaining for instance *how* personal-level perception, cognition or intentionality is for action or develops from practical coping or embodiment, without endorsing a metaphysical picture such as the inner/outer dichotomy, or copy-correspondence models of truth (see for instance Wheeler 2005; and Clark 1997, chap. 8).

This possibility of a philosophical rejection of representationalism coupled with an acceptance of cognitive representationalism can actually be found in some of Rorty's early writings. Let us consider for instance *Philosophy and the Mirror of Nature:* Rorty here defends representationalism (in cognitive psychology) against some of its early critiques, namely Gilbert Ryle and Norman Malcolm. As long as we posit and define mental representations apart from Cartesian and foundationalist epistemology, and thus in order to study exclusively the *causal* conditions of knowledge (that is, cognition), nothing should lead us to be intolerant towards the hypothetical constructs of the psychologist:

> the notion of psychological states as inner representations is unobjectionable but fairly uninteresting. To say that psychological states are states postulated to explain behavior, ones which we do not yet know how to identify with physiological states, is not to discover the true nature of the mind; it is only to reemphasize that there is no "nature" to be known (Rorty 1979, 242).

From an anti-realist point of view, representationalism in cognitive science is thus a way to criticize philosophical models of the mind: there is no essence or nature of mind or knowledge; there are just some vocabularies (as Rorty will later say), including the vocabulary of cognitive science and its concept of "mental representation". It would be a mistake to confuse this psychological concept with what essentialist or foundationalist philosophers mean by "mental representation".

Rorty's position rests on a basic distinction between what Ray Jackendoff later called "the psychological view" and the "philosophical view" (Jackendoff 1992, chap. 8). The *philosophical view* of the mind is concerned with a question such as "What is the relation between the mind and the world, such that we are able to know reality, such that our beliefs and desires relate to the world, and such that our statements may be true or false?". The *psychological view* attempts rather to answer a different question such as "How does the brain function as a physical machine, in such a way that the world can appear to us and that we can move in it?". Both these sorts of questions may presuppose and/or postulate mental representations in order to answer the questions. However in one case, the representationalist solution is weighed down by costly postulates, corresponding notably to the assumption on which the initial problems rest (separation mind/world; the idea of "making true"); in the other case, representations are posited in order to resolve the problem without making any *explicit* philosophical commitments. Of course one can and should question the plausibility and the relevance of this distinction as it was posed by Jackendoff in 1992: whether it is a question of the computational theory of mind or the 4E approaches to cognition, research programmes in cognitive science have been constituted on the basis of philosophical commitments which orient their various empirical studies. However, certain exceptions exist, and can illustrate—in accessory fashion—the distinction posed by Jackendorff, or at least one of its declensions when we address the question of representations.

For example, according to Jackendoff himself, psychological representations may not have referential or semantic properties: “a representation is not necessarily *about* anything; if you like, it does not strictly *represent* anything” (Jackendoff 1992, 160).

Since they do not represent anything, it is absurd, for Jackendoff, to ask whether these representations can be “true” or “false”—whether, for example, the representations that my auditory system builds and through which I auditively perceive the *Eroica* symphony are *true* or *false* (Jackendoff 1992, 165). The only “truth” or “falsehood” is relative to the conditions of the functioning of the system, and not in its relation to reality. Since they cannot be true or false, these representations cannot either, *a fortiori*, posses an epistemic or philosophic role in the justification of our beliefs, for example. For a long time now, Chomsky has also been convinced that folk and philosophical notions like “content”, “intentionality” and “reference” have no place at all in a naturalistic framework dedicated to the understanding of cognitive faculties. For Chomsky, mental representations are individuated by means of their role in cognitive processing. The functional roles of mental representations are here related to properties that have nothing to do with content, truth conditions, reference, or intentionality. Their important properties are formal or syntactic. These formal and syntactical properties are sufficient for individuating and studying the causal role of these representations in cognitive processing. These representations do not mean or represent anything; defining their reference is of no scientific interest:

> The internalist study of language also speaks of “representations” of various kinds, including phonetic and semantic representations at the “interface” with other systems. But here too we need not ponder what is represented, seeking some objective construction from sounds or things. [...] Accessed by performance systems, the internal representations of language enter into interpretation, thought, and action, *but there is no reason to seek any other relation to the world, as might be suggested by a well-known philosophical tradition and inappropriate analogies from informal usage.* (Chomsky 1995, 53, my emphasis)

“Informal usage”, here, means the very widespread tendency to embrace a linguistic model of mental representations, assuming they have semantic content or truth conditions, like daily linguistic products. Chomsky’s representationalism, here, wants to sever the ties between the explanatory posits of cognitive science and our folk understanding of what representations are, but considers that the very term of “representation” can still be applied for describing the natural, intracranial and subpersonal reality he is studying (even though its meaning, in scientific inquiry, is very different from our daily conceptions).

Let us take stock. The arguments put forward by philosophers that attack representationalist accounts of knowledge, language or perception from considerations on what knowledge, language or perception are do not seem to entail a rejection of representationalism as a scientific hypothesis on subpersonal cognitive processing. It is not a surprise that Rorty's first *Meisterwerk* includes both a criticism of representationalist-Cartesian epistemology *and* a minimal defence of representationalism as a scientific hypothesis. But this combination requires a distinction between philosophical views on mind, knowledge and representations, and scientific views on those phenomena. This distinction might be hard to find in the contemporary literature, but there are exceptions, most notably Chomsky. Up to now, we have not found any argument put forward by pragmatists against the existence and/or against the theoretical plausibility of mental representations, if representationalism—as a scientific hypothesis (ontological and/or methodological)—is kept distinct from philosophical issues.

# 3 Dewey's anti-representationalism: instrumentalism and meaning

Dewey's instrumentalist theory of knowledge was supposed to "give the *coup de grace* to *representationalism*", as Dewey wrote in a often quoted letter to C.A. Strong in 1905[10]—representationalism being here the idea that knowledge is a representation of some ready-made reality. But Dewey's instrumentalism should by no means be reduced to a claim on the nature of scientific theories. Dewey's criticism of what he named "spectatorial theories of knowledge" (LW4: 18) led him to defend some precursory version of extended cognition. Knowledge is here, first and foremost, an activity, "in which linguistic artefacts transact business with physical artefacts, tools, implements, apparatus".[11] This activity is thus by no means restricted to what happens in the head. On the contrary:

> Pendulums, lenses, prisms, yard sticks, and pound weights and multiplication and logarithmic tables have a great deal more to do with valid knowing, since they enable the organism to partake with other things in the effecting of consequences, than have bare consciousness or brain and nerves. (LW1: 261)

The tools of inquiry are, according to Dewey, all the items that contribute to the controlling of a situation and to the restoration of a transactional continuity between the organism and its environment. *With* idealism, Dewey emphasizes

10 Quoted in Menand 2001, 361.
11 Dewey, letter to Bentley, April, 9, 1951; quoted in Hickman 1990, 4.

the *creative* nature of intelligence;[12] but *against* idealism, he emphasizes that intelligence is embodied, practical and technical, and that experience is not limited to knowledge, ideas and judgements.[13] *With* realism, Dewey recognizes that inquiry involves relations of response with facts whose constraining dimensions are not the pure product of our imagination or our ideas.[14] An object is something that resists us; it can *object.*[15] The object is also what inquiries bear upon; the object is one of the objectives of inquiry.[16] But *against* realism, Dewey denies that the function of knowledge and, more generally, of the mind, is to represent a reality which is supposedly independent of experience, and that certain types of discourse (such as scientific theories) would be able to describe facts and reality *as they are in themselves*, by providing a privileged description of the world. Dewey also denies that objects exist independently of our operations, offering themselves up to our discovery. The subject of inquiry is a subject because he/she *subjects* her environment.[17] It is because the primary function of knowledge and intelligence is to make a difference in the world that it is vain to make reference to a reality that supposedly lies beyond experience, the latter being our "inclusive frame of reference"[18]—a reality whose access would supposedly guarantee the acceptability and the success of our propositions and our hypotheses. It is not only our material actions which change the world; judgements and hypotheses also help us to do something in the realm of experience. What they refer to renders us capable of doing something new, so that the world is constantly modified and constantly modifying itself.

In inquiry, there are no representations, there are rather things with *representative capacities or meaning* (LW12: 52). In order to function as a representation, a thing must become a mediator towards some purported result or some conclusion in the context of inquiry. This requires the thing to be public, objective and accessible. This does not require conventions or mental ideas, but agreement in action, as we will see below. Ideas do not have intrinsic representative capacities: they gain them in virtue of their operative functions in inquiry (LW12: 516). Cognitivist functionalism starts with mental representations, and defines mental states from the relations between these representations, perceptual entries and behavioural exits. Deweyan functionalism, in its critique of representa-

**12** LW4: 111.
**13** MW6: 86; LW1: 28.
**14** MW10: 341.
**15** LW1: 184.
**16** LW12: 122.
**17** LW1: 184.
**18** Dewey 2012, 329.

tive realism, derives the representative powers of ideas from transactional functions in which the organism and the environment are intertwined. Ideas are not intellectual entities, but anticipated consequences of certain operations (LW12: 113). Sensations and ideas are not distinct psychical existences or entities; they are functioning factors "within a single concrete whole", to use the terminology of Dewey's "The reflex arc concept in psychology".

But what about the existence or at least the explanatory relevance of mental representations, if we base ourselves on what contemporary scientific theories tell us about perception or reasoning? According to Svend Brinkmann,

> Dewey's psychology has space for the physiological and for the social but not for individual mental representations in between. An inner world of mental representations is mythical for Dewey, and he would thus be skeptical of the tendency to speak of inner representations of an outer world, for what value could such static representations have in a world undergoing constant change? (Brinkmann 2013, 71).

But here too, we must not mistake the part for the whole: how much does the rejection of internalism, of the inner/outer dichotomy and of spectatorial epistemology entail the rejection of every kind of representationalism? Concerning our best contemporary scientific theories, the anti-realistic commitments of Dewey are very clear:

> There is something both ridiculous and disconcerting in the way in which men have let themselves be imposed upon, so as to infer that scientific ways of thinking of objects give the inner reality of things, and that they put a mark of spuriousness upon all other ways of thinking of them, and of perceiving and enjoying them. It is ludicrous because these scientific conceptions, like other instruments, are hand-made by man in pursuit of realization of a certain interest–that of the maximum convertibility of every object of thought into any and every other. (LW4: 153)

The entities that are posited by our best theories should not be given an ontological value. Ontological representationalism would be a violation of this principle. But what about methodological representationalism? Would it be reasonable to entertain skepticism about an explicitly anti-realist appeal to the necessity of mental representations in successful psychological and neuroscientific theories? But do the predictive and explanatory successes of these theories entail that representationalism—only as a methodological hypothesis—is itself successful? The positing of hypothetical constructs does not require a defence of the existence of some entities, but it requires some further explanation of the processes from which these constructs are supposed to work. And it is here that we meet the issue of the problematic relations between mental representations and mental content, even if one endorses methodological representationalism.

If we follow Dewey, representational content is first and foremost a product of interpersonal and normative practices. If external representations are meaningful, it is not because of the intrinsic meaningfulness of some interpretative mental acts—on pain of regress or of falling into the traps of a mentalistic and foundationalist mythology:

> Failure to place the representative capacity of immediate qualities as signs and of meanings as possible significances in their context in inquiry leads to their being supposed to be psychical or mental existences, which are then endowed with the miraculous power of standing for and pointing to existences of a different order. (LW12: 516)

"The miraculous power of standing for and point to existences of a different order". Indeed. And this is a problem that even proponents of methodological representationalism should take at face value. It does not seem that representational content[19] exists at a natural and subpersonal level, so that—by implication —it would be reasonable to do *as if* (naturally) *contentful* physical structures exist. Can we use a (inter-)personal-level entity as the general blueprint for explanations or descriptions at the subpersonal level? It becomes a category mistake to think that natural and subpersonal phenomena of covariation or causal dependency could harbour or amount to representational or semantic properties. The latter properties are necessarily derived from linguistic and social practices, in which shareable structures (pictures, sentences, models, ...) can acquire a representational status in virtue of what rule-following agents do and must do with them. The methodological appeal to mental representations as made out of natural content is here *a priori* excluded from the stipulation of necessary conditions for the existence of semantic content. These conditions will never be satisfied by a brain (even if having a brain is a necessary causal condition for being able to participate to linguistic and social practices from which representational properties emerge).[20]

---

**19** In what follows, I will use "content", "representational content" and "semantic content" interchangeably. (Representational) content is the content of mental representations. In the representationalist literature (including the one that is discussed and criticized by anti-representationalism), this content has semantic properties: it can be true or false, correct or incorrect; it carries meaning (and not only information).

**20** In the introduction to his *Perspectives on Pragmatism* (Brandom 2011, 12–13), Brandom argues that the very idea of "subpersonal representations" should not be problematic to pragmatists, as long as these representations are not conceived as *propositional* representations (like personal-level representations), and are merely postulated for causally explaining practical abilities. By contrast, from Dewey's perspective, "subpersonal representations", even non-propositionally conceived, are problematic, since they are supposed to be *naturally contentful.*

But wait a minute. Are we not violating here the basic distinction between ontology and methodology? Why should concerns about the *nature* of content or meaning have some implication for the rejection of representationalism as a purely *methodological* stance? For Dewey's philosophy of science, the preference for some theoretical posits is never totally determined by predictive or explanatory successes as they would be defined in *autonomous* or *insulated* scientific practices. Prediction, control and explanation are embedded in human interests and, more broadly, in *experience*. Inquiry—including scientific inquiry—contributes to the development, to the transformation and to the enrichment of experience from the inside,[21] with the melioristic aim of turning us into better creatures—actors, citizens and knowers.[22] *If* experience fundamentally involves the construction, the modification and the sharing of meanings and values, and *if* we want to understand how it works so that we might make it better, how is it *enlightening* to refer to some basic and intrinsically contentful entities, even if they are not taken as literally existing? Are we just not displacing the problem? The type of living creatures we *are* (or want to be), and the inclusion of scientific inquiry in experience may here influence our rejection of some theoretical posits when we want to understand *how* we think, interact and reason from meanings and values.

For Dewey, what makes an entity become a linguistic vehicle (or a vehicle of meaning), is its use to point at something, to refer *to*..., on the basis of rules of interpretation that are public.[23] In truth, anything which has the potentialities of referral (and hence use) which are publicly recognized in a community of action can be said to be meaningful,[24] and is thus included one way or another in a language; a meaning never emerges without language,[25] because, more precisely, a quality can only become meaningful on the basis of an allocentered and normative relation between at least two persons (these are essential characteristic of the linguistic relation). "Meaning" is not limited to speaking or writing; everything which is employed as a sign, or better a symbol, is language:[26] gestures, images, artifacts, ceremonies, monuments, but also tools.[27] "Meaning" is the property of a transaction between two persons and an object, before subsequent-

---

**21** LW1: 128–131; LW1: 269; LW10: 90–91; LW12: 71–72.
**22** MW9: 236.
**23** MW9: 34.
**24** LW1: 143–148.
**25** LW1: 227.
**26** LW12: 27.
**27** LW1: 145–147; LW12: 48 sqq.

ly being a property carried by the object—linguistic understanding is fundamentally based on an intersubjective agreement in action.[28]

Thought, meaning and language cannot be dissociated for at least two reasons:

(1) The referential and inferential capacities of human thought arise and develop necessarily by the purport and manipulation of *external* symbols. If language is identified with speech or discourse, then, says Dewey, there is assuredly thought without language. But if language also includes all manners of signs or symbols, then there is no thought without language.[29] However thought cannot be identified or reduced to language:[30] thought and its qualitative dimensions can present a richness and a singularity which cannot always be translated by symbols.[31] To say that language is necessary for thought, is above all to insist on the fact that signs are necessary for thoughts to exist, because our thoughts do not deal with things, but with what those things mean, by virtue of their meaningful qualities: "The concern of thinking is with things as they carry the mind beyond themselves; they are vehicles not terminal stations" (LW6: 5). It is by means of these meanings that our thoughts can be related to what is *absent* and *abstract*. It is the public language which structures and enriches—directly, but also by allowing for the emergence of new capacities for thought—*our* capacities for thought, and not the inverse. The function of language is not to *express* a thought which would supposedly pre-exist (or to represent reality): language is rather that through which thought can have an object; it is what institutes the external meanings in virtue of which thought and thoughts can develop themselves. Language, as a regime of meanings, forms the objects of thought, but also makes certain forms of thought (certain capacities for using meanings) possible: it is a *mode of action* whose advantages, once they *can* be systematically exploited by certain creatures, are numerous for thought (silent judgements, capacities for memory, for inference, for abstraction, for anticipation, for meta-representation, for transmitting data, counterfactual reasoning concerning possible situations, remembering and anticipating[32] ...). The use of language as a means to carry and to acquire

**28** LW1: 141.
**29** LW6: 4–5; LW1: 134, 147.
**30** LW1: 134.
**31** LW5: 250.
**32** LW2: 331.

ideas is only an extension and a refinement of its initial capacity to make meanings emerge.[33]

(2) Thought, as we have seen, is not an internal semiotic commerce; the signs that it employs are external, as are the rules which govern their use (their meanings), which are defined institutionally. Consequently, the conditions of possibility and the exercise of the referentiality of thought are not to be sought in the internal acts which, for example, would supposedly have a foundational precedence over the intentionality of public language. It is rather the inverse: for Dewey, the nature of the referential properties of mental states is derived from semantic properties which are public, both for the definitional and genetic reasons mentioned above, and for reasons relative to the criteria by which our thoughts may be judged as correct (or not).

Because of the fact that meanings are public and shared, every use of the meanings must be accompanied by criteria as to whether they are correct, criteria which are also institutionally defined. Thought and reference, since they mobilize meanings, are normative activities because, fundamentally, "meanings are rules for using and interpreting things; interpretation being always an imputation of potentiality for some consequence" (LW1: 147).

A meaning *is* a rule for public use—which implies, in addition, that the acquisition of these norms of use which allow for the emergence of thought can only be relative to a social learning. The public norms which define the linguistic space are not magic: language, for Dewey, is a *mode of action* among others. Language emerges from previous forms of action and cooperation, which it has in return enriched[34] by transforming the objects of these forms of action into meaningful things. The order of meaning is in genetic continuity with physical and biological phenomena, given that it is, in fact, "these phenomena taken up and incorporated in a larger field of interactions".[35] This is the emergentism of Dewey.

I have just suggested here that Deweyan philosophy of language and, more broadly, meaning, would invite us to see the methodological appeal to meaningful intracranial entities as dubious. The linchpin of this critique of representationalism is based on considerations concerning the nature of meaning and content. Still, as I have argued in the previous section, there are some minor versions of representationalism that do not define mental representations in re-

33 MW9: 18 – 19.
34 LW12: 62.
35 LW3: 50.

lation with content. They consider that mental representations should not be modeled on cases of external representations, and thus do not entertain referential relations with their objects in virtue of some content (be it natural or not). A proponent of anti-representationalism might of course reply that one does not see the point of calling intracranial entities 'representations' if they do not have content, reference or intentionality. But the friend of "contentless" representationalism might answer that there may be scientific changes by which we *retain* terms even though the meanings of the latter ones are changing. An invitation to drop the term 'mental representation' just because of massive change of meaning presupposes that the term "mental representation" *has to* be connected to our daily, pre-scientific concept of "representation" (which essentially includes "content" in its intension), so that giving up this connection is renouncing to some intelligible term of "mental representation". But there is no *a priori reason* to think that concepts proposed in the context of scientific inquiry have to be derived or must respect the bounds of common sense (where representations have content). On the contrary : one may think that the autonomy of these concepts is the best guarantee for avoiding the fallacies correctly diagnosed and criticized by Dewey : using concepts such as "content" with their common sense meaning for describing and explaining intracranial and subpersonal processes, as so many forms of representationalism do.

The debate between cognitive representationalism and cognitive anti-representationalism now meets a broader issue in philosophy of science, dealing with the conditions in virtue of which some term does not refer to anything, rather than referring to something which is very different from what the descriptions associated with it prescribe. For reasons of space, I will not consider this debate here. I will rather have a look at James' philosophy of psychology in order to see whether it can offer us another argument for cognitive anti-representationalism.

# 4 James and the pragmatics of representationalist descriptions

Representationalism can be a methodological position, not an ontological thesis. But what do scientists (such as psychologists or neuroscientists) *do* when they posit mental representations in order to explain certain phenomena? In William James' *Principles of Psychology* one can find beautiful descriptions of the representationalist position adopted by the psychologist… and of its ambiguities. In particular, the description of what James calls the *psychologist's fallacy* finds its place in Chapter VII of the *Principles*, entitled "The methods and the traps of psychology".

For James, the question of method, or the *how*, is inseparable from an examination of the particularity of the posture of the psychologist, i.e. the *who*. The minds studied by the psychologist are *objects* which exist in a world containing other *objects*. Mental phenomena are, however, rather particular objects: they are indeed *about* other objects (James 1890, vol. 1, 190). Now the psychologist herself is also related to a world of objects which, in the context of a laboratory, are individuated and defined on the basis of the psychologist's knowledge and theoretical project. Her viewpoint is thus a very particular one: it bears both on subjective phenomena (thoughts, sensations) which are about objects, and on worldly objects, which are specific to her perspective. The objectivity of thought is obviously not *of the same order* as the objectivity deployed when the psychologist objectifies a mental state on the basis of her own perspective: without for all that making the objective of her thought a *subjective* fact, it is to be noted that the aspectuality of the objects of a thought is not necessarily identical to the aspectuality of these "same" objects as seen and described by the observer. Given that the psychologist has as an object the description of a mental state and the object of the latter, and that she also entertains *a* relation with "*that*" object, she can surreptitiously and involuntarily introduce into the mental state of the subject her own knowledge of the object, and thus describe the mental state as being conscious of the object in the same way that she herself (the psychologist) is. To commit the psychologist's fallacy is to confuse the viewpoint of the mental state concerning its object (and, by extension, the viewpoint that the subject under study has about the world) with the viewpoint of the psychologist about the "same" object and, more generally, about the mental state under study. What the psychologist knows about the substance of thought, about its genesis, about its object, about the relation of thought to its object as she observes it, about the causal origin of thought, and its role in the mental life of the subject, all this is included in the content of thought and in particular in its mode of intentionally aiming at an object.

James describes this drama as follows:

> The *great* snare of the psychologist is the *confusion of his own standpoint with that of the mental fact* about which he is making his report. I shall hereafter call this the 'psychologist's fallacy' *par excellence*. For some of the mischief, here too, language is to blame. The psychologist, as we remarked above, stands outside of the mental state he speaks of. Both itself and its object are objects for him. Now when it is a *cognitive* state (percept, thought, concept, etc.), he ordinarily has no other way of naming it than as the thought, percept, etc., *of that object*. He himself, meanwhile, knowing the self-same object in *his* way, gets easily led to suppose that the thought, which is *of* it, knows it in the same

> way in which he knows it, although this is often very far from being the case. (James 1890, vol. 1, 196, author's italics).

The psychologist's fallacy, as James diagnoses it here, has a number of major variants (James 1890, vol. 1, 196–197), but hereafter we will concentrate on the following one: The fallacy can occur when one considers that the knowledge carried by mental states about the states of the world, is identical to the knowledge that the psychologist can have about these states of the world; and in particular, that the objects of the experience of the subject are identical to the "same" objects as they are defined and individuated by the psychologist. Because the subject and the psychologist use the same word to speak about an object, or because the psychologist names the object of the mental state of the subject by using the name that she herself gives to that object, one ends up supposing that the object of the two mental states is the same, and thus that the experienced relation, the mode of access or the way of intentionally aiming at the object are likewise identical.

The classical theory concerning the perception of space is, for James, a prime example of this sort of fallacy. For James, our relation to space is a direct relation, a sort of familiarity that is non-conceptual (*knowledge by acquaintance*); this relation does not imply, in order to it to be effective, the deployment of conceptual theoretical knowledge *about* this space (*knowledge about*), an unconscious exercise which may for example take the form of operations of synthetic association of parts, of inferences from the simple to the complex by way of algorithms (James 1890, vol. 2, 281–282). To suppose that the direct experience of an ordinary subject concerning space and the vague whole which constitutes it, mobilizes a theoretical knowledge having as its object the assembly of parts and positions in this space, is to confuse what a scientist knows about space by means of abstraction and theoretical analysis, which the way in which ordinary everyday thought relates directly to space.

The psychologist's fallacy can thus consist in supposing that a thought must have the properties of the object it is aiming at. By naming a thought, by referring to the object(s) of that thought, we are inclined to think that the thought itself must greatly resemble its object; in elementary terms, then, that the thought must be *something like an object*. Like the environment, consciousness is considered as being made up of objects, of elements, of mental cogs or atoms, whose relational properties are artificial (they are postulated by a synthetic activity) and/or purely intellectual. Thought comes to be considered as a succession of units of thought, built on the basis of a set of sensations or ideas which are discrete, independent and stable, which can be repeated and recombined indefinitely in all sorts of complex combinations on the basis of an elementary principle of association (James 1890, vol. 1, 553–554). In this way one ends up with the *myth of mental objects* (James 1890, vol. 1, 236). In his criticism of this myth, James denounces in particular the empiricist psychologies of Locke and Hume, and the analytical and elementary psychology of Herbart, which all adhere to this atomistic conception of thought. All that is at stake in the chapter "The stream of thought" consists, as is well known, in overcoming this conception; it is a question of giving its rightful place to an image of thought as a stream, as a flow, where relations and continuity between moments of thought are real and primary.

There are several ways of defining the scope of James's denunciation of the psychologist's fallacy. An advocate of representationalism might accept that it is important, when postulating the existence of mental representations, to avoid defining their object or their content on the basis of the confusions identified by James. The vehicle/content confusion is a contemporary variant of James' fallacy (Millikan 1993). But the observations of James can also be used to ground an anti-representational stance: the psychologist's fallacy is already at work when one postulates the existence of mental representations. Because the observer mistakes the perspective of the brain for his very own perspective, he puts in the brain contents that only exist at some linguistic and (inter)personal level, *and turns these contents into natural and/or explanatory entities.*

This anti-representationalist reading of James echoes with a critique of representationalism developed by Maturana and Varela in the early 80's. Maturana and Varela strongly criticized the conflation which is too often made between the *operational* or *mechanistic* perspective of the Central Nervous System (inside which mental representations are supposed to be found, be it at a neural level or at a functional level) and the *observer-dependent* perspective of the theorist, considering from the outside (and in its relations with the environment) the organism in which the brain takes place. These three perspectives (the brain, the organism in which the brain is located, the observer) are not all congruent in their relations with the environment, and in their causal mechanisms. True, the observer can see that a set *Y* of neurons enables some organism (*O*) to deal with *X* (external state of affairs). But, for one thing, *O*'s brain (in which *Y* occurs) does not have with *X* the same relations as the ones the observer and *O* have with *X*. The nervous system does not have access to the correlation the observer constructs between parts of itself and the world. *We* make semantic projection onto neural signals by looking at *O*, *X* and *Y* from *our* linguistic and perceptual perspective—neither from God's eye point of view, nor from the brain's point of view; it is dubious whether brains (not to mention God) might have *point of views* (for brains do not *see*—only organisms do). For Maturana and Varela, the brain is not a solipsistic engine: it is an interacting part of the organism, and plays a role in the structural coupling of the organism with the environment. But it is not a *representational* engine: it does not relate to the environment as we do, or as machines that we have conceived do, in virtue of information and instructions (Maturana and Varela 1987, 169; Maturana 1978, 50; Maturana 1972, 23)

Pragmatically, what is happening when one observer (a scientist) is using representationalism? We have a human person, the observer, facing an organ (the brain), or representations or models of the brain (images, data, ...). The observer entertains relations with the environment: he sees, feels, touches or converses *about* objects or states of affairs in the environment. When he converses about objects with his colleagues, he is producing and acting from linguistic contents, said or written in utterances. These linguistic contents notably exist in virtue of linguistic rules and conventions. The observer may believe that the brain he is observing entertains the same type of relation to the environment as the relations that he has, so that the brain would have cognitive relations with the objects the observer interacts with, by using or producing contents. But this is an illusion: unless one is under the grip of the prejudice that he/she is what his brain is or is doing, there is no reason to think that the operations of the brain—a subpersonal organ—, and its relations to the environment are *like* our semantic operations and our cognitive relations with that environment. True, the brain plays a crucial causal role in the production of our cognitive and semantic behaviours, but that does not entail that it *harbours* all the dimensions of this cognitive and semantic behaviour, including its objects (as represented) and the linguistic contents from which one may think about something. Representationalism consists here in the mistake of thinking that the representationalist idiom (and especially the notion of "content") is and must be something more than an idiom that can be trivially used for very metaphorically (and grossly) making sense of the brain: it would be the only or the best way of doing so, because the brain really harbours the units of this idiom, or would otherwise be unintelligible to us without the use of this idiom; and these intracranial units we posit would represent the environment as *we* know or interact with it.

## Conclusion

The readings and uses that can be made, in the philosophy of cognitive science, of the anti-representationalist pragmatists are diverse. One might think that the adoption of a global anti-representationalism concerning language or knowledge naturally obliges one to adopt an anti-representationalist stance, ontological or methodological, in the philosophy of cognitive science. But this in no way implies that this global anti-representationalism can *justify* the rejection of a methodological representationalism, for example. The case of Rorty is exemplary on this point. Notwithstanding, one can find in Dewey and in James elements for criticizing the adoption of a methodological representationalism: on the basis of considerations concerning the nature of meaning (in Dewey), or considerations concerning the singular posture of the scientist who observes and models the mind and the brain (James). Unfortunately, these critical elements are not often evoked today, including in the context of debates within the 4E cognitive science community. What pragmatism brings to us concerning action, sociality or the technical nature of cognition is not sufficient to produce a precise criticism of non-simplified versions of cognitive representationalism, but that should not worry us: as I have sketched out, there are other resources, *within pragmatism itself*, for addressing the question of mental representation.

## References

Brandom, R. (2011). *Perspectives on Pragmatism. Classical, Recent, and Contemporary*, Cambridge, MA/London: Harvard University Press.

Brinkmann, S. (2013). *John Dewey: Science for a Changing World*. Transaction Publisher.

Calvo Garzon, F. (2008). "Towards a general theory of antirepresentationalism." In: *British Journal for the Philosophy of Science* 59, 259 – 292.

Chemero, A. (2009). *Radical Embodied Cognitive Science*. Cambridge, MA/London: MIT Press.

Chomsky, N. (1995). "Language and nature." In: *Mind* 104 (413), 1 – 61.

Clark, A. (1997). *Being There. Putting Brain, Body and World Together Again*. Cambridge, MA: MIT Press.

Clark, A. (2008). *Supersizing the Mind: Embodiment, Action, and Cognitive Extension*. New York: Oxford University Press.

Dewey, J. *The Middle Works: 1899 – 1924*. ed. by J.A. Boydston. 15 vol. Carbondale & Edwardsville: Southern Illinois University Press, 1976 – 1983 (quoted "MW" followed by the volume and the pages numbers).

Dewey, J. *The Later Works: 1925 – 1953*. ed. by J.A. Boydston. 17 vol. Carbondale & Edwardsville: Southern Illinois University Press, 1981 – 1990 (quoted "LW" followed by the volume and the pages numbers).

Dewey, J. (2012). *Unmodern Philosophy and Modern Philosophy*. ed. by Ph. Deen. Carbondale: Southern Illinois University Press.

Gallagher, S. (2008). "Philosophical antecedents to situated cognition." In: P. Robbins & Aydede, M. (Eds). *Cambridge Handbook of Situated Cognition*. Cambridge, Cambridge University Press, 35 – 51.

Gallagher, S. (2014). "Pragmatic interventions into enactive and extended conceptions of cognition." In: *Philosophical Issues* 24, 110 – 126.

Hickman, L. (1990). *John Dewey's Pragmatic Technology*. Bloomington: Indiana University Press.

Hutto, D. & Myin, E. (2013). *Radicalizing Enactivism. Basic Minds Without Content.* Cambridge, MA/London: MIT Press.
Jackendoff, R. (1992). *Languages of the Mind.* Cambridge, MA: MIT Press.
James, W. (1890). *The Principles of Psychology.* Volumes 1 and 2. New York: Dover.
Johnson, M. (2006). "Cognitive science." In: J. Shook & J. Margolis (Eds.). *A Companion to Pragmatism.* Malden/Oxford: Blackwell, 369–377.
Madzia, R. (2013). "Chicago pragmatism and the extended mind theory: Mead and Dewey on the nature of cognition." In: *European Journal of Pragmatism and American Philosophy* V, 1, 193–211.
Margolis, J. (2007). *Pragmatism without Foundations.* second edition. New York/London: Continuum.
Maturana, H. (1972). "Biology of cognition." In: H. Maturana & F. Varela. *Autopoiesis and Cognition. The Realization of the Living.* Dordrecht/Boston: Reidel, 1–58.
Maturana,H. (1978). "Biology of language: the epistemology of reality." In: G. Miller & E. Lenneberg (Eds.). *Psychology and Biology of Language and Thought: Essays in Honor of Eric Lenneberg.* New York: Academic Press, 27–63.
Maturana, H. & Varela, F. (1987). *The Tree of Knowledge. The Biological Roots of Human Understanding.* Boston and London: Shambhala.
Menand, L. (2001). *The Metaphysical Club. A Story of Ideas in America.* New York: Farrar, Straus and Giroux.
Millikan, R. (1993). "Content and vehicle." In: N. Eilan, R. McCarthy & B. Brewer (Eds.). *Spatial Representation.* Oxford: Blackwell, 256–268.
Price, H. (2014). *Expressivism, Pragmatism and Representationalism.* Cambridge University Press.
Putnam, H. (1999). *The Threefold Cord: Mind, Body, and World.* New York: Columbia University Press.
Rockwell, T. 2005. *Neither Brain nor Ghost. A Nondualist Alternative to the Mind-Brain Identity Theory.* Cambridge (MA): MIT Press.
Rockwell, T. (2013). "Representation and radical empiricism." In: *Intellectica* 60, 219–240.
Rockwell, T. (2014). "How computational neuroscience revealed that the pragmatists were right." In: T. Solymosi & J. Shook (Eds.). *Neuroscience, Neurophilosophy and Pragmatism. Brains at Work with the World.* Palgrave MacMillan, 57–69.
Rorty, R. (1979). *Philosophy and the Mirror of Nature.* Princeton University Press.
Rorty, R. (1990a). *Objectivity, Relativism and Truth. Philosophical Papers vol. 1.* Cambridge, UK/New York: Cambridge University Press.
Rorty, R. (1990b). "Pragmatism as anti-representationalism." Introduction to: J.P. Murphy. *Pragmatism. From Peirce to Davidson.* Boulder/San Francisco/Oxford: Westview Press, 1–6.
Rorty, R. (1991a). *Essays on Heidegger and Others. Philosophical Papers vol. 2.* Cambridge, UK/New York: Cambridge University Press.
Rorty, R. (1991b). "Pragmatismo." In: M. Carrilho (Ed.). *Dicionario do Pensamento Contemporâneo.* Lisbon: Don Quixote, 265.
Rorty, R. (2000a). "Universality and truth." In: R. Brandom (Ed.). *Rorty and his Critics.* Malden/Oxford: Blackwell, 1–30.
Rorty, R. (2000b). "Response to Michael Williams." In: R. Brandom. *Rorty and his Critics.* Malden/Oxford: Blackwell, 213–219.

Rorty, R. (2007). *Philosophy as Cultural Politics. Philosophical Papers vol. 4.* Cambridge, UK/ New York: Cambridge University Press.
Rowlands, M. (2010). *The New Science of the Mind. From Extended Mind to Embodied Phenomenology.* Cambridge, MA/London: MIT Press.
Schulkin, J. (2014). "Pragmatism, cognitive capacity and brain function." In: T. Solymosi & J. Shook (Eds.). *Neuroscience, Neurophilosophy and Pragmatism. Brains at Work with the World.* Palgrave MacMillan, 71–102.
Sellars, W. (1956). "Empiricism and the philosophy of mind." In: H. Feigl & M. Schriven (Eds.). *Minnesota Studies in the Philosophy of Science, Volume I: The Foundations of Science and the Concepts of Psychology and Psychoanalysis.* University of Minnesota Press, 253–329.
Shook, J. & Solymosi, T. (2014). "Introduction." In: J. Shook & T. Solymosi, T. (Eds.). *Pragmatist Neurophilosophy. American Philosophy and the Brain.* London/New Delhi/New York/Sydney: Bloomsbury, 1–9.
Solymosi, T. (2011). "Neuropragmatism, old and new." In: *Phenomenology and the Cognitive Sciences* 10, 347–368.
Steiner, P. (2010). "The bounds of representation. A non-representationalist use of the resources of the model of extended cognition." In: *Pragmatics & Cognition* 18 (2), 235–272.
Steiner, P. (2014a). "Enacting anti-representationalism. The scope and the limits of enactive critiques of representationalism." In: *Avant. Trends in Interdisciplinary Studies* 5 (2), 43–86.
Steiner, P. (2014b). "The delocalized mind. Judgments, vehicles and persons." In: *Phenomenology and the Cognitive Sciences* 13 (3), 437–460.
Steiner, P. (2015). "A problem for representationalist versions of extended cognition." In: *Philosophical Psychology* 28 (2), 184–202.
Von Eckhardt, B. (1995). *What is Cognitive Science?* Cambridge, MA: MIT Press.
Wheeler, M. (2005). *Reconstructing the Cognitive World. The Next Step.* Cambridge, MA: MIT Press.
Wilson, R. (2004). *Boundaries of the Mind: The Individual in the Fragile Sciences.* Cambridge, UK: Cambridge University Press.

# Part II: **Embedding Embodied Cognitive Science: A Larger Picture**

Mark Johnson

# Pragmatism, Cognitive Science, and Embodied Mind

**Abstract:** Most of the first-generation cognitive science that emerged in the second half of the twentieth century was founded on assumptions about mind, meaning, thought, and language that were at odds with a classical Pragmatist philosophical perspective. However, today there is a rapidly developing second-generation cognitive science that is providing ample scientific support for some of the central tenets of pragmatist conceptions of cognition and language. These shared views start with a naturalistic, anti-dualistic, and yet non-reductive conception of mind. From this basis they develop an account of experience, meaning, thought, and communication that recognizes the central role of the body in shaping everything we experience, think, and do. The sciences of the embodied mind are allowing us to give detailed accounts of the bodily basis of meaning and thought, while a general pragmatist orientation allows us to identify both the limitations and important implications of this mind research.

Pragmatism and cognitive science have had a rocky, but steadily improving relationship. They were like two characters in the standard plot line of so many romantic comedies, who start off intensely disliking each other, noticing each other's faults and disagreeable traits. Eventually, however, fate brings them together, the sparks begin to fly, and they realize that they had a deep connection all along, though they were both blind to it. The early days of their relationship were fraught with suspicion, misunderstanding, and even hostility. The cognitive scientists ignored pragmatist philosophy, and the pragmatists either ignored or actively criticized the sciences of mind. The reason for this was that, in its early stages of development—its adolescence—cognitive science was grounded on assumptions that were mostly incompatible with classical pragmatism. The situation was made even worse by the fact that many pragmatist philosophers were wary of scientism, fearing that scientific theories of mind were bound to be overly reductionist in spirit, thereby failing to capture the depth and fullness of mind, thought, and language. Consequently, until recently there has been very little recognition of the great potential for a mutually beneficial relationship between pragmatists and cognitive scientists. Or, to put it in today's somewhat pathetic romantic vernacular, it has only been in the last few years that the two have been able to "hook up".

Such a suspicion of cognitive science among pragmatists is ironic, given that Charles Sanders Peirce, William James, and John Dewey each championed the integration and co-evolution of philosophy and the sciences. These classical pragmatists made extensive use of the best available biological and cognitive science of their day in developing their views of mind, knowledge, and value. If they were alive today, I have no doubt that they would appreciate the importance of cognitive science research for the philosophy of mind and language. I propose to explore some of the shared views that reveal the promise for a creative relationship between pragmatism and contemporary cognitive science.

## 1 Different conceptions of cognitive science

So, what was it that kept the future lovers feuding for so long? What kept pragmatists and cognitive scientists apart until now? The key to the answer is that what George Lakoff and I (1999) have called "first-generation cognitive science", which arose in the 1940's and developed rapidly through the 1960's and down to the present day, was founded on a mostly disembodied notion of mind and thought. This disembodied view of cognition was incompatible with pragmatism's anti-dualist, embodied, and emergentist perspective. It was not until the 1970's that an alternative, second-generation "embodied" cognitive science began to take shape, providing experimental support for a view of experience, consciousness, mind, meaning, and thought that was compatible with mainstream pragmatist philosophy.

First-generation cognitive science was a blending of analytic philosophy of mind, information-processing psychology, generative linguistics, model theory, computer science, and research in artificial intelligence. The orientation was functionalist, insofar as it took mental operations to be formal algorithmic operations on abstract meaningless symbols that could supposedly be run on any of a number of suitable hardwares (machines) or wetwares (biological organisms). Many functionalists are materialists who do not think "mind" and "body" are two separate ontological realities, but their view can still be labeled "disembodied", insofar as they think that an account of mental operations does not depend on the particular makeup of the body or other material that instantiates the functional program of mind. Consequently, they do not think that the very structure of our thought and reasoning are determined by the nature of our bodies and brains. Furthermore, most functionalists assume that sentence-like propositions are the core of meaning and thought, and so they employ formal logic, propositional attitude theory, and speech act theory to represent the range of possible mental states and operations.

It is no surprise, therefore, that pragmatists tended to dismiss first-generation cognitive science, mostly because of its reductionist and disembodied character, while cognitive scientists of this persuasion returned the favor by either ignoring or rejecting what they perceived to be just one more speculative and scientifically unsupported philosophical world-view. To return to my romantic comedy analogy, we would say that, in their mutual ignorance and disdain, the two partners were all too ready to point out each other's manifold faults, while remaining blind to their respective virtues.

Fortunately, by the mid-1970's, the plot took an unexpected turn. One of the characters began to grow and change. Cognitive science underwent a make-over of substantial proportions. This new second-generation, embodied cognitive science was naturalistic, anti-dualist, emergentist, and aware of the need for multi-level explanatory frameworks—just like pragmatism! A number of scientific research programs began to discover the importance of the body in the constitution of mind and thought. Research in cognitive psychology was revealing how human conceptual systems were dependent both on the nature of our bodies and brains, and also on the nature of our environments (Lakoff 1987; Barsalou 1999). The new orientation known as Cognitive Linguistics was showing how meaning is grounded in our sensory, motor, and affective processes and how most of our abstract concepts are metaphorical (Lakoff & Johnson 1999; Gibbs 2006; Feldman 2006). Cognitive neuroscience was growing up, producing computational models of embodied cognitive operations, whereby sensory-motor processes are recruited for so-called "higher" cognitive functions. (Feldman 2006; Lakoff & Narayanan, in press). But the growth and change was not just one-sided. Pragmatism, too, has grown over the last decade to appreciate the need to base our accounts of mind, thought, and language on the best empirical scientific work available (Solymosi & Shook 2014). Although most of the people doing the scientific research and modeling are still today largely ignorant of pragmatist philosophy, the convergences between second-generation cognitive science and pragmatism are quite striking.

Let us explore some of the more important shared perspectives that speak for the possibility of a rich and very constructive ongoing dialogue between embodied cognitive science and pragmatism.

## 2 Areas of convergence between second-generation cognitive sciences and classical Pragmatism

### Naturalistic, multi-level methods for the study of mind

By a naturalistic perspective on mind, I mean the view that what we commonly call "mental" phenomena, as well as all cultural phenomena, are part of nature. Naturalistic approaches typically share a materialist ontology, and so they deny the existence of supernatural and transcendent entities, agencies, and causes. Consequently, the empirical methods of inquiry utilized by the natural sciences are deemed appropriate and necessary for the study of all aspects of human cognition and symbolic interaction.

However, all naturalisms are not the same. There are differences concerning which methods of explanation are able to provide an adequately rich account of mind, thought, and language. For example, *strong reductionist naturalism* claims that in a suitably mature science of mind (which does not yet exist) all of the relevant phenomena will be explained using only assumptions and methods employed in a single discipline, such as bio-chemistry, or, perhaps ultimately, physics. Eventually, so the story goes, all our "higher level" functions and processes will be reduced to events and causes of the sort investigated in the physical sciences.

There is an alternative, less reductionist, conception of naturalism that recognizes that a fully satisfactory theory will ultimately employ multiple levels of explanation, where each form of explanation is appropriate for a particular emergent level of functional organization within a creature in ongoing interaction with its environments. This second version might be called a *pluralistic naturalism*. For example, William Bechtel (2009) argues that mechanistic explanations in the sciences of mind will be generally reductionist in spirit, but they will, where necessary, recognize the need for multiple levels of explanation. He observes that, since mechanisms consist of parts working together to produce whole functional units, reduction to the operations of the parts will be a key element for explaining the behavior of those wholes. However, these various functional wholes are sometimes part of larger, more extensive systems, and therefore each of those larger wholes will be caught up in broader causal patterns of interaction that may not be operative at the lower level of the individual parts. Consequently, although there will be some inter- and intra-level reduction possible, an adequate explanation of complex cognitive and affective phenomena will involve multiple levels of organization and will require different explanatory models and methods to capture the fullness of the phenomena being explained. Bechtel summarizes:

> [...] [A] final feature of mechanistic explanation that should be noted is that, insofar as it emphasizes the contributions made by parts of a mechanism to its operation, a mechanistic analysis is in an important sense, reductionist. However, insofar as it also recognizes the importance of the organization in which the parts are embedded and the context in which the whole mechanism is functioning, it not only sanctions looking to lower levels but also upward to higher levels. [...] Thus, far from sanctioning only a focus downward on the components, a mechanistic perspective as well requires an account of engagements with other systems at the same level and, potentially, of the constraints imposed by being incorporated into a higher-level mechanism. In fact, therefore, the mechanistic perspective is inherently multilevel (Bechtel 2009, 21–22).

Bechtel's focus is primarily on the nature and operations of mechanistic models, and the methods he countenances are mostly those of various sciences (e.g., physics, chemistry, molecular biology, dynamical systems theory, etc.). Pragmatists would insist that the natural sciences are not the *only* appropriate methods of inquiry, because other critical and interpretive methods may also prove necessary to explain the full range of cognitive phenomena at different levels of organization and complex interaction. Consequently, we would expect some methods to be appropriate at the cellular level, others required for an adequate explanation of the autonomic nervous system, others fitting for the whole-body organization of sub-systems, and still different methods necessary for a complete explanation of social and cultural phenomena.

This broader pluralistic conception of naturalism is shared by many cognitive scientists and nearly all pragmatists. It is naturalistic in the sense that it understands humans to be complex, highly evolved, embodied dynamical systems in ongoing interaction with their environments. It is pluralistic insofar as it sees increased complexity of organism-environment interactions as giving rise to emergent qualities and functions, and thus requiring multiple explanatory models that are adequate to each of the emerging levels of organization.

## The central importance of continuous organism-environment interactions

The view I am developing here is fundamentally dependent on the idea that the locus of all experience, meaning, thought, communication, value, and action is the ongoing interaction between an organism and its environments. *You can only think what your brain, body, and environments (physical and cultural) allow you to think. You can only mean and communicate what your brain, body, and environments (physical and cultural) allow you to mean and communicate. You can only know what your brain, body, and environments (physical and cultural) allow you to know.* Consequently, any view that assumes a fundamental ontological dualism that locates mind, thought, meaning, or value as merely *within the subject* and places "the world" (or any subset thereof) as *outside* mind is already misguided from the start.

It is very tempting to regard mind as consisting of a set of pre-given capacities whose structures and formal operations are not determined by the environments in which they operate. According to this view, the environment might supply the content, but not the structures, of cognitive operations. In *The Principles of Psychology* (1890), William James presciently warned against this temptation as follows:

> Mental facts cannot be properly studied apart from the physical environment of which they take cognizance. The great fault of the older rational psychology was to set up the soul as an absolute spiritual being with certain faculties of its own by which the several activities of remembering, imagining, reasoning, willing, etc. were explained, almost without reference to the peculiarities of the world with which these activities deal. But the richer insight of modern days perceives that our inner faculties are adapted in advance to the features of the world in which we dwell, adapted, I mean, so as to secure our safety and prosperity in its midst. (Psychology, Briefer Course, 3)

Those scientists and philosophers who began to realize that organisms are clusters of interrelated habits of perception, movement, feeling, and action, soon also saw that *habits incorporate aspects of our environments*. This means that habits are not just structures or dispositions internal to organisms, but instead stretch out into the world and embody aspects of our environments. Dewey focused on this fact as the key to developing a genuinely non-dualistic philosophy. He observes that habits are internal and external all at once: "[...] habits are arts. They involve skill of sensory and motor organs, cunning or craft, and objective materials. They assimilate objective energies and eventuate in command of environment" (Dewey 1922, 15–16). He illustrates this fundamental ontological fact (i.e., that habits embody and project into aspects of their environment) by observing that

> Habits are like functions in many respects, and especially in requiring the cooperation of organism and environment. Breathing is an affair of air as truly as of the lungs; digesting an affair of food as truly as of tissues of stomach. Seeing involves light just as certainly as it does the eye and optic nerve. Walking implicates the ground as well as the legs; [...][N]atural operations like breathing and digesting, acquired ones like speech and honesty, are func-

> tions of the surroundings as truly as of a person. They are things done *by* the environment by means of organic structures or acquired dispositions. (Dewey 1922, 15).

The primacy of ongoing organism-environment interaction means that all of the discriminations we make within our experience are just that, namely, they are simply recurring patterns of that interactive process. Dewey (1925; 1938) emphasized that our reflective processes of thought always involve selecting qualities, patterns, and relations from our primary experience. We need to be careful when using a word like "selection", for it can mislead us into positing some inner homunculus that does the selecting, when, in fact, the vast majority of the "selecting" that goes on is non-conscious and automatic.

We also need to be careful not to regard the products of selective attention as somehow capturing everything that matters in any given situation. Dewey thought that this was such a major temptation that it he called it The Philosopher's Fallacy (Dewey 1925). The fact that we cannot help but select parts or aspects of a particular situation (relative to our needs, perceptual makeup, and motor capacities), does not sanction the conclusion that only those selected dimensions are important for a full account of an experience.

Another extremely important point is that, as long as a creature lives, the organism-environment transactions that make it what it is are *always in process.* Mind is not a fixed thing or static structure. As psychologist and neuroscientist Don Tucker has argued, "the mind's organization emerges through a developmental process of controlling neural activity, beginning in the womb and continuing throughout life" (Tucker, in press, 93). Learning is not just a matter of adding new content to a fixed, prestructured mind; rather, cognition and learning *change the brain itself*, by changing neuronal connections. Tucker summarizes his key insight: "Cognition is neural development, continued in the present moment. The neural connections that are formed in the process of forming a concept are then the scaffolding for further brain development that allows mental activity" (Tucker, in press, 94).

## The rich, irreducibly complex character of experience

So far, I have suggested that the ontological and epistemic starting point for an analysis of mind, thought, language, and value must always be the totality of the situation of the ongoing organism-in-interaction-with-its-environment. Any analysis or explanation that starts with anything less than this entire arc and process of organism-environment coupling leads to a narrowing and distorting of experience, in a way that is bound to leave out important factors. Obviously, we cannot articulate the entire environing situation, because understanding and thought are necessarily selective. Any particular method of inquiry will therefore be partial and perspectival. It will circumscribe only what it takes to be the relevant phenomena within a situation, and it will focus only on those qualities, relations, and events that it deems significant for its purposes. This limited, perspectival character is not some catastrophic human failing, but rather merely a fact about the intrinsic limitations of human understanding.

Pragmatist treatments of inquiry, meaning, thought, knowledge, and truth can be extremely helpful in (1) revealing and, when appropriate, criticizing the basic limiting assumptions of particular methodologies and technologies of experiment and inquiry, and (2) providing a broader framework for incorporating multiple modes of inquiry and explanation, and integrating them with the cognitive sciences, in search of the fullest possible account of the depths of experience.

James and Dewey were especially insistent that both philosophy and common sense suffer from the mistaken view that experience consists in our "inner" feeling of how we are affected by our world. Dewey in particular observed that focusing only on the "how" of experience—how we feel ourselves being impacted by our perception of our surroundings—leads to a subjectivizing of experience that has been especially prominent since the Enlightenment. This exclusive focus on experience as an "inner" perceiving and processing of mental representations creates an ontological and epistemic gap between the inner (the mental) and the outer (the physical), with no way to connect the two. Consequently, what "mind" knows, on this subjectivist view, is only its own inner ideas or representations, and we are then left with the skeptical problem of how we could ever be sure that our inner representations somehow correspond to objects and events in the outer world.

James and Dewey saw that such a skepticism is unnecessary and misleading, once we understand the unity of the organism-environment reciprocity. They struck at the heart of any dualism of inner-outer, subject-object, and mind-body by insisting that our world comes to us as what James called a "much-at-once-ness"—an integrated, complex, multi-dimensional whole—from which we later select (mostly unconsciously) objects, properties, and relations upon which we focus our attention and thinking. James and Dewey repeatedly emphasized that, if philosophy is going to be grounded in our concrete situation and reconstruct the world for the better, then we have to start with a non-reductive, suitably rich and complex account of experience. Experience, they argued, is neither inner nor outer, but both at once, and it far exceeds, in its richness, any particular theoretical account of it. In Dewey's words:

> Experience is *of* as well as *in* nature. It is not experience which is experienced, but nature—stones, plants, animals, diseases, health, temperature, electricity, and so on. Things interacting in certain ways *are* experience; they are what is experienced. Linked in certain other ways with another natural object—the human organism—they are *how* things are experienced as well. Experience thus reaches down into nature; it has depth. It also has breadth and to an indefinitely elastic extent. It stretches. That stretch constitutes inference (Dewey 1925, 12–13).

Humans are not merely spectators of a shadow-play of inner representations. We are not locked up within an inner space of "mind" or "consciousness." Rather, *we are in and of the world,* always active within it in an ongoing process through which, by virtue of our activity and the activity of things within our environment, we are in at least partial touch with our world and not alienated from it. Inner/outer, mind/world, and subject/object are abstractions from the unified flow of experience.

One important implication of the James/Dewey non-dualistic view of experience is that it is not correct to say that the mind is merely the brain, since experience encompasses the entire arc of organism/environment engagement (which is an enactive process) (Varela, Thompson, and Rosch 1991). Sometimes neuroscientists are criticized, and rightly so, for claiming that all experience and thought takes place *within* the brain. What they should say, according to a pragmatist non-dualist ontology and according to good cognitive science, is that thought takes place via structures and processes operating at many levels: in neurons, in a cortex, in a brain, in chemicals in the blood, in an active body, in social interactions, within cultural institutions, and thus in a multi-dimensional environment. In other words, any satisfactory account of cognition will have to include the whole creative process of organism-environment engagement.

That said, there is still an important truth in the claim that the mind is the brain. That truth is this—everything that we call experience, thought, conceptualization, feeling, valuing, and acting has to be enacted through neural and chemical activity within the brain. The brain needs a body and a world in order to survive and function, but without neural activity, none of the processes in the list of "mental" operations just mentioned can occur. So, even if one balks at the bald "the mind is the brain" claim, the brain is surely the center of the action. Most neuroscientists realize that the mind is not just the brain (even though there is no mind without a brain), but since they spend their entire lives struggling to understand how brains work, it is understandable that they are prone to regard the brain as the locus of the action of cognition. When they are suitably reflective, however, they of course know that there is no mind, thought, valuing, feeling, or acting without a brain in a body in an environment, with each dimension being absolutely indispensable. Even if there could someday be a brain in a vat, there could never be a mind in a vat. Consequently, it is best to avoid locutions such as "Fear happens in the amygdala (and other organs)," or "Planning and inference occur in the prefrontal cortical areas," because the fear, inference, and planning each involve the person's ongoing engagement with their environment, and so they are not merely brain episodes.

## A non-dualistic, non-representational theory of mind

Once we recognize the primacy of organism-environment interactions and the depth and breadth of experience, then all ontological dualisms are out. Distinctions like inner/outer, mind/body, subject/object, and even organism/environment, are not ontological primitives, but rather dimensions of the continuous processes of experience that we happen to mark out as important in our inquiries, for one reason or another. Discarding such dualisms (i. e., denying them status as ontological absolutes) requires us to concomitantly discard the Representational Theory of Mind, by which I mean the view that thinking is conceiving and operating on inner mental representations or symbols that are somehow supposed to be connectable to an objective world outside the mind. As mentioned earlier, such a view holds that minds know their "ideas", "concepts", or "representations", which are attributes of the mental. Thinking is taken to be inner episodes of quasi-perception of entities (ideas/images/concepts) whereby we discern conceptual relations and draw inferences.

The classical pragmatists saw, what was later confirmed by cognitive neuroscience, that there is no mind-homunculus that processes mysterious inner representations. There is no need for an inner perceiving agent, because there is nothing "inner" to perceive or think about, since there are no inner representations at all, in any significant sense of that term. There is no need for a synthesizing transcendent ego that supposedly unifies our discrete inner sensations, ideas, or thoughts, because these were never atomistic representations in the first place. Since "inner" and "outer" are not two separate and distinct metaphysical realities, there is no need for a third thing—an inner representation—to mediate between the assumed two types or dimensions of reality.

Of course, neuroscientists are in the business of identifying patterns of neural connection and activation associated with cognitive and affective processes, and they are prone to call these "representations". These patterns of connection are, indeed, "in" the brain. However, these are not representations in some inner *mental* realm, because there is no such realm and no homunculus or executive center in the brain that could somehow experience these representations, note their relations, or see how they map onto some aspect(s) of the world. Instead, there are just patterns, qualities, and structures of organism-environment couplings—realized neurally—that shape our experience, both non-conscious and conscious. There is the much-at-once-ness of experience, within which there are discriminative selections that direct the organism's engagement with its world. All of this discrimination and discernment is realized through activated functional clusters of neurons, but none of this requires the existence of some inner interpreter whose job it is to "read" the meaning of the patterns and strengths of neural connections.

Some will protest by observing that even the best neuroscientists routinely employ the language of representations in their accounts of cognitive, conative, and affective phenomena. And so they do. Such talk is especially common in accounts of neural "maps" (e.g., Edelman 1992; Damasio 1999; Feldman 2006). By a "map" is meant a spatial relation of neurons or clusters of neurons in cortical sheets that preserves the topological relations of objects, events, and qualities in our environment as it interacts with our sensory systems. For example, there are auditory maps, in which what we experience as "adjacent" pitch tones (A to B, or B to C, etc.) are "represented" in the spatial relation of neurons in the auditory cortex (above the ears). Thus, neurons activated when we hear middle-A are located in auditory cortex next to those activated in hearing the pitch middle-B, and those for middle-C are located next to those for B. However, this use of the term "representations" does *not* support the Representational Theory of Mind, because it does not say that hearing adjacent notes or pitches consists in some inner locus of awareness that interprets aspects of particular neural maps as adjacent pitches. We do not perceive the neural maps. Instead, we perceive the *pitches*—partly because of the nature of our auditory environment, partly because of the structure of our inner ear, and partly because of maps in the activated neuronal clusters in our auditory cortex. The firings of those adjacent neuronal assemblies in some temporal sequence *just is* our hearing of the successive pitches middle-A, B, C. We are not possessed of some mysterious inner ego that *interprets* the meaning of alleged inner auditory representations (maps) that have been given rise to by environmental cues. There is no third thing—the mental representation—needed to mediate between self and world.

Notice that, on this view, humans were never separated, either ontologically or epistemically, from their world, because "mind", consciousness, meaning, thought, feeling, and inference are all just emergent phenomena and functions based on the structures and activities of the organism engaging the organized energies of its environment.

## Concepts are embodied, and they go beyond their neural dimensions

If there are no ghostly inner entities (representations) to serve as the content of our thoughts, then what *is* there to serve as the stuff of conceptualization and reasoning? Eventually, I began to ask myself, "Where ARE these ideas or concepts that nearly all philosophers seem to regard as the objects and content of our thoughts?" The more cognitive science and neuroscience I became familiar with, the more I realized that concepts could not be representations in some inner mental realm, nor could they be Fregean objective senses lodged in some sort of Platonic heaven accessible to all rational beings. Concepts are not abstract entities that could possibly be contained in some mental arena or processing plant. If they are not located *in* something called "the mind," then what are they *in?* Perhaps one will say that concepts are in the brain, in the body, and in the world, all at once. However, these metaphors of containment simply reinforce the very *inner/outer* dualism we are trying to overcome. It would probably be less misleading if we said that the whole *inner/outer* dichotomy is just not appropriate for explaining how we feel, conceptualize, reason, value, and act. Instead, we should say that concepts are realized via functionally bound clusters of neurons that are connected and co-activated as part of our embodied interactions with aspects of our environment.

One way to explore the nature of what we call "concepts" is to ask what it means to have an idea of, say, a chair. What Lakoff (1987) called the "classical objectivist theory of categories" holds that a concept of a kind, such as *chair*, would be a representation of a list of properties and relations that jointly constitute some perceived object as a chair. However, as Lakoff has pointed out in considerable detail, that does not seem to be the way human brains work. Instead, having a concept of a chair amounts to possessing the ability to simulate neurally certain specific kinds of experiences associated with chairs. According to what has come to be known as the *Simulation Theory of Meaning*, our understanding of a concept of some particular object or event consists in our ability to simulate a series of perceptions, body movements, feelings, emotions, and appraisals that are involved in our bodily and social engagement with those particular kinds of objects and events. Ben Bergen (2012) has surveyed a large number of experimental studies showing how, as we hear or read certain words and sentences that involve objects, actions, and scenes, clusters of functionally integrated neuronal assemblies are activated, as though we were actually seeing, touching, and interacting with that object or person within a scene. For example, in thinking of chairs, parts of our brain that would be involved in seeing various kinds of chairs would be activated, such as simulating possible views and perspectives I might have as I walk around different chairs. There would also be parts of our motor and pre-motor cortex involved as they would be if we actually sit on, lift, and move chairs. So, part of the meaning of "chair" would be the range of visual, tactile, auditory, and motor actions we have, or might possibly have, with chairs. Besides sensory and motor simulation, there would also be some activation of the sorts of feelings and emotions associated with our experience of chairs. Finally, there would be a simulation of possible social and cultural settings, practices, and events in which chairs play an important role. Our repeated experiences with various chairs would lead us to develop cognitive prototypes for central members of the category *chair* (Rosch 1977), so that certain kinds of possible simulated interactions with chairs would be more likely to occur in certain cultural settings. Consequently, whenever I see a chair, imagine a chair, or read or hear the word "chair", there would be complex simulative activity in my brain and body of these sensory, motor, affective, and cultural dimensions (Feldman 2006; Bergen 2012).

## An embodied view of meaning

The Simulation Theory of Meaning attempts to give an account of how meaning is embodied in organisms that are embedded in and acting in the world. It thus provides a radical alternative to disembodied views that claim that an account of the operations of mind can be given without any serious inclusion of how our bodies and brains work to generate, constrain, and open up meaning. For the pragmatist, who denies mind-body dualism, then, the central problem about meaning is to explain how organisms with our types of bodies and brains could have evolved systems capable of performing the full range of cognitive processes, from the ability to experience objects, qualities, and events all the way up to our most creative acts of thought and communication!

The view of meaning that seems to be developing in recent cognitive science research is roughly pragmatist: The meaning of any thing, event, or symbol is what it points to by way of experiences—experiences past, present, and future. The meaning of a given term is not some ghostly inner conceptual entity, but is instead realized as a set of related activations of functional neuronal clusters that are associated with that term (where the term picks out an object, person, or event). As Charles Sanders Peirce argued in his celebrated theory of semiotics, signs do not have meaning in themselves, or even merely as standing in relation to some state of affairs in the world. Rather, to get a meaningful sign you must also have a subject or person who understands the sign as indicating some state of affairs.

As a result of this organism-sign-referent connectedness, the meaning of anything is as much a matter of the nature of the perceiver/conceiver (i.e., her brain and body) as it is a matter of how the relevant environment is structured. Besides implicating some structured environment, meaning thus depends on the bodily makeup of the organism for whom things and events are meaningful. This is the reality behind my earlier claim that "you can only think what your brain, body, and environment (physical and cultural) allow you to think." One may substitute for the word "think" in this sentence any of a number of terms such as "understand," "grasp the meaning of something," "perceive," "conceive," "feel," "act," "know," etc.

In *Philosophy in the Flesh* (1999), Lakoff and I surveyed some of the ways our bodily makeup and our environments give rise to our spatial relations concepts. Spatial relations like front-back, near-far, right-left, up-down, center-periphery, in-out, and over-under are based on the patterns that emerge given the ways our particular types of bodies interact with their surroundings. For example, *Up* and *Down* are important concepts to us, because we are bodily creatures who can sometimes stand erect in a gravitational field in which we experience different degrees of verticality. There is no *Up* and *Down* intrinsically in the nature of the cosmos. These ideas are defined relative to our embodied being in the world. Cultures the world over project body-part relations onto objects and scenes, such as when we speak of the head and foot of a bed, a mountain, or a sheet of paper.

Another set of body-based meaning structures are what Lakoff and I (Lakoff 1987; Johnson 1987) have called *image schemas*. Given the ways human bodies are typically (but not universally) put together, we are likely to be able to discriminate a large number of recurring patterns of organism-environment interaction. Consider, for example, how people who have use of their legs routinely move around in their environment. There is a recurrent pattern to these acts of locomotion: A mover (person or animal) moves certain body parts (e.g. legs, torso, arms) using certain physiologically-constrained gaits (such as walking, skipping, or running), with varying speeds (e.g., slow, medium, fast), and exerting varying levels of energy (e.g., trudging along vs. scampering). There thus emerges what we might call a *Locomotion* schema, consisting of the following structural elements that can be realized via different parameterizations: Mover, Gait (e.g., walk, run, skip, hop), Speed (slow, medium, fast), Effort (on a scale of exertion), and Body Part (e.g., leg, hand, foot, knee) (Dodge & Lakoff 2005). Different actions of locomotion can be characterized by inserting the relevant values into the parameters of the schema. The neural systems governing such parameters of action have been fairly well mapped out, such as how the different gaits of a cat or a horse are generated.

In a similar fashion, a large number of body-based image schemas for perception and action have been described, such as: Container, Source-Path-Goal, Object, Action, Verticality, Balance, Compulsive Force, Iteration, Straight, and Center-Periphery. The Embodied Meaning Hypothesis entails that all languages around the world will make use of most of the available image schemas to express meaning relations. There is ample evidence that languages vary regarding which image schemas they use and how they are specifically articulated, but the appropriation of such image schemas appears to be a universal phenomenon. As Ellen Dodge and George Lakoff (2005) conclude:

> In sum, we have seen that there is great cross-linguistic diversity in spatial-relations terms and their use of schematic structure. Languages differ as to the basic spatial distinctions they make, the combinations of distinctions they "package" together in their spatial-relations terms, and the grammatical class membership of these spatial terms. [...] However, we have also seen that these spatial-relations terms can instead be analyzed as complex combinations of more primitive image-schematic structures. (Dodge & Lakoff 2005, 67)

One simple example of this combination of image-schematic structures is the semantics of the English word "into". In English, *into* is understood via a blending of a *Container* schema (with a Boundary, Exterior, and Interior) and a *Source-Path-Goal* schema (with a source point, path, and destination). The Source of the *Source-Path-Goal* schema is mapped onto the Exterior within the *Container* schema, and the Goal onto the Interior, with the Path of motion running from outside the bounded area to the interior. Thus, the meaning of "into" involves motion starting from outside a container, across its boundary, and ending within the bounded area.

What we traditionally think of as abstract thought and concepts are no less embodied than these more obviously body-based patterns and felt contours of experience. The orientation now known as Conceptual Metaphor Theory (Lakoff & Johnson 1980; Kovecses 2010; Dancygier & Sweetser 2014) has supplied over thirty years of research into the ways our abstract concepts are defined by vast systems of metaphors, in which we recruit the meaning and inference patterns from a body-based or social domain of experience for understanding and reasoning about an abstract, or non-concrete, domain of experience. To cite just one example, consider the meaning of the sentence *The presidential candidate came out of the closet.* The use of "out of" activates a body-based Container schema, plus a Source-Path-Goal schema, as we just saw with the semantics of "into", only this time the Source maps into the Interior of the Bounded Region and the Goal onto the Exterior. One interpretation of this sentence has the candidate literally existing in his or her physical closet and then exiting that enclosure. But in the metaphorical interpretation, the closet is a metaphorical container that "hides" the knowledge of the candidate's sexual orientation. The Interior is a space of the unknown, private, and non-accessible, whereas the Exterior space is metaphorically the known, public, and accessible. There exists a large and rapidly growing literature on the operation of large-scale conceptual metaphors in different cultures, practices, behaviors, and forms of communication (Lakoff & Johnson 1980; Gibbs 2008; Kovecses 2010; Dancygier & Sweetser 2014).

Image schemas, spatial relations, conceptual metaphors, and other types of body-based meaning are not linguistic; instead, they are conceptual and are operative both before children acquire language and in meaningful activities that extend far beyond language in adult life. This shows that meaning is not just linguistic in nature, since there are many forms of meaning-making that are not dependent on language. Embodied meaning processes like these lie at the heart of various mostly non-linguistic forms of expression and communication, such as visual art, visual advertising, film, music, dance, architecture, spontaneous gesture, ritual acts, and dramatic performances. In film, for example, in addition to spoken dialogue, there are all sorts of ways meanings are communicated through juxtaposition of images, temporal flow of action, sound effects, music, lighting, camera angle, and so forth (Tikka 2008). In other words, embodied meaning reaches deep into our visceral connection with everything that surrounds us and makes up the stuff of our lives, and it is not dependent on language, but rather underlies language and makes it possible.

## Feeling and emotion in meaning and thought

The classical pragmatists knew well that feeling and emotion were an essential part of our capacity to experience meaning, to understand, to conceptualize, to reason, and to value something. However, when pragmatism was eclipsed in professional philosophy by so-called Analytic Philosophy, this core insight was neglected and even denied. Analytic philosophy's exclusive focus on concepts, propositions, and inferences reinstated the thought/feeling and cognition/emotion dichotomies that have plagued philosophy throughout most of its history. Fortunately, with the emergence of second-generation cognitive science, research on the role of emotions in thought processes has recovered many of the lost insights from earlier pragmatist works.

In his *Principles of Psychology*, William James famously employed the term *thought-feeling* to emphasize the presence of feeling in all of our thought processes. He even extended this to the claim that logical relations are *felt* just as much as they are thought. We feel, he boldly asserted, the direction, tendency, and connection of our patterns of thinking:

> If there be such things as feelings at all, then so surely as relations between objects exist in rerum natura, so surely, and more surely, do feelings exist to which these relations are known. There is not a conjunction or a preposition, and hardly an adverbial phrase, syntactic form, or inflection of voice, in human speech, that does not express some shading or other of relations which we at some moment actually feel to exist between the larger objects of our thought. (James 1890, 238)

Every thought of a particular object or event will be surrounded by a "halo" or "fringe" of felt tendencies and connections.

Dewey went even further than James when he argued that our very capacity to think and reason intelligently depends upon our felt sense of the entire situation in which our thinking emerges: "the selective determination and relation of objects in thought is controlled by a pervasive and internally integrating quality, so that failure to acknowledge the situation leaves, in the end, the logical force of objects and their relations inexplicable" (Dewey 1930, 246). His point was that the felt qualitative unity of a given situation is what gives us a sense of which relations and connections are relevant for our thought. Without this grounding in the sense of a situation, our thoughts would float free from our embodied engagement with the particular affordances of our world. Our ability to "make sense" of any object, act, or event requires us to grasp—at a felt level—what is going on:

> This state of things in which qualitatively different feelings are not just had but are significant of objective differences, is mind. Feelings are no longer just felt. They have and they make *sense*, record and prophesy. That is to say, difference in qualities (feelings) of acts when employed as indications of acts performed and to be performed and as signs of their consequences, *mean* something. And they mean it directly; [...] Feelings make sense [...]. (Dewey 1925, *198)*

This classical pragmatist emphasis on the crucial role of feelings and emotions in meaning and thought was discarded and even ridiculed over the last three-quarters of the twentieth century during which Analytic Philosophy rose to prominence. However, things have changed for the better over the past three decades with the rapid growth of cognitive neuroscience and its rediscovery of the importance of emotion for thought. Most decisively, Antonio Damasio (1994; 1999; 2003; 2010) has shown the indispensability of emotion for intelligent social reasoning. He defines emotions as more or less automatic processes of bodily response, based on the body's ongoing monitoring of how it is being affected by its engagement with aspects of its environment. Emotions operate mostly beneath our awareness, but they are sometimes felt by us, when we perceive those changes in our body-state as it interacts with its surroundings: "Emotions are complex, largely automated programs of *actions* concocted by evolution. [...]

Feelings of emotion, on the other hand, are composite *perceptions* of what happens in our body and mind when we are emoting" (Damasio, 2010, 109). Following James, Dewey, and Damasio, I (Johnson 2007) have argued that emotions are central to meaning, insofar as emotions are the body's response to its continuous assessment and evaluation of "how things are going" for it. What could be more meaningful to you than your ability to take the measure of the affordances of our environment—to feel how your body is being changed to adapt to, or to reconfigure, parts of your environment?

## The aesthetic dimensions of meaning

As soon as we begin to appreciate the embodied character of meaning and thought, it becomes clear that when we focus on patterns, images, qualities, relations, feelings, and emotions, we are dealing with what has traditionally been the subject matter of aesthetic theory. Unfortunately, aesthetics has long been denigrated as dealing only with subjective aspects of experience and judgment, and therefore as not worthy of a place in theories of meaning, thinking, knowing, and communicating. Following Dewey's (1925; 1934; 1938) lead, in *The Meaning of the Body: Aesthetics of Human Understanding* (2007), I argued that any adequate theory of meaning will have to focus on those dimensions that we typically regard as central to our experience, appreciation, and creation of the various arts.

An account of meaning that recognizes the central role of aesthetic dimensions is precisely what is required if we hope to explain how we make and experience meaning in a broad range of arts. It finally allows us to overcome the pervasive prejudice that only words and sentences have meaning, and that we cannot really talk about meaning in the arts, beyond any linguistic component they might have. In sharp contrast, once you see that meaning is about how objects, events, and qualities provide meaningful affordances for various experiences, you find non-linguistic meaning everywhere and you can then begin to understand how the arts can be so meaningful to us (Shusterman 2000). Aesthetic theories that assume that only language has meaning and therefore end up trying to reduce our experience of the arts to language-like structures miss most of what is really going on in art. In contrast, when you start with the aesthetics of embodied meaning, you have the resources to explain a wide range of more-than-linguistic meaningful experiences in spontaneous gesture, painting, sculpture, architecture, dance, the theater arts, film, and ritual practice. Even better, you then have the resources to give a more experientially nuanced account of linguistic meaning that reveals its grounding in aspects of our sensory, motor, and affective experiences.

## Experience is not just linguistic

Just as you have to pick your cognitive science (namely, second-generation cognitive science) if you hope to get an adequate account of experience, mind, concepts, thought, values, and virtually everything else, likewise you have to pick your Pragmatism carefully. Not all pragmatisms are the same. Recently there has emerged a movement that has come to be known as *Linguistic* (or *Analytic*) *Pragmatism*. This orientation owes its origins mostly to the influential work of Richard Rorty, an analytically trained philosopher who later began to appreciate the merits of certain selected aspects of pragmatist thought and continental philosophy (Rorty 1979). Rorty liked Dewey's critique of all forms of dualism and his undermining of any form of foundationalism concerning knowledge, truth, or values. What Rorty didn't like in Dewey was the latter's insistence on starting from and returning to "experience", in the sense I have articulated above (Rorty 1982). Rorty thought that Dewey used his notion of experience as a way to get the very sort of ontological grounding that Dewey had taken such pains to deconstruct throughout his career. Consequently, Rorty argued that Dewey's "empirical metaphysics" was just another foundationalist metaphysics (Rorty 1982). In other words, for Rorty there is no good metaphysics, and philosophy should confine itself to critique of ontological, epistemological, moral, and political foundationalism.

Rorty's rejection of metaphysics rested on his claim that everything we can experience, think, know, and communicate is mediated by our language (or what he called our Vocabularies), so that our only legitimate analytic focus should be on language practices, since we have no unmediated access to the world beyond language. Alan Malachowski approvingly sums up this suspicion of "experience" when he writes: "New Pragmatism has not bound itself by the sorts of commitments that were always going to hold classic pragmatism back. Of these, empiricism, with the accent on experience, is the principal factor. New Pragmatists are able to discard much philosophical baggage by shifting its focus to language" (Malachowski 2010, 31).

Students of Rorty, such as Robert Brandom (2011), have subsequently developed this position into an extensive corpus of writings on how to understand the conditions for justified assertions and reason-giving in making claims to knowledge. This work has plenty of insights about speech act and argument conditions, but you will not find any account of either the depth and qualitative richness of experience or the central role of embodiment in the experience, making, and communication of meaning. Not surprisingly, since the body is basically ignored, you will not find any adequate treatment of the importance of the embodied aesthetic dimensions of experience.

As soon as you put the modifying adjective "Linguistic" or "Analytic" in front of the term "Pragmatism", the deed is done. Analytic Pragmatism is what you get when you keep the founding assumptions of Analytic philosophy of mind and language and simply add the pragmatist insight that speaking and communicating are *actions*, and so subject to social and institutional constraints. In like fashion, when you add the modifier "Linguistic" to "Pragmatism", you are left with meaning being confined to language, which leaves out both the prelinguistic embodiment of meaning and the role of qualities, feelings, and emotions in unifying and defining the character of a situation.

What this means is that, once pragmatism is eviscerated and narrowed into Analytic/Linguistic Pragmatism, most of the cooperative relations between embodied cognitive science and classical pragmatism are lost. Take away the embodiment of mind, meaning, and thought, and you are right back in first-generation cognitive science, with its view of meaning as propositional and constrained only by speech act and epistemic rules.

## Meaning and the *six* E's

In those cognitive science and philosophy of mind orientations that appreciate the pragmatist insight about the primacy of ongoing organism-environment interaction, it has become quite popular over the past decade to speak of "the 4 E's" to indicate that consciousness and mind are *embodied, embedded, enactive, and extended.* To the traditional 4 E's, I am thus urging the addition of a 5$^{th}$ E—*emotional*—and a 6$^{th}$ E—*evolutionary.* The 6 E's are, indeed, a convenient way to summarize some of the key tenets of an embodied cognition approach, in the following manner: (1) Our capacities for meaning making, understanding, and reasoning are evolved structures that emerged over our ongoing *evolutionary* history through the appropriation of prior capacities for perceiving, doing, and feeling. (2) Within this evolutionary framework, consciousness, mind, meaning, thought, and value are grounded in our *embodied* engagement with our world, and they are emergent phenomena of the increasing functional complexity of the organism. (3) These phenomena exist only because mind is, through our embodiment, *embedded* within our surroundings that offer affordances for our perceptions and actions. (4) The term *enactive* captures the way our ongoing transactions with aspects of our environment reconfigure experience. Meaning and value are enacted in the very process by which the organism engages its environment. (5) *Extended* mind recognizes the many ways we off-load cognitive processes, resources, and products onto structures in our environment. (6) Finally, the *emotional* character of consciousness, meaning, and mind marks the felt, qualitative dimensions of our enactive coupling with, and remaking of, parts of our surroundings.

## Embedded meaning requires embodied meaning

Recently, a lively in-house debate has arisen within the 6-E (previously 4-E) community about whether there is a fundamental and irresolvable tension between the embodied and extended aspects of mind (Clark 2012). If meaning, thought, and mind can actually be deposited into the world beyond the organism, then it would appear that mind must go beyond its embodied dimensions. If mind, consciousness, and thought can (at least in part) be lodged "out there", then they cannot be only phenomena "in here" (i.e., within the embodied mind). This is an interesting and potentially fruitful debate that can help us sharpen our understanding of what it means to ground everything on organism-environment interactions. Here I want only to note that, since meaning requires a sign, its referent (i.e., the experiences it indicates), and an embodied person to whom the sign is meaningful, it is not correct to assert that any of the objects, patterns, or actions that make up the world of extended mind outside and beyond our skins could have meaning in themselves. *Nothing in the "extended" world has any intrinsic meaning in itself, because in order to have a meaning relation, there must be someone for or to whom that object, relation, quality, or event is meaningful.* Meaning has to be enacted in the brain and body of a more or less functioning organism that possesses certain specific capacities (Tucker, in press). What we call "pens", "clouds", and "quarks" do not *mean* anything in themselves; instead, creatures like us experience them as meaningful patterns in our environment, because of our bodily makeup, the affordances provided us by our environment, and our values and interests.

This crucially important point is easily overlooked by extended-mind theorists, because when they talk about how elements, procedures, and products of mental operations are off-loaded onto the environment, they have already *assumed* that those off-loaded structures are meaningful, forgetting most of what was required to make them meaningful in the first place (namely, an at least partially functioning brain, in an at least partially functioning body, operating within an environment that is physical, interpersonal, and cultural). Those who speak of externalized aspects of mind, consciousness, and meaning as *entirely* "out there" in the "external" world, independent of people, are falling into the same error as that committed by objectivists who think that sentences could have meaning merely by standing in relation to some external state of affairs, completely independent of the being for whom the sentence is meaningful.

I am thus urging that view that an adequate account of mind and thought will have to involve *all* of the six E's, with the acknowledgment that none of them alone adequately captures what goes into meaning, experience, and action.

## It's values all the way down

If everything starts from organism-environment interaction, then we begin with values, so we simply do not need some special source of values or norms. To live is to have values, which means that, as active organisms, we are preferentially directed toward specific states of affairs in the course of our interactions with our surroundings. As Damasio (2010) has pointed out, the continued existence of any organism requires a permeable boundary within which the conditions of life are maintained, and this takes the form of homeostasis. The organism must secure a dynamic balance of energies, of food taken in and waste excreted, of internal chemical balance, of maintenance of a temperature range, and so forth. If we understand homeostasis to be restoring a disrupted balance to some preferable set point, then perhaps a better word for the experiential process would be allostasis, which is a dynamic process by which an animal can create a new set-point or dynamic equilibrium. The very conditions of life thus give rise to what Damasio calls an Ur-Value: "For whole organisms, then, the primitive of value is *the physiological state of living tissue within a survivable, homeostatic range*" (Damasio 2010, 49; italics in original).

> My hypothesis is that objects and processes we confront in our daily lives acquire their assigned value by reference to this primitive of naturally selected organism value. The values that humans attribute to objects and activities would bear some relation, no matter how indirect or remote, to the two following conditions: first, the general maintenance of living tissue within the homeostatic range suitable to its current context; second, the particular regulation required for the process to operate within the sector of the homeostatic range associated with well-being relative to the current context. (Damasio 2010, 49)

Since people are not just biological organisms, they will have a number of other values that go beyond their need for sustaining the conditions of life. I (Johnson 2014) have argued that besides values arising from our biological makeup, there would also be values tied to our intimate interpersonal relations with other people (e.g., our need for nurturance, affection, care), values emerging from our role as members of larger social groups (e.g., our sense of justice, rights, and group responsibilities), and values pertinent to our question for meaningful and fulfilled lives.

All of these values are natural, which is to say that they arise for us given the kinds of bodily, interpersonal, and cultural beings we are. There is therefore no need to justify values as coming from some absolute or transcendent source. It is simply an experimental issue of determining what kinds of activities we value (both non-consciously and consciously), what helps us live harmoniously with others, and what tends to give our lives meaning and a sense of purpose.

The previous list presents just some of the areas of overlap between classical pragmatism and recent cognitive science that suggest potentially productive avenues of dialogue and ongoing engagement between the two orientations. I have argued that the cognitive science of the embodied mind gives us considerable evidence to support many of the defining assumptions of the pragmatism of Peirce, James, and Dewey, though not of Linguistic or Analytic Pragmatism.

So, I am hoping that the romance between Pragmatism and Cognitive Science will grow in depth and meaning. In spite of the fact that they appear to have gotten off on the wrong foot in their early days of courtship, they now appear to have fallen for one another. Let us hope we have the pleasure of being able to grow to know their lovely offspring, like young babies who have learned to walk and are now discovering the wonders of human language and other forms of creative communication. May the union of classical pragmatism and second-generation cognitive science grow and prosper, just like in our favorite romantic comedies.

# References

Brandom, R. (2011). *Perspectives on Pragmatism: Classical, Recent, and Contemporary.* Cambridge, MA: Harvard University Press.

Clark, A. (2012). "Embodied, Embedded, and Extended Cognition." In: *Cambridge Handbook of Cognitive Science.* Cambridge: Cambridge University Press, 275–291.

Damasio, A. (1994). *Descartes' error: emotion, reason, and the human brain.* New York: G.P. Putnam's Sons.

Damasio, A. (1999). *The Feeling of what happens: body and emotion in the making of consciousness.* New York: Harcourt, Brace & Co.

Damasio, A. (2003). *Looking for Spinoza: joy, sorrow, and the feeling brain.* New York: Harcourt, Inc.

Damasio, A. (2010). *Self Comes to Mind: Constructing the Conscious Brain.* New York: Pantheon Books.

Dancygier, B. & Sweetser, E. (2014). *Figurative Language*. Cambridge: Cambridge University Press.
Dewey, J. (1925). Experience and Nature. Vol. 1 of *The Later Works: 1925–1953*. ed. by J. A. Boydston. Carbondale: Southern Illinois University Press, 1981.
Dewey, J. (1930). "Qualitative Thought". In: Vol. 5 of *The Later Works: 1925–1953*. ed. by J. A. Boydston. Carbondale: Southern Illinois University Press, 1988.
Dewey, J. (1934). Art as Experience. Vol. 10 of *The Later Works: 1925–1953*. ed. by J. A. Boydston. Carbondale: Southern Illinois University Press, 1987.
Dewey, J. (1938). Logic: The Theory of Inquiry. Vol. 12 of *The Later Works: 1925–1953*. ed. by J. A. Boydston. Carbondale: Southern Illinois University Press, 1991.
Edleman, G. M. & Tononi, G. (2000). *A Universe of consciousness: how matter becomes mind.* New York: Basic Books.
Flanagan, O. (1992). *Consciousness reconsidered.* Cambridge, MA: MIT Press.
Gendlin, E. (1997). "How philosophy cannot appeal to experience, and how it can." In: D. Levin (Ed.) (1997). *Language beyond postmodernism: saying and thinking in Gendlin's philosophy.* Evanston: Northwestern University Press, 3–41.
Gibbs, R. (2008). *The Cambridge Handbook of Metaphor and Thought.* Cambridge: Cambridge University Press.
Jackson, F. (1986). "What Mary didn't know." In: *Journal of Philosophy* 83, 291–295.
James, W. (1890). *The Principles of Psychology.* New York: Dover, 1950.
James, W. (1900). *Psychology.* American Science Series, Briefer Course. New York: Henry Holt & Company.
Johnson, M. (2007). *The Meaning of the Body: Aesthetics of Human Understanding.* Chicago: University of Chicago Press.
Johnson, M. (2014). *Morality for Humans: Ethical Understanding from the Perspective of Cognitive Science.* Chicago: University of Chicago Press.
Johnson, M. & Lakoff, G. (2002). "Why Cognitive Linguistics Requires Embodied Realism." In: *Cognitive Linguistics* 13(3), 245–263.
Kovecses, Z. (2010). *Metaphor.* Oxford: Oxford University Press.
Lakoff, G. & Johnson, M.(1999). *Philosophy in the flesh: the embodied mind and its challenge to western thought.* New York: Basic Books.
LeDoux, J. (2002). Synaptic self: how our brains become who we are. New York: Viking.
Malachowski, A. (2010). *The New Pragmatism.* Durham, Great Britain: Acumen Publishing Limited.
Rorty, R. (1979). *Philosophy and the Mirror of Nature.* Princeton: Princeton University Press.
Rorty, R. (1982). *Consequences of Pragmatism.* Minneapolis: University of Minnesota Press.
Shusterman, R. (2000). *Pragmatist Aesthetics: Living Beauty, Rethinking Art.* Lantham, Maryland: Rowman & Littlefield.
Tikka, P. (2008). *Enactive Cinema: Simulatorium Eisenteinense.* Jyvaskyla, Finland: University of Art and Design Helsinki.
Tucker, D. (1992). "Developing Emotions and Cortical Networks." In: M. Gunnar & C. Nelson (Eds*.). Minnesots Symposium on Child Psychology*, Vol. 24: Developmental Behavioral Neuroscience. Hillsdale, NJ: Erlbaum, 75–128.
Tucker, D. (in press). *Right Wing, Left Wing: The Dialectical Development of Identity.* Oxford: Oxford University Press.

Varela, F., Rosch, E. & Thompson, E. (1991). *The Embodied mind: cognitive science and human experience.* Cambridge, Mass.: MIT Press.

Erkki Kilpinen

# Why It's Better Be Pragmatism: Assembling Some Philosophical Foundations for Future Cognitive Science

**Abstract:** A difficulty in assessing the relation between pragmatist philosophy and cognitive science, particularly embodied cognitive science, is due to the fact that pragmatism is a minority position in contemporary philosophy. It has much fewer adherents than analytic philosophy or phenomenology. Its picture does not get any clearer by the efforts of contemporary 'neo-pragmatists'. In fact, pragmatism does not belittle the importance of epistemic issues. Accordingly, this article argues that *classic* pragmatism goes best together with the embodied interpretation of cognitive science. The contact surface between these two bodies of thought is in their similar views about human action. Furthermore, this shared view about empirical description of action enables pragmatism to add even something new to cognitive science. This new ingredient is its interpretation of logic. According to pragmatism, the major interest in logic (deduction included) is in the advancement of knowledge, rather than in demonstration.

## 1 Philosophy as a house divided, and pragmatism as a minority position therein

To consider what the relation between pragmatist philosophy and cognitive science (and its philosophy) might be like, is a little bit complicated affair, because there is no unanimity, even among pragmatism scholars, about what this philosophy actually is. The issue is complicated further because of the general situation in contemporary philosophy. Pragmatism is a minority position in comparison with positivist-analytic[1] philosophy and phenomenology. At the turn of the millennium, a sociological survey (Gross 2002) reported that only some five percent of professional philosophers in the United States identified themselves as pragmatists. That number may have increased meanwhile, and pragmatism

1 The usual tern 'analytic philosophy' would be anachronistic, as it refers to post-WWII, mostly English-speaking philosophy, influenced by Wittgenstein (Glock 2008). Therefore the clumsier expression, 'positivist-analytic'. Phenomenology, positivism and pragmatism all date back to the last decade of the 19th and the first decade of the 20th century.

has advanced in Europe as well, but it is obvious that it has not caught up with the other movements in popularity. However, in current discussions it has often been conflated with them, and this has blurred its outlines even further.

The curious history of pragmatism also has played its role in this. Let me come back to the contemporary situation after a brief review of that history. Pragmatism first came to wider cognizance through a book under this title, by William James, published in 1907. It was a popularizing volume (originally a series of popular lectures), where James seemed to take a too careless attitude toward truth and its position in human knowledge. Thus originated the reputation that pragmatism still has, in some quarters, that it represents dubious epistemological relativism. "These disparagements are all boomerangs," answered George Herbert Mead (1938, 97) in his time, for example, but such rejoinders did not suffice to convince outsiders. Because of its reputation, pragmatism was for decades in near eclipse in academic philosophy, until it began to make recovery in the early 1980s. This was mostly due to the efforts of Richard Rorty (1931–2007) who called his own philosophy 'neo-pragmatism'. Assessing Rorty's accomplishment, one can certainly say that he made academic philosophy safe for pragmatism again—to paraphrase his own sayings (Rorty 1961; 1982). His achievements beyond this are of a bit more ambiguous nature. Regarding pragmatism's reputation as epistemological relativism, Rorty tried to make a virtue out of this supposed necessity, by belittling epistemology altogether. "Take care of freedom, and truth will take care of itself" (Rorty 2006), and so forth. One may wonder whether proclamations like this suffice to deliver a philosopher from the snares of "epistemocentrism" (Pierre Bourdieu's expression, 1997/2000), but this is not the place to go into details on that. Instead, Rorty's curious use of the term 'representation' demands more attention, because it has influenced contemporary discussion. Rorty made a name for himself with the book *Philosophy and the Mirror of Nature* (1979), which already in its title asserted that philosophy should not treat the human mind as "mirroring" nature, or outside reality. The proper name for this stance and its consequences (denial of the correspondence theory of truth, etc.) was for Rorty "anti-representational philosophy". In this he claimed to be going in the footsteps of the original pragmatists.

I called Rorty's use of the term 'representation' curious, because from an alternative viewpoint pragmatism appears as a representational philosophy *par excellence*. It is established knowledge that pragmatism and the general theory of semiotic have a founding father in common, in Charles S. Peirce. It is only slightly less established knowledge that these two bodies of thought converge also later on, particularly in the work of Josiah Royce, G. H. Mead, and C. W. Morris.[2] This pertains to the interpretation of pragmatism, because it reminds about its meaning-theoretic position, the triadic theory of meaning, which Peirce and Mead held explicitly, and others more implicitly. Here we meet the first touching point between pragmatism and cognitive science, as Dan Sperber (1996, 78 *et passim*) today adheres to this theory almost with the very same words, but without mentioning his pragmatist predecessors. This suggests that *classic* pragmatism has pertinence to cognitive science, but neo-pragmatism apparently less. This is not any news, of course. The excellent accounts by Teed Rockwell (2005) and Anthony Chemero (2009) have already traced core ideas of (embodied) cognitive science to the classic pragmatists, although Chemero (2009) refers to them as 'American naturalists', and talks about their psychological rather than philosophical contributions. No objection against that, but the leading idea of this paper is that the philosophical contributions of pragmatism seem to have pertinence as well. Regarding its psychology, Edward Slingerland has called the original pragmatist interpretation of human cognition "the mother tongue of thought" (Slingerland 2008, 222), but he adds that current neo-pragmatism makes only an "almost pragmatist turn" in its study (Slingerland 2008, 88).

Of course there are such authors who would like pragmatism to have nothing at all to do with cognitive science. Jerry Fodor (2008) uses the term, but merely as a dirty word for everything that he dislikes. It is curious, however, that the anti-pragmatist Fodor and the neo-pragmatist Rorty almost meet each other in their common fascination with language. For both of them, it is a universal frame of reference, almost a prison-house (Fredric Jameson), out of which it is impossible to escape. However, one of the contributions of classic pragmatism is precisely the conclusion that language can be transcended, without inflicting any harm on its important features. This is made possible by the more comprehensive theory of meaning that pragmatism has, the semiotic theory[3] mentioned above. I shall return to these issues later on.

At the moment, we are not yet done with the question about pragmatism's position in contemporary philosophy. Due to its minority position, it tends to get defined in terms of the more wide-spread approaches. Academic philosophy

2 About Mead's semiotic contributions, see Kilpinen (2002). Colapietro (2004) discusses John Dewey as a semiotician.

3 Much confusion is added in that semiotic and de Saussure's language-theory, called *sémiologie*, are regrettably often conflated. Short (Short 2007, 15 f.) puts forward a strong argument about their incompatibility. Briefly: *sémiologie* is a dyadic, pragmatist semiotic a triadic theory of meaning.

in our time has been called *A House Divided* (the title of an anthology: Prado 2003). The opposites in that divide are analytic philosophy and phenomenology. From their respective viewpoints, pragmatism usually appears as a junior version of either. Prado (2003), for example, situates it together with contemporary 'Continental' philosophy, meaning, of course, contemporary neo-pragmatism and forgetting the original variant.

Picture 1 attempts to show the general situation in philosophy graphically.

Figure 1

The line in the picture describes a continuum between analytic philosophy and phenomenology. If we ask where pragmatism falls on that continuum, one can say that all logically possible answers have been suggested at some time. Pragmatism has been taken as a forerunner of logical positivism (e.g., Morris 1937; Kolakowski 1972), or fused together with the Continental tradition influenced by phenomenology, as just noted. For some, in turn, it offers a mediating position in the 'Great Divide', by avoiding the extremities of the two opposites. Perhaps such interpretations are of some value, but this, to repeat, is to define pragmatism in terms of other philosophies, not in its own right. If it, for a change, may speak from its own viewpoint, the situation looks different. Picture 2 attempts to capture the difference.

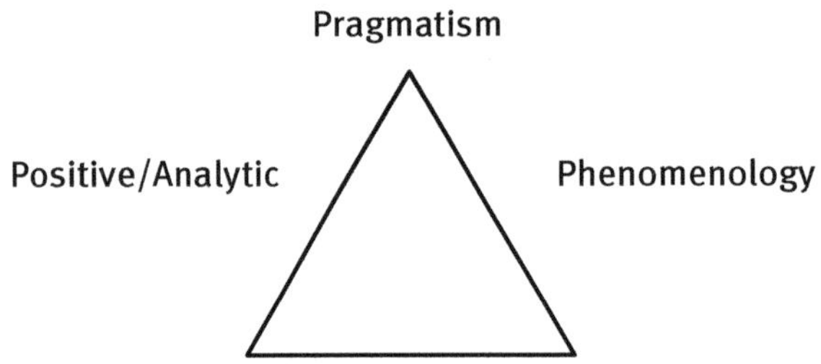


Figure 2

The regular triangle is meant to express that those major philosophies can quite well be taken as equally close to, or equally far from each other—you can choose your own descriptive adjective. By this I mean that a pragmatist is often struck by *similarities* between phenomenology and positivism, rather than by their differences.

Before going into those similarities and differences in closer detail, I wish to make the point that borders between philosophical creeds and schools are not, of course, watertight. Both the analytical and the phenomenological tradition have produced thinkers who in some respects might come closer to pragmatism than their own original traditions. Gilbert Ryle (1949/1970) is one such case in the analytic camp; Maurice Merleau-Ponty (1945/1962) is such a phenomenologist.[4] These cases remind that we are not adding apples and oranges in these comparisons, but the affinities and differences between traditions also demand a few more explaining words. I first refer to what I see pragmatism having in common with its rivals. After that I concentrate on pragmatism itself. Regarding its overlap with others, it turns out that pragmatism has occasionally been able to beat those others in their own games, colloquially speaking. What then is the 'own game' of analytic philosophy, and what is that of phenomenology?

Philosophers, both inside and outside the analytic movement, would probably answer that it has advanced the study of *logic* in particular. All right, but pragmatism in a sense has advanced it even further. The conclusion that logic, even deductive logic, can have an *ampliative* knowledge-interest was first reached in pragmatism. The task of logic (deduction included) is not confined to explication and proof. The term 'ampliative' refers to the interest of widening and enlarging knowledge, and the first to specify this explicitly, in regard to logical theory, was pragmatism's founding father Peirce. He specified it in his distinction between the 'corollarial' and 'theorematic' modes of deduction, of which the former answers to the traditional idea of deductive inference, whereas the latter is a more dynamic mode. It suggests also new truths, it does not just preserve the truth of the premises of the inference, as is the usual understanding about deduction. This was Peirce's original insight,[5] but it has eventually been achieved (and developed further) in the analytic tradition as well. Jaakko Hintikka (2007) in particular has drawn some new conclusions on its basis. As he writes, regarding its importance to human life,

> A theory of information (knowledge) acquisition is both philosophically and humanly much more important than a theory of whether or not already achieved information amounts to knowledge. Discovery is more important than the defence of what you already know (Hintikka 2007, 18).

---

4 Sandra Rosenthal and Patrick Bourgeois (1991) provide an interesting comparison between Merleau-Ponty and the pragmatist Mead.

5 Peirce was explicit on this distinction and its consequences, but the basic idea was approximated also by other pragmatists, Josiah Royce and C. I. Lewis. Dewey drew on Lewis, and intimated about the ampliative function of deduction, in his 1938 *Logic*.

According to Hintikka, a further conclusion from this insight is that a 'logic of discovery' is quite possible, after all, Karl Popper and others to the contrary notwithstanding. But the first modern philosophers to suggest this came from the pragmatic movement.

So much about analytic philosophy. What in turn is the forte of phenomenology? Or, let me specify and talk about phenomenological-*cum*-hermeneutic philosophy, because the relevance of philosophy for human studies is what most interests me here. Both adherents and outsiders probably agree that *interpretive study of meaning* is the strong suit of phenomenological hermeneutics. A pragmatist will agree, but will also note that this interpretive study needs not be restricted to the study of *human* meaning, as hermeneutics restricts it. Pragmatism studies meaning by interpretive means as well, but specifically by means of the aforementioned *semiotic*, which is a more comprehensive approach than hermeneutics. It is more comprehensive in the subject matter: Semiotic study in the pragmatist sense is not limited to humans and their language. Consequently, for hermeneutics, the interpretive study of meaning has traditionally been what separates human studies from natural sciences. For pragmatism, interpretive study of meaning rather *joins human studies together* with the study of other living nature. This is what pragmatists meant as they took meaning and intelligence as "almost co-extensive with life," as Mead (1938, 68) once expressed the idea (for its more recent and more comprehensive treatment, see Deacon 2012). Regarding the question, which approach goes better together with recent cognitive science, my bet is that pragmatism beats phenomenology here.

## 2 Pragmatism: overcoming the inaction fiction by habit-concept and fallibilism

Enough of other philosophies; where does pragmatism itself come out strong? The short answer is that it does so in the study of action. Others might reply that their approaches study action as well; 'philosophy of action' is an established genre in the analytic tradition, for example. A pragmatist's rejoinder will be that nominally, yes, but those other approaches are confined to a specific, narrow viewpoint. They treat action from the viewpoint for which I borrow an ironic name from the philosopher Dennett (1995): the 'mind-first explanation' of action. Writing about the theory of evolution, Dennett (1995) means by 'mind-first explanation' what is elsewhere known as 'intelligent design'. In other words, he uses it as a polite euphemism for creationism. Accordingly, from the pragmatist viewpoint one might suggest that other philosophies entertain 'creationist' ideas about action, but let us not get impolite.

In more serious terms, the sore point in the 'mind-first explanation' of action is not so much its reference to mind, but the implicit assumption that action (always) needs an explanation in the first place. Action is thus taken as a contingent empirical phenomenon, in this understanding. In other words, many philosophers are still in the confines of the "inaction fiction," as the psychologist Daniel Wegner (Wegner 2002, 218) called that position. Pragmatism, instead, takes action as the way in which human beings exist in the world (a phrase originally coined by Hans Joas), and if an addendum is needed, it would be that much of the same goes for other living beings as well.

I shall have more to say on this, but first I wish to settle accounts about the pragmatist position in epistemic questions (see above). Contrary to its denigrators (and misguided adherents), pragmatism does not disregard epistemology *tout court*. It rather suggests that the *job-description* of epistemology needs a fresh look. To put it laconically: Where other philosophies have theories of knowledge, pragmatism has the theory of inquiry.

In classic pragmatism, *inquiry is an epistemic concept*, and this is a point that too often gets missed. Let's not forget that this philosophy originated in Peirce's 'doubt/belief model of inquiry' in the 1870s, and in a sense reached its high point in Dewey's *Logic: The Theory of Inquiry* (1938). By 'inquiry' pragmatism does not mean mere empirical procedures of investigation; for it the term has epistemic connotation as well. We humans are all inquirers, though not all of us are inquirers in the academic sense.

If we take, for the sake of argument, 'inquiry' in an epistemic sense and as a basic notion, this will suggest some reformulation of traditional epistemological questions. They have usually been of the form: On what grounds can we be sure that the beliefs that we have amount to genuine knowledge? Pragmatism asks instead:[6] (1) What is knowledge good for; why do we need it in the first place? (2) Secondly, supposing that knowledge is good for us, how can we get it?

General answers to these questions are in turn of the form: (1) We need knowledge to guide us in action; and (2) we obtain knowledge by conducting inquiries. Taken as premises for further conclusions, these principles then suggest that "The criteria of knowledge concern the conditions on which the results of epistemological inquiry can be relied as a basis of action," as Hintikka (Hintikka 2007, 30) concludes today, carrying on the original argument of pragmatism.

---

6 Implicitly so. For these formulations I have drawn on Hintikka's recent work rather than old-time pragmatists, but I argue that the sayings do justice to their intention as well.

To return to the pragmatist conception of action specifically, the following important point escapes commentators regrettably often: In pragmatism, the basic concept to refer to action-phenomena is *habit*, rather than 'action' simply and solely. Many languages have proverbs that call 'habit' the 'second nature' of us humans. Not quite, pragmatists replied (James 1890 in particular): Habit is actually not our second nature; it rather is our *first* nature. We can do voluntarily only what we have learned to do habitually. Above we made the point that the idea of 'inaction' is foreign to pragmatism. This hints that pragmatists may have something specific in mind, as they take 'habit' as their basic action-concept. For them, it does not refer to "mere slothful repetition of what has been done", as Peirce once pithily put it (NEM 4, 143; ca. 1898). Instead, by 'habit' they mean the *sui generis process character* of action (Kilpinen 2009a; 2012; 2015). In so thinking they have adumbrated the position of Alva Noë (Noë 2009, 119) today, according to whom "we are always already in full stream". We are in the full stream of an already ongoing process of action, and for this idea pragmatists redefined the traditional term 'habit'. By this they wanted to bring out the process character of action, but there are also other points involved. One of them is that for which Noë today has found a nice expression, by calling habits "world-involving" modes of action. John Dewey (particularly in 1922) was nearly as explicit on this, as he explained that habitual action ties us inherently together with the outside world. Acceptance of these two principles then suggests a third, quite radical one, *viz.* that "only a being with habits could have a mind like ours," as Noë has put the matter (Noë 2009, 97–98).

Classic pragmatists were not quite as explicit on this, but they did relate 'habit' to mental and cognitive phenomena. Peirce, whose passion was logic, stated that the logical significance of habit is in that "habit is generalization, and from a logical point of view, generality is continuity" (Peirce RLT, 189; 1898). A famous passage from Dewey expresses the idea in a more down-to-earth manner:

> The reason why a baby can know little and a experienced adult know much when confronting the same things is not because the latter has a 'mind' which the former has not, but because one has already formed habits which the other has still to acquire. The scientific man and the philosopher like the carpenter, the physician and the politician know with their habits, not with their 'consciousness.' The latter is eventual, not a source. (Dewey 1922, 182–183).

To some (perhaps most) people it seems counter-intuitive to think that 'habit' might have logical significance, or anything to do with logic, in the first place. To a pragmatist it is not so. One more observation, in a sense a consequence of the above, "world-involving" character of habitual action, remains to be made explicit.

For pragmatism, action is an already ongoing process, but so is the world (reality, if you like), the arena of our action, as well. Pragmatists assume a process world, in other words, subscribe to process ontology (Kilpinen 2009b, 166 f.).[7] This pertains to the analysis of action in the sense that the main problem then is to make the small-scale process (action) go on with the same rhythm with the large-scale process (the world). This, however, can be established only for a time being, because those two processes have different dynamics. Consequently, sooner or later action will fail and falter, in other words, it is *inherently fallible.* But if so, then the following conclusion appears quite natural, *viz.* that "The whole function of the brain is summed up in: error correction," as W. Ross Ashby laconically put the idea, later on in the 20th century (as cited by Clark 2013, 181). If one prefers to talk about mind or consciousness instead of brain, the same principle holds even then. Of course, the phrase is to be taken as a programmatic slogan. The *entire* function of either the brain or the mind is not just error correction, but it can quite well be called their *original* function.

## 3 Pragmatism as non-reductive naturalism

The last few paragraphs have shown some contact surface between pragmatism and (the philosophy of) cognitive science, so that it now is ripe time to enquire whether these are just happy coincidences, or whether there is a closer relation between the two bodies of thought. While taking up this issue, I avail myself of the list of questions, that the organizers of this conference have prepared in advance. Three of them in particular suggest to me ideas that might have more general relevance. As I take up the questions, I also suggest a preliminary answer to each of them. After this first enumeration, I attempt to give those answers a fuller form. Here are the three topics that I find inviting:

(1) Does the pragmatist paradigm shift in contemporary cognitive science mean a revival of behaviourism? – No, but it might mean a *limited acceptance* of it. Behaviourism is slightly better than its reputation, though only slightly. I do not suggest its complete acceptance, but I do think that it has got one important point right. It reminds about the inalienable role of the outer world in

7 On process ontology generally, see Rescher (1996; 2000).

our action, the "world-involving" character of habitual action that we met above.

(2) Is there a way of reconciling the notion of representation with anti-Cartesianism?—Yes, there is, if some conceptual underbrush around the term 'representation' is removed. Above we have noted that it has more than one meaning. The history of pragmatism begins with an explicit attack on Cartesianism—and the weapon in that attack is the notion of representation (or sign, as the pragmatists often preferred to call it).

(3) What new methodological principles for cognitive sciences can be drawn from the work of the classical pragmatists?—A methodological principle, with obvious pertinence to cognitive science, is forthcoming from the pragmatist interpretation of logic. Perhaps Peirce was right, as he suggested that his distinction between 'corollarial' and 'theorematic' modes of deductive inference "is of extreme importance for the theory of cognition" (Peirce NEM 4, 56; c. 1902).

Let us now consider these putative answers in closer detail.

# 4 Who's afraid of behaviourism?—At least, not a pragmatist

Thinking about behaviourism nowadays, I am reminded of a joke by Mark Twain. He said that everyone should have at least one friend, but Satan apparently doesn't have any. So, let me be Satan's friend, Twain proposed. I propose to be taken as a sort of friend of behaviourism, though not as a particularly close friend. I am not inviting anyone to embrace that doctrine, that would lead to a dead end, and we know it by now. Today, however, behaviourism has so few friends that its modest contribution tends to get forgotten. That contribution was mentioned above, it is the inherent role of the outer world in our doings. I am tempted to paraphrase Mark Twain again. As he said, we ought to be a little wiser than the proverbial unlucky cat who happened to sit down on a hot stove lid and burned her tail. "That cat will not sit down on a hot stove lid again," said Twain, "and that is well. But she won't sit down on a cold lid, either." To translate this into more academic language, we ought to take the behaviourist *aspect* in action seriously (though not the entire doctrine), lest we inadvertently hurt ourselves like the poor cat did, and are unable to find out why.

Behaviourism is today attacked from two sides. For traditional humanists it has always been an ogre, but today it has gained enemies on the naturalist side as well. Many of those who call themselves 'evolutionary psychologists' today dislike behaviourism just as much. It may be that both of these anti-behaviourist positions search for support in Noam Chomsky's famous critique (1959) on B. F. Skinner's behaviourist language-theory. Chomsky's critique is still taken as a paradigmatic verdict on behaviourism at some quarters.[8] However, Chomsky's own interpretation about the origin and function of language does not appear as plausible today as it did only a short while ago (Evans & Levinson 2009; Tomasello 2009).

Leaving the verdict on that question for professional linguists, one can say instead that, of all philosophies, pragmatism does not need to be afraid of behaviourism. As is relatively widely known, behaviourism could be jocularly called a prodigal son of classic pragmatism. Not merely jocularly, however, because the founder of psychological behaviourism, John B. Watson, originally studied philosophy under Dewey and Mead.

Furthermore, pragmatism, in a sense, has from early on taken an insurance policy against excessive, radical behaviourism. It was taken in Dewey's classic article "The Reflex Arc Concept in Psychology" (1896), where he explained that a physiological stimulus does not appear "*ex abrupto*". It rather is something that happens *to* an already ongoing action-process and stalls its course. To bring home his point, he said, quite aware of the seeming paradox, that a stimulus does not actually explain the ensuing response, it rather is the other way round. The apparent paradox is in that an effect cannot precede its cause in time, but Dewey's point was meant as rhetorical. Concisely expressed it was that stimulus and response together constitute a *unit* in the ongoing flow of behaviour, a hindrance that is to be removed.[9] The final conclusion of this idea was later on drawn by Mead, who noted that our sense reactions to the outside world "depend upon what we are doing, not the reverse" (Mead 1903, 37).

Later on, in his lectures on social psychology, posthumously published as the book, *Mind, Self, and Society* (1934/2015), Mead commented on Watson's behaviourism explicitly. In those comments he blamed Watson for unnecessary reductionism, he does not explain consciousness, but explains it away. Mead, however, also gave Watson some credit for taking consciousness as a problem to be investigated. Interestingly, the same point recurs a generation or two later, in Dennett's (1978a, b) assessment of B. F. Skinner's behaviourism. I find Dennett's treatment of Skinner more to the point than Chomsky's famous on-

---

**8** For evolutionary psychologists' reliance on Chomsky's 'universal grammar' hypothesis, see Wallace (2010).

**9** An excellent summary of Dewey's classic paper is provided by Chemero (Chemero 2009, 19–20).

slaught, as Dennett gave Skinner some modest credit for taking consciousness as a problem to be explained, and keeping the role of the outside world in focus. “The role of effect [i.e., consequences of behavioural conditioning] will not go away,” as Dennett put it in the title of the latter of those papers (Dennett 1978b). In this he carried on the *Auseinandersetzung* with behaviourism begun in pragmatism by Dewey and Mead.

## 5 What do representations represent?

Some people maintain that from the pragmatist viewpoint, behaviourism is merely the other side of Cartesian dualism. Cartesian dualism means the separation of mind and body, usually traced, in philosophy, to René Descartes in the 17th century. Compared with the behaviourist hypothesis, this supposition has had a much greater number of supporters. This is actually no wonder. It has been found that in normal conduct of life, “we all act like dualists,” as suggests a chapter title in a recent book on cognitive neuroscience (Gazzaniga 2008, ch. 7). We act like dualists in the sense that we as a matter of course take living beings in general, human beings in particular, as equipped with intentionality and agency. This is so pervasive a feature in our mind that the cognitive processes with which we handle information from animate and inanimate objects are different; that is, they involve different perceptions and inferences (Gazzaniga 2008, 268f.). However, although we all act like dualists, this does not justify Fodor’s (Fodor 2008, 14) claim that “Descartes was right,” and pragmatism was and still is wrong, in the study of cognition. But if not this, what course then should we take?

More specifically put, how should we actually understand the idea of representation, without falling back to Cartesianism, but without accepting the ‘anti-representational’ stand of the neo-pragmatists, either. (Theirs, incidentally, is a contradictory position, in that they declare an ‘anti-representational’ general program, yet encapsulate themselves inside language, which of course from its semantic side is a system of representations.) I believe that a notion of representation is needful for making sense of cognition. Otherwise, Popper’s good piece of advice, that we’d better let our hypotheses die, instead of ourselves, would be impossible to follow. Neuroscientist Antonio Damasio has put the same idea into words in a more exact manner:

> If actions are at the root of survival and if their power is tied to the availability of guiding images, it follows that a device capable of maximizing the effective manipulation of images in the service of the interests of a particular organism would have given enormous advan-

> tages to the organisms that possessed the device and would probably have prevailed in evolution. Consciousness is precisely such a device. (Damasio 2000, 24)

For images, we can read: representations, because the function of images is to guide action by representing the world. They do not have to represent it exactly, it suffices if they make action easier. Some living creatures apparently do not have them at all. What about those creatures? Even they have to act, but their action is most vulnerable to failure, and they seldom get a second chance in their action, as do we who use images (representations) to facilitate our action.

Accordingly, and to sum up: Some notion of representation is needed in making sense of our relation to the world. However, we can now answer the question posed in the above section-title: Do representations depict the world? Yes, they do, but not in any old manner, not in an action-neutral manner, so to speak. Instead, they depict the world in terms of action-possibilities. In recent cognitive science, this is often related to J. J. Gibson's (1979) classic hypothesis about what he called "affordances". That is all right, but thereby not all has yet been said. We must pay attention, not only to how we may act, but also to how some other agents might act upon us. It often is so that their action is not benevolent, as is reminded in today's cognitive science (Sterelny 2003) and as Dewey already reminded in his *Experience and Nature* (1929). To juxtapose Dewey's and Sterelny's characteristic expressions, a process world is a hostile world. In closer terms it is "precarious and perilous," as Dewey said in the book in question. Action is a simple affair only in a neutral or benevolent environment, in a precarious environment is a much more complicated project. We must be alert toward two directions, both regarding our own action-possibilities (affordances), as well as regarding possible sinister threats from the outside.

# 6 Implications of fallibilism

Above, I made the point that the dynamic pragmatist notion of logic, coming out most explicitly in Peirce's logic, might be of systematic relevance for cognitive science. In order to demonstrate why it is so, I need to return to the fallibilist conception of action that classic pragmatism maintained. In the previous section, we made some observations that give support to the idea that the world is inherently 'involved' in our action. But precisely because of this, in order to have the initiative, we'd better deal with the world by means of representations (in the language of pragmatism: signs), instead of burning our fingers at every attempt. Sooner or later, we will lose the initiative, the world will answer with negative feedback, our action will fail, and our expectations will be disappointed. I have suggested that all this is, in a sense, an *intrinsic* feature in our action, in our relation to the world. The idea that it is intrinsic has recently received some support from cognitive science literature, and now I will first briefly review this support, and then spell out the concluding thesis of this paper as concisely as I can.

The item from recent literature that I'm referring to is Andy Clark's 2013 paper, from which I already quoted the illustrious maxim by W. Ross Ashby. However, there is more to come. Judging by the more recent findings that Clark reviews, it indeed appears to be so that our mind is connected to the world (outside reality) in a fallibilist manner.

As appears to be the practice in recent (embodied) cognitive science (Noë 2005; Rockwell 2005; Rizzolatti & Sinigaglia 2008; Chemero 2009) and as already was the case with classical pragmatism, Clark sets out from the assumption that perception and action are intrinsically linked together. Whatever else perception might be, it is always perception of action-possibilities, "enactive", as Noë (2005) calls it. However, as the possibility of failure is always lurking around in our course of action, it obviously follows that this is reflected in the workings of our mind and sense organs; though, it seems, only below the threshold of consciousness. Clark's examples are about visual perception and he cites findings that tell that already in the operations of our retina there is a 'sieving operation' taking place. In his words, "Perception is here depicted as a process that attempts to match incoming 'driving' signals with a cascade of top-down predictions [...] that aim to cancel it out" (Clark 2013, 186). To simplify just a little bit, in dealing with the outside world, our sense organs handle such perceptions that *appear* anomalous, counter to expectations. The more specific joint task of those organs and the brain is to show which of the perceptions actually are genuinely anomalous and which ones only seemingly so. "Explaining away" is the term that Clark uses in this context, the point being that we do such 'sieving' or 'explaining away' all the time in our dealings with the world. Very nice, and to me it seems that the underlying logic in this 'explaining away' is very similar to the 'logic of abduction' in Peirce's philosophy, as he took (like today's cognitive scientists do) perception and logical inference as shading in with each other, without a sharp line of demarcation.

But if we accept ideas like these, at least to an extent, and connect them with those conclusions reached above, we can forge together the argument of this paper, about the possible relevance of pragmatism for cognitive science. Let me spell it out stepwise.

1. We need not be afraid of behaviourism, because pragmatism is insured against its ill effects. If we are not afraid of it, we can grant it its modest contribution, its reminder that our life and doings are intrinsically connected to

the outside world. The traditional philosophical notion of 'externalization of subjectivity' will not do as a master-key to our actions, it may even be a downright myth.

2. Even though we are connected to the outside world, we, thank goodness, are not connected to it quite immediately. That would be too risky, and we would not have enough leeway to guide our action. Instead, Mother Nature has enabled us to deal with the world in a mediate manner, by means of representations, or signs, as the pragmatists sometimes called them. We do orient to the world by their means, but not in an action-neutral way, as the classic computational theory of mind suggested. It rather is so that we orient to the world, via representations, in an *action-intensive* way, to coin a phrase.
3. In our orientation to the world, we strive to anticipate both the success of our actions, and also intrinsic alterations in the world, those not brought about by us. This reminds that process ontology, that the pragmatists held, is not mere metaphysical ballast, but belongs to the systematic treatment of our activities. Because our doings take place in a 'precarious and perilous world' (Dewey), it follows that they are inherently fallible, the possibility of failure is built-in. This insight receives support and explanation from the cognitive science finding (Clark) that 'explaining away' spurious anomalies from genuine anomalies (the latter thus demanding specific attention) is built in our perceptual activities. So far so good.
4. But just because the possibility of error is always lurking around, so that our mind attempts to anticipate its appearance and develop strategies for its avoidance, this mind needs to be equipped with a logic that recognizes the ampliative knowledge-interest in the use of reason. This is the genuine addendum that pragmatism brings into the discussion. Points (1)–(3) above are such where classic pragmatism already has seen a great deal, but perhaps not finished the job, in view of later developments, although it only rarely is so, as far as I can see, that pragmatist conclusions and cognitive science findings actually contradict each other. However, the ampliative knowledge interest in logic is a genuine original achievement of pragmatism (although later recognized and augmented in analytic philosophy). It thus proves my point that pragmatism not only goes well together with that picture about the human mind that cognitive science is today presenting. In addition, pragmatism is even able to offer something new and needful. In case cognitive science needs philosophical founding (according to

Popper everything that we say or do reflects some philosophy), best ingredients for this purpose are forthcoming from pragmatism.[10]

## References

Bourdieu, P. (2000). *Pascalian Meditations.* Transl. R. Nice. Cambridge: Polity Press. (The French original 1997).

Chemero, A. (2009). *Radical Embodied Cognitive Science.* Cambridge, MA & London: MIT Press.

Chomsky, N. (1959). "A Review of B. F. Skinner's 'Verbal Behavior'." In: *Language* 35 (1), 26–58. Republished, e.g., in L. A. Jakobovits & M. S. Miron (Eds.). *Readings in the Psychology of Language.* New York: Prentice-Hall, 1967.

Clark, A. (2013). "Whatever next? Predictive brains, situated agents, and the future of cognitive science." In: *Behavioral and Brain Sciences* 36 (3), 181–204.

Colapietro, V. (2004). "Toward a truly pragmatic theory of signs: Reading Peirce's semeiotic in light of Dewey's gloss." In: E. L. Khalil (Ed.). *Dewey, Pragmatism, and Economic Methodology.* London & New York: Routledge, 102–129.

Damasio, A. (2000). *The Feeling of What Happens: Body, Emotion and the Making of Consciousness.* London: Vintage Books.

Deacon, T. W. (2012). *Incomplete Nature: How Mind Emerged from Matter.* New York & London: W. W. Norton.

Dennett, D. C. (1997 [1978a]). "Skinner Skinned." In: *Brainstorms: Philosophical Essays on Mind and Psychology.* Harmondsworth: Penguin, 53–70.

Dennett, D. C. (1997 [1978b]). "Why the Law of Effect Will Not Go Away." In: *Brainstorms: Philosophical Essays on Mind and Psychology.* Harmondsworth: Penguin, 71–89.

Dennett, D. C. (1995). *Darwin's Dangerous Idea: Evolution and the Meanings of Life.* New York: Simon & Schuster.

Dewey, J. (1896). "The Reflex Arc Concept in Psychology." In: *The Psychological Review* 3(4), 357–370. Republished in the edition: *John Dewey. The Early Works, 1882–1898,* vol. 5. ed. by J. A. Boydston. Carbondale: Southern Illinois University Press (1975), 96–109.

Dewey, J. (1922). *Human Nature and Conduct.* New York: Holt. Republished in the edition: *John Dewey. The Middle Works, 1899–1924,* vol. 14. ed. by J. A. Boydston. Carbondale: Southern Illinois University Press (1983).

Dewey, J. (1958 [1929]). *Experience and Nature.* 2nd ed. New York: Dover. Republished in the edition: *John Dewey. The Later Works, 1925–1953,* vol. 1. ed. by J. A. Boydston. Carbondale: Southern Illinois University Press (1981).

Dewey, J. (1938). *Logic: The Theory of Inquiry.* New York: Holt. Republished in the edition: *John Dewey. The Later Works, 1925–1953,* vol. 12. ed. by J. A. Boydston. Carbondale: Southern Illinois University Press (1991).

---

**10** Though written years later, this article bears fruits of those researches that I conducted during my Fellowship at the Swedish Collegium for Advanced Study (Uppsala), in 2009–10. I wish to dedicate this article to that wonderful institution.

Evans, N. & Levinson, S. C. (2009). "The myth of language universals: Language diversity and its importance for cognitive science." In: *Behavioral and Brain Sciences* 32 (5), 429–448.

Fodor, J. (2008). *LOT2: The Language of Thought Revisited.* Oxford: Clarendon Press.

Gazzaniga, M. (2008). *Human: The science behind what makes us unique.* New York: Harper/Collins.

Gibson, J. J. (1986 [1979]). *The Ecological Approach to Visual Perception.* New York, NY & Hove, East Sussex: Psychology Press.

Glock, H.-J. (2008). *What Is Analytic Philosophy?* Cambridge: Cambridge University Press.

Gross, N. (2002). "Becoming a Pragmatist Philosopher." In: *American Sociological Review* 67 (1), 52–76.

Hintikka, J. (2007). *Socratic Epistemology: Explorations of Knowledge-Seeking by Questioning.* Cambridge: Cambridge University Press.

James, W. (1890). *The Principles of Psychology*, I-II. New York: Holt. Republished in three volumes in the critical edition: *The Works of William James.* Cambridge, MA & London: Harvard University Press (1981).

James, W. (1907). *Pragmatism: A New Name for Some Old Ways of Thinking.* New York: Longman's, Green & Co. Republished as the first volume of the critical edition: *The Works of William James*, Cambridge, MA & London, England: Harvard University Press (1975).

Kilpinen, E. (2002). "A neglected classic vindicated: The place of George Herbert Mead in the general tradition of semiotics." In: *Semiotica* 142 (1/4), 1–30.

Kilpinen, E. (2009a). "The habitual conception of action and social theory." In: *Semiotica* 173 (1/4), 99–128.

Kilpinen, E. (2009b). "Pragmatism as a Philosophy of Action." In: S. Pihlström & H. Rydenfelt (Eds.). *Pragmatist Perspectives.* Helsinki: Societas Philosophica Fennica. Vol. 86 in the series *Acta Philosophica Fennica*, 163–179.

Kilpinen, E. (2012). "Human Beings as Creatures of Habit." In: A. Warde & D. Southerton (Eds.). *The Habits of Consumption.* Helsinki: Helsinki Collegium for Advanced Study. Vol. 12 in the series *Collegium: Studies across Disciplines in the Humanities and Social Sciences*, 45–69.

Kilpinen, E. (2015). "Habit, Action, and Knowledge, from the Pragmatist Perspective." In: U. Zackariasson (Ed.). *Action, Belief and Inquiry: Pragmatist Perspectives on Science, Society and Religion.* Helsinki: Nordic Pragmatism Network. Vol. 3 in the series *Nordic Studies in Pragmatism*, 157–173.

Kolakowski, L. (1972). *Positivist Philosophy: From Hume to the Vienna Circle.* Harmondsworth: Penguin.

Mead, G. H. (1903). "The Definition of the Psychical." Abridged in Mead 1964: 25–59. Available unabridged at: http://www.brocku.ca/MeadProject.

Mead, G. H. (1938). *The Philosophy of the Act.* C. W. Morris et al. (Eds.). Chicago: University of Chicago Press.

Mead, G. H. (1964). *Selected Writings.* ed. by A. J. Reck. Indianapolis: Bobbs-Merrill. Reprint, Chicago: University of Chicago Press, 1981.

Mead, G. H. (2015 [1934]). *Mind, Self, and Society.* The Definitive Edition. ed. by C. W. Morris. Annotated edition by D. R. Huebner & Hans Joas. Chicago & London: University of Chicago Press.

Merleau-Ponty, M. (1962 [1945]). *Phenomenology of Perception.* Transl. C. Smith. London: Routledge & Kegan Paul.

Morris, C. W. (1937). *Logical Positivism, Pragmatism, and Scientific Empiricism.* Paris: Hermann et Cie.
Noë, A. (2005). *Action in Perception.* Cambridge, MA & London: MIT Press.
Noë, A. (2009). *Out of Our Heads: Why You Are Not Your Brain, and Other Lessons from the Biology of Consciousness.* New York: Hill & Wang.
Peirce, C. S. (1976). *New Elements of Mathematics.* 4 vols. C. Eisele (Ed.). The Hague & Paris: Mouton; Atlantic Highlands, NJ: Humanities Press. (Cited as NEM).
Peirce, C. S. (1992). *Reasoning and the Logic of Things: The Cambridge Conferences Lectures of 1898.* K. L. Ketner (Ed.). Cambridge, MA & London: Harvard University Press. (Cited as RLT).
Peirce, C. S. (1992–98). *The Essential Peirce: Selected Philosophical Writings.* 2 vols. N. Houser, C.J.W. Kloesel & the Peirce Edition Project (Eds.). Bloomington & Indianapolis: Indiana University Press. (Cited as EP).
Prado, C. G. (Ed.) (2003). *A House Divided: Comparing Analytical and Continental Philosophy.* Amherst, NY: Humanity Books.
Rescher, N. (1996). *Process Metaphysics: An Introduction to Process Philosophy.* Albany: State University of New York Press.
Rescher, N. (2000). *Process Philosophy: A Survey of Basic Issues.* Pittsburgh: University of Pittsburgh Press.
Rizzolatti, G. & C. Sinigaglia (2008). *Mirrors in the Brain: How Our Mind Shares Actions and Emotions.* Transl. F. Anderson. Oxford: Oxford University Press. (The Italian original published in 2006).
Rockwell, W. T. (2005). *Neither Brain nor Ghost: A Nondualist Alternative to the Mind-Brain Identity Theory.* Cambridge MA & London: MIT Press.
Rorty, R. (1961). "Pragmatism, Categories, and Language." In: *The Philosophical Review* 70 (2), 197–223.
Rorty, R. (1980). *Philosophy and the Mirror of Nature.* Oxford: Blackwell.
Rorty, R. (1982). *Consequences of Pragmatism: Essays 1972–80.* Brighton: Harvester Press.
Rorty, R. (2006). *Take Care of Freedom, and Truth Will Take Care of Itself: Interviews with Richard Rorty.* Ed., with an introduction by E. Mendieta. Stanford, CA: Stanford University Press.
Rosenthal S. & P. Bourgeois (1991). *Mead and Merleau-Ponty.* Albany: State University of New York Press.
Ryle, G. (1970 [1949]). *The Concept of Mind.* Paperback edition, London: Hutchinson.
Short, T. L. (2007). *Peirce's Theory of Signs.* Cambridge: Cambridge University Press.
Slingerland, E. (2008). *What Science Offers the Humanities: Integrating Body and Culture.* Cambridge: Cambridge University Press.
Sperber, D. (1996). *Explaining Culture: A Naturalistic Approach.* Oxford: Blackwell.
Sterelny, K. (2003). *Thought in a Hostile World: The Evolution of Human Cognition.* Oxford: Blackwell.
Tomasello, M. (2009). "Universal grammar is dead." (Comment on Evans & Levinson 2009). In: *Behavioral and Brain Sciences* 32 (5), 470–471.
Wallace, B. (2010). *Getting Darwin Wrong: Why Evolutionary Psychology Won't Work.* Exeter: Imprint Academic.
Wegner, D. M. (2002). *The Illusion of the Conscious Will.* Cambridge, MA & London: MIT Press.

Tibor Solymosi

# Recovering Philosophy from Cognitive Science

**Abstract:** I outline a reconstruction of philosophy via cognitive science in order to restore philosophy's traditional aspiration toward justice. I constructively criticize Putnam's metaphilosophical reading of the history of thought as a learning process to note that philosophy has an intricate relationship with democracy and science. From this perspective, I advocate for philosophy as a project of reconstruction and not a project of reconciliation as many neopragmatists and neurophilosophers now see it, thanks to their Sellarsian inheritance. In rejecting the Humean-Sellarsian conception of experience, I take up and modify Dewey's transactional conception of experience and situate it within dual process/system theories of behavior in moral psychology and behavioral economics. In developing a third process or system of cultural affordances, I also elaborate on what Tschaepe has called *philosophical allostasis*, which conceives of philosophy as a creative enterprise that seeks to navigate amelioratively the tension between the old and the new.

Philosophy is notoriously difficult to define. One common pattern of philosophers since Socrates and Plato is that it is the cultural practice of negotiating the traditional ideas and beliefs of one's culture in light of new ideas produced by the cutting edge of creative inquiry. For my purposes, I consider three ways of characterizing philosophy. The first is Hilary Putnam's historical view of three philosophical enlightenments. The second is Wilfrid Sellars's conceptual framework of the scientific and manifest images of humanity in the world. The third way I introduce draws from recent work in cognitive science that conceives of human cognition as a dual-process system, one fast, one slow.

My aim is to articulate not only the need for thinking about a third system but also to recover and reconstruct an old view of philosophy that moves away from Locke's narrow view of philosophy as a handmaiden to the sciences. I contend that this myopia is detrimental to the intelligent use of insights coming from cognitive science, especially with regard to their ethical, social, and political consequences. We must recover philosophy from cognitive science in order to appreciate and utilize cognitive science to transform the lived experience of a greater number of people. I call for the criticism of democratic culture via tools (ideas, methods, data) developed by or drawn from cognitive science.

I draw together what may seem like very disparate lines of thought in order to thread together a cable (to borrow a metaphor for consilience from Peirce). Primarily, I am seeking patterns of reflective activity. The difference that makes a difference is that in considering the consequences of an embodied, embedded, extended and enactive conception of experience there comes a demand for reconsidering many of our cultural practices, most especially those having to deal with social justice. This demand, as Dewey envisaged, goes beyond science and thus beyond knowledge: it is creative, imaginative, and hopeful. It is a democratic demand, of which Dewey is an early and passionate prophet (1939). Given the evolutionarily enlightened view of pragmatism, the cash value of not only embodied cognitive science but also the recovery of philosophy lies in new possibilities for justice and the flourishing of life, human and non-human alike.

# 1 Putnam's three enlightenments

Putnam describes the history of philosophy in terms of enlightenment. He conceives of philosophy as

> reflective transcendence, that is, standing back from conventional opinion, on the one hand, and the authority of revelation (i.e., of literally and uncritically accepted religious texts or myths) on the other, and asking 'Why?' Philosophy [...] thus combines two aspirations: the aspiration to justice, and the aspiration to critical thinking (Putnam 2004, 92).

These aspirations Putnam uses to philosophize about three eras in philosophy's history.

The first Putnam calls the Platonic enlightenment. Putnam takes Plato's *Euthyphro* and Book V of the *Republic* as examples of asking *why* with an eye toward justice. In the former, both Socrates and Euthyphro are interested in justice: Euthyphro because he wishes to sue his father for murder; Socrates because he's on trial for impiety. For both men, the question of why one act is pious and another not leads to an examination of the relationship between piety and justice—which notably ends with Euthyphro's acting differently than he originally planned. Book V of the *Republic* is where Plato gives his early feminist argument for the equality of the sexes, likening the differences between men and women as nothing more than the differences between bald men and long-haired men. In this case, Socrates asks why reproductive differences should matter for anything other than reproduction, especially with regard to education and politics. In both cases, critical thought led to a difference in action.

Putnam's second enlightenment is the Enlightenment of 17th and 18th century Europe. During this period, Putnam sees two great forces characterizing the time. The first is the shift from the Divine Right of Kings to social contract theories of justice. Putnam writes that

> we can say that the lasting effect of the social contract conception—one that we tend to take for granted—is the widespread acceptance of the idea that governments derive their legitimacy from the consent of the governed; while the lasting effect of the Enlightenment's talk of natural rights is the prevalence of the idea that every human being should have the opportunity to develop certain capabilities (particularly those capabilities needed to play the role of an autonomous citizen in a democratic polity) (Putnam 2004, 93).[1]

The new science of modern physics is the second great force Putnam sees characterizing the Enlightenment. Newton's physics was on the minds of many people regardless of whether they were capable of understanding the mathematics at work.

Before introducing his third enlightenment, Putnam compares and contrasts these first two. "On the side of similarities," he writes, "there is the same aspiration to reflective transcendence, the same willingness to criticize conventional beliefs and institutions, and to propose radical reforms" (Putnam 2004, 94). Putnam goes on:

> there is the same enthusiasm for the new science (in Plato's case, enthusiasm for Euclidean geometry), and there is the same refusal to allow questions of ethics and political philosophy to be decided by an appeal to religious texts and/or myths. Yet there is also a very large difference (Putnam 2004, 95–96).

The difference is that those who have knowledge are not only the few, according to Plato, but are also the only ones worthy of ruling, of having political power. Plato's enlightenment is narrowly meritocratic: only those who have proven themselves knowledgeable are legitimate rulers, whereas the Enlightenment distinguishes itself by making both political power and knowledge available to one and all. This democratic orientation is continued in the third enlightenment.

This enlightenment, Putnam says, is "one that hasn't happened yet, or hasn't at any rate fully happened, but one that I hope *will* happen, and one worth struggling for" (Putnam 2004, 96). The best philosopher of this enlightenment is John Dewey, so Putnam labels it the *pragmatist* enlightenment. As with

---

**1** Of course, this idea of legitimacy from the consent of the governed is still very much an ideal as many non-white and non-male people have never given consent or only have very recently been able to have their voices heard.

the first two enlightenments, the main characteristic is the valorization of "reflective transcendence, or, to use an expression Dewey himself once used, *criticism of criticisms*" (Putnam 2004, 96). As with the previous enlightenments, there is the concern for democracy. However, there is a crucial difference that marks the pragmatist enlightenment from its predecessors.

The first two enlightenments valorized reflective transcendence by valuing reason over blind faith or unchecked emotional impulses. But the idea of reason became, as Dewey noted, too divorced from lived experience and the problems of human beings. The reason of philosophers led to the epistemology industry Dewey so lamented. As Putnam discusses, Dewey preferred "to speak of the application of *intelligence* to problems" (Putnam 2004, 97). In advocating a terminological change, Dewey also pushed for a rejection of traditional rationalism and traditional empiricism. Both, Dewey recognized, were aprioristic and based on faculty psychology. Neither traditional rationalism nor traditional empiricism, however, is supported by our lived experience or our best experimental science.

In his comparison of the three enlightenments, Putnam first notes that "Pragmatism in general [...] is characterized by being simultaneously *fallibilist* and *anti-skeptical*" (Putnam 2004, 99). He notes secondly that for Dewey social-scientific research must be informed by the problems of people, not by the curiosities of intellectuals. And, thirdly, Putnam notes that for Dewey communal life precedes ethics.

I take the first of Putnam's points for granted. The other two points about Dewey are two main themes I address with particular regard for embodied cognitive science. As a first pass toward that end, I now constructively criticize Putnam's criticism of the history of philosophy. I want to creatively expand on his main themes by both elaborating and adding to the story. What I find lacking and in need of consideration is that in each of these enlightenments there is a complex relationship between a new science, a crisis in democracy, and the emergence of new philosophy.

The Platonic enlightenment comes not only with the new science of what is now called Euclidean geometry but also at a time of serious democratic crisis. Plato grew up in the shadow of the Golden Age of Athenian democracy under Pericles. He came of age in the aftermath of Athens' loss to Sparta in the Peloponnesian War and the Tyranny of the Thirty, which culminated in the trial and death of Socrates. But Plato also did something that Socrates didn't do. He embraced a new technology that Socrates himself found dangerous: writing things down (Allen 2012). Plato wrote that Socrates thought writing things down made our memories weaker and thus made us more reliant on our environments and not on oneself. The precariousness of the ancient world may have warranted such skepticism of doing things with the dynamic and unruly environment. Humans' interactions with their environments, especially in the Mediterranean, were precarious, not stable. Many things could and did go wrong, so far better to have your wits about you than be dependent upon instruments that could be broken, stolen or unreliable. Writing was no simple task for the ancients; the need for servants and for not-readily available resources (most especially if you did not have the wealth for such resources, let alone the servants to employ to use the resources) was far more immediately demanding than it is today (which is to say, as easy as it is for any member of the First World to pick up a pen or type on a keyboard, the industrial and human investments and costs to afford such opportunities are great, just not as immediately apparent to its user).

By the time of the Enlightenment, the situation was quite different. The democratic crisis was not the loss of democracy as it was in ancient Athens. Rather, it was the birth of modern democracy that came hand-in-hand with the rejection of the authority of the Church. This shift to both science and democracy, however, was not strictly one of mental attitude: it was itself a technological shift. Without specific tools, such as the printing press, or beakers and telescopes, human engagement with the world would not have been as successful or as far-reaching.

A common story, if not the orthodox view, of modern science is that it has provided us with great technological innovations through the application of the theories science discovers. But, as Larry Hickman argues, this view puts the cart before the horse. The experimental framework that made modern science possible would not have been possible were it not for the tools and techniques already present, from beakers and telescopes to standardized units of measurement. Hickman argues that "Technology in its most robust sense [...] involves *the invention, development, and cognitive deployment of tools and other artifacts, brought to bear on raw materials and intermediate stock parts, with a view to the resolution of perceived problems*" (Hickman 2001, 12, italics in original).[2] This cultural matrix of modern Europe was thoroughly technical before scientific. As such, Hickman argues, we are better off talking in terms of *technoscience* instead of science and technology. He writes,

> The term 'technoscience' signals the fact that the 'scientific revolution' of the seventeenth century initiated the systematic use of instrumentation for experimental purposes, thus forever conflating putatively 'pure' theory and putatively 'applied' instrumental practice. In

2 On the similarities and differences between Dewey's philosophy of technology and Martin Heidegger's, see Hickman 2007.

> the seventeenth century, science transcended its role as 'knowledge' and became forever 'technoscience' or instrumentally experimental knowledge (Hickman 2001, 43).

This point similarly extends to the greater availability of information to people generally that afforded greater democratic practices. This availability is not just a matter of the printing press making more books, pamphlets, journals, and newspapers available; it is also a matter of the quality of life improving for enough people to have the leisure time to read, write, and converse—and for some to actively experiment.

With the pragmatist enlightenment, matters are more complex. There has been the advance of industrialization that has expedited leisure and information accessibility. But there was also the democratic crisis of the American Civil War. As industrialization and American Reconstruction go on after the war, American intellectuals come into their own, most especially in response to the new science —itself an intellectual crisis. As Dewey argued (1910), Darwin's theory of evolution revolutionized—and continues to do so today—both science and philosophy in the turn away from immaterial fixity to concrete change. This forward-looking shift is at the heart of Dewey's democratic faith as much as it is integral for understanding experimentalism in science and society.

To appreciate the dramatic change in the pragmatist enlightenment, we need only to consider the differences in culture and lived experience during the whole of Dewey's life, from 1859 to 1952. From horses and buggies to airplanes, from wood and stone buildings to steel skyscrapers, from pen and paper to telegrams and telephones, from water wells and outhouses to indoor plumbing, Dewey was born into a world drastically different from the one in which he died. All the while, from 1859 on, Darwin's influence continued to spread.

Yet Putnam does not emphasize this influence of Darwin. Perhaps he's reluctant for fear of the all-too-common accusation of scientism. Regardless, Dewey saw the biological import as crucial for appreciating the shift in conception of experience from the sensationalistic spectator view of modern empiricism to the transactional and relational view of radical empiricism. Consider the following by Dewey:

> Suppose we take seriously the contribution made to our idea of experience by biology,—not that recent biological science discovered the facts, but that it has so emphasized them that there is no longer an excuse for ignoring them or treating them as negligible. Any account of experience must now fit into the consideration that experience means living; and that living goes on in and because of an environing medium, not in a vacuum. Where there is experience, there is a living being. Where there is life, there is a double connexion maintained with the environment. In part, environmental energies constitute organic functions; they enter into them. Life is not possible without such direct support by the environment.

> But while all organic changes depend upon the natural energies of the environment for their origination and occurrence, the natural energies sometimes carry the organic functions prosperously forward, and sometimes act counter to their continuance. Growth and decay, health and disease, are alike continuous with the activities of the natural surroundings. The difference lies in the bearing of what happens upon future life-activity. From the standpoint of this future reference environmental incidents fall into groups: those favorable to life-activities, and those hostile. (Dewey 1917, 48)

Among the insights Dewey makes in this passage is the connection between experience as organism-environment transaction (henceforth Œ)[3] and living by which he means the living being's doing things with the surrounding medium. Experience is thus embodied and embedded. But this embodiment and embeddedness lack meaning, carry no significance, if experience is not also, always and already, extended and enactive. With the dynamic transaction, there is the opportunity for growth and health; without properly functioning metabolic regulatory mechanisms modulating this transaction, there is disease, decay, and death.

From this transactional view, the continuity between nature and culture is easily conceived. The oppositional relationship historically held between culture and nature does not follow from this evolutionary view. Rather, as Dewey noted and Mark Johnson continues to remind us, culture is a part or phase of nature. This intellectual shift in our orientation is perhaps the central insight of the pragmatist enlightenment. One of the greatest road blocks, however, to completing this shift and all it affords us is an attractive but ultimately flawed view of philosophy, science, and humans.

## 2 Reconsidering Sellars's scientific and manifest images

A popular characterization of philosophy comes from Sellars's "Philosophy and the Scientific Image of Man" (1963). It begins with the famous statement that "The aim of philosophy, abstractly formulated, is to understand how things in the broadest possible sense of the term hang together in the broadest possible sense of the term." That first paragraph closes with a criterion of success:

3 Following Dewey's view of experience as organism-environment transaction, I argue the diphthong of *O* and *E*, for *organism* and *environment*, respectively, captures their ontological inseparability in a way that words like *transaction* do not. See Solymosi 2012, and Solymosi & Shook 2013.

> To achieve success in philosophy would be [...] to 'know one's way around' with respect to all these things, not in that unreflective way in which the centipede of the story knew its way around before it faced the question, 'how do I walk?', but in that reflective way which means no intellectual holds are barred. (Sellars 1963, 1)

In other words, we have Putnam's reflective transcendence, that standing back and asking *why*. For Sellars, his reflections produced two distinct images to account for "man-in-the-world": the manifest and the scientific.

These two images of humanity's place in the world have influenced many leading philosophers of mind, not least because the images conveniently capture what is at stake. Due to the limits of this essay, I am restricted to focusing primarily on Sellars's view but will draw on Owen Flanagan's recent extensions of Sellars for illustration.[4]

According to Sellars, human beings held an original image[5] of the world as thoroughly personified. Accounts of the world worked by attributing personality to various parts or aspects of the world. From rivers and rocks to trees and bees, humans acted as if these things were not inanimate or mechanical, but as fully animate: as things with souls. But as time went on, Sellars argued, humans found it more useful to account for the non-human world in terms of natural regularities. So the manifest image of humans as immaterial souls, with specific mental capacities or faculties, distinguished itself from the original image. That is, humans were distinct from nature in that empirical explanation identifying natural regularities accounted for all that was non-human, whereas extra-empirical accounts were found suitable for accounting for all that is uniquely human. Early Western philosophy was occupied with articulating and maintaining the manifest image.[6]

The scientific image eventually became detailed and powerful enough in its predictions to set itself in competition with and ultimately in opposition to the manifest image. This opposition was no mere theoretical construct: it was and continues to be a felt difficulty for many people. This difficulty, as set forth in Sellars's terms, is a conflict that arises from the dichotomous terms in which it is set. Sellars himself put the question for philosophy, asking, "How, then, are we to evaluate the conflicting claims of the manifest image and the scientific image thus provisionally interpreted to constitute *the* true, and, in principle,

4 See Solymosi 2014.
5 This image should not be confused with the manifest or scientific images. The manifest image, according to Sellars, develops out of the original as nature becomes depersonalized.
6 I note that Sellars's anthropology is very narrow and may strike a non-Western perspective as exclusionary. By no means do I endorse it.

*complete* account of man-in-the-world?" (Sellars 1963, 25) Given the mutually exclusive terms, there is no clear way to reconcile the two. For Sellars, philosophy's job is to hold these two irreconcilable images together in a synoptic vision.

Flanagan modifies this conflict by setting the Sellarsian dyad into his project of eudaimonics, the science of human flourishing. For Flanagan, there is not only the hard problem of scientifically understanding phenomenal consciousness but also what he calls "the really hard problem." The original hard problem is couched in contrasting realisms, what Flanagan calls an objective realism and a subjective realism. This contrast—despite Flanagan's efforts to turn Sellars's dyad into a sextet of art, science, technology, ethics, politics, and spirituality (Flanagan 2007, 7)—recapitulates the scientific and manifest images. Flanagan sees this seemingly unbridgeable explanatory gap as leaving our metaphysics incomplete. But this gap also carries an existential threat, his *really hard problem* of finding meaning in a material world. Eudaimonics is intended to solve this problem. The problem with this problem is the same as the problem with Sellars's initial statement of the conflict. Namely, the bifurcation of images, a bifurcation seen between the subjective and the objective, value and fact, mind and body—a bifurcation between culture and nature. Or, to put it in the terms of neopragmatism, it is a gap between the space of reasons and the causal world described by natural science, wherein reasons are not causes.

The conflict as Sellars frames it ignores the influence of Darwin. Despite Sellars's notable efforts in ridding the given from accounts of experience, he nevertheless maintains a Kantian split. This split carries through the contemporary use of Sellars's manifest and scientific images. This split presumes a scientific realism ultimately based in what Lakoff and Johnson (1999) describe as *Objectivism*. On this view, there are substances first and foremost that are discovered by science. Relations between these objects are not as ontologically real as the substances themselves. To redescribe yet again what is at work here, consider Putnam's observation that we are still in the midst of the pragmatist enlightenment. In which case, Sellars is a complicated, transitional figure who stands with one foot in the Enlightenment, with its faculty psychology and objectivist metaphysics, and the other foot in pragmatism. The next step forward is toward a pragmatist metaphysics of relations and processes: a step toward radical empiricism.

Following my critique of Putnam, I sketch a criticism of Sellars's scientific realism as one of discovery and not creativity. Elsewhere, Putnam describes pragmatism's advantage over a correspondence theory of truth and inquiry that discovers and a coherence theory of truth and inquiry that creates as a method of truth and inquiry in which humans create strategies for solving their problems and subsequently discover which of them work better in the world (Putnam 2002, 97). This view of experimentalism is not only fallible and anti-skeptical but thoroughly technoscientific. For the means of creation and discovery are contingent on the particular problematic situation. Depending on what tools and resources inquirers have at hand, myriad resolutions of exigencies are available. But this view of inquiry does not fit with the Sellarsian view of awareness as a linguistic affair, of a hard divide between culture and nature. So long as the scientific realist view is held, the conflict between the images will go on without end. What this pragmatist view of inquiry proposes—from the replacement of objects with relations, of oppositions with continuities, of stasis with dynamics, of representations with affordances—will become apparent as I first redescribe philosophy yet again, now in terms of contemporary cognitive science, after which I turn to the relationship between cognitive science and philosophy.

## 3 On creating a morally imaginative culture

Johnson notes and challenges the popular dual-process view of cognition put forth in moral psychology. Similarly, John Shook and I challenge the popular dual-process view of cognition set forth in behavioral economics. For present purposes and the limits of this chapter, I present an imaginative synthesis of Johnson's processes and Shook's and my systems.

A prefatory note is in order. I contend that the Sellarsian conflict is implicit in this dual-process/system view. As I go through the fast process or system 1, keep in mind the scientific image; likewise for the manifest image and the slow process or system 2. Lastly, as is the case with Putnam's three enlightenments, I see the processes/systems as potential heuristics for inquiry.

The first process/system is fast, unconscious, automatic, effortless, emotional and intuitive, and always operative (Johnson 2014, 2; Solymosi & Shook 2013, 144 – 145). The second process/system is in near perfect dichotomy to the first: it is slow, conscious, deliberative, difficult, rational and principled, and rarely at work (Johnson 2014, 2; Solymosi & Shook 2013, 144 – 145). Johnson, Shook and I recognize the experimental data on which these two processes/systems are based. However, we disagree with the proponents of these views (e. g., Jonathan Haidt, and Daniel Kahneman) because we contend there is more to the story than just these two processes or systems. Rather, Johnson, Shook and I suggest a third process or system that better accounts for the richness of human lived experience. For the three of us, a core aspect of this third process/system is the role of intelligence and imagination (Johnson 2014, 90; Solymosi & Shook 2013, 155).

Intelligence is Dewey's alternative to reason. The historical connotations of words like reason and experience became problematic for philosophers, including Dewey who later expressed regret at using the word *experience* and considered *culture* a better term. Johnson's third process continues Putnam's valorization of reason by taking up the Darwinian insight discussed earlier. This insight rejects the faculty psychological conception of reason as opposed to emotion. In taking up intelligence in lieu of reason, Johnson argues that this third process is a process that recognizes and operates on the continuity between rational deliberation and affective impulses. This process takes effort and time, but, in the end, is transformative of experience. Since Johnson recognizes experience as Œ, he is able to situate his third process of moral imaginative deliberation in terms of Dewey's distinction between valuing and valuation, which correspond to the first and second processes, respectively.

Valuing is the habitual and intuitive process of immediate response. Values are determined evolutionarily, historically, and personally. The production of values is the result of experience so-conceived as Œ. Valuings as habits are useful and suitable for organisms in specific environments. The problem is that things change. "Consequently," Johnson argues, "we cannot simply assume that our intuitive, non-reflective appraisals and valuings are adequate when we encounter new conditions and complexities" (Johnson 2014, 92).

If we are to adjust to change, and if progress is possible, then there must be another process for the critical reflection upon our valuings, their suitability to new circumstances, etc. Such a process is valuation. It requires a *conscientiousness* and a *fallibilism*. It is a deliberate effort not only to see whether specific values hold to any standard we may value but to be willing to revise and modify any moral standards we hold.

Given his naturalism, Johnson looks to the metabolic regulatory mechanisms of homeostasis and allostasis to further ground his case for moral imaginative deliberation (2014, 93). Shook and I similarly ground our case for system 3 (following Kahneman's popular and influential sketch of the fast system 1 and the slow system 2) in these metabolic regulatory mechanisms (Solymosi & Shook 2013, 142ff.). Leaving the details to Johnson (2014) and Solymosi and Shook (2013), for present purposes the difference between homeostasis and allostasis is expressed in their different prefixes and identical suffixes. Both regulatory mechanisms seek stability as the suffix suggests. How stability is attained, however, differs accordingly. In homeostasis, the mechanisms seek a return to a previous stable set point, such as internal body temperature as achieved through shivering or sweating. In allostasis, a new stability is attained through anticipation of conditions yet to come. A classic example is the body's production of stress hormones, such as cortisol, adrenaline, and/or testosterone prior to battle and/or sex.

In recognizing the difference between homeostasis and allostasis, Johnson, Shook and I not only find a parallel with the fast process/system and the slow process/system; we also find a resolution of the tension between the fast and the slow. This tension is at the heart of the challenge Haidt believes his research presents for moral psychology—namely that human behavior is primarily the result of unconscious affective intuitions that are only later, if at all, rationalized by ad hoc principles—and at the heart of the challenge Kahneman believes his research presents for behavioral economics—namely that human decision making is more error prone because more emotional and reactionary than rational and deliberative. The error is that both the fast process/system 1 and slow process/system 2 are conceived in homeostatic terms: the equilibrium sought is a return to a previously established state.

Allostasis provides a resolution because it points a way forward for modifying Œ without a concern for returning to the old state. Critical to this shift in perspective are Johnson's appeal to Dewey's distinction between valuing/valuation and Shook and my appeal to Dewey's conception of the *situation*, the cultural milieu in which Œ occurs. Our system 3 is, in a sense, culture, as Dewey conceived it.

Prior to human culture, Œ was valuing: the habitual responses to exigencies in life. Slowly but surely the value of conscious intervention (the second process or system) evolved to better deal with such exigencies. What is insufficiently discussed is the allostatic roots of conscious activity. Johnson, Shook and I are effectively arguing that without recognizing the sensationalistic conception of experience underlying their work, researchers like Haidt and Kahneman are recapitulating the old dichotomies pragmatists like Dewey have been working to overcome: body/mind, world/mind, fact/value, scientific/manifest. The reconstruction we pragmatists are after rejects these oppositions in favor of the continuities between each term. When we recognize that Œ is continuous with culture—that the natural and the cultural are not at odds with one another but that the latter grows out of and transforms the former—we claim the relationship between these processes/systems must be reconsidered in light of the cultural situation in which such moral imaginative deliberation occurs.

Some cultures promote such moral imaginative deliberation better than others. A way to consider such differences would be to draw out contemporary parallels with Plato's devolution of constitutions of city and soul in the *Republic*. Another would be to consider, as Shook and I suggest (Solymosi & Shook 2013, 144–145), an overweight man trying to lose weight in an average American city. Given the fast-food culture of such a city, the odds of this man's success depend on his ability not only to morally and imaginatively deliberate in general but also on the specifics of his life situation. If his route to work takes him past several fast-food restaurants, his system 1 and system 2 are in conflict. But if he is able to find an alternative route, then he may not face such distress. But such alternatives are not always, regularly, or easily available to many people: their Œ is delimited by their cultures.

Johnson recognizes this in his criticism of Marc Hauser's moral-faculty approach. Morality, for Hauser, is delimited to a specific module of the brain (despite there being no such place in the brain where it all comes together for morality). Johnson argues that what the cumulating evidence is beginning to show

> is that *what we are describing are some of the many interwoven abilities that constitute a whole human person. These are capacities necessary for most of our larger-scope cognitive operations, and they depend on a more-or-less functioning brain, operating in a more-or-less functioning body (consisting of multiple interdependent bodily systems), permeated with affective valence, and inextricably intertwined with a physical, interpersonal, and cultural environment* (Johnson 2014, 148, italics in original).

Embodied cognitive science not only points to the mechanisms within the body and brain that are necessary but insufficient for moral deliberation but also opens the scope of inquiry to the medium in which the organism lives. For humans, the engagement with the environment affords better and worse ways of acting and inquiring. This transactional view demands that we find better and varied ways of inquiring into the cortex in the brain in the body in the environment; but "not as marbles are in a box but as events are in history, in a moving, growing never finished process" (Dewey 1925, 224).

## 4 Philosophical allostasis

Flanagan modifies Sellars's dyad of the scientific and manifest by expanding it into the sextet of art, science, technology, ethics, politics, and spirituality. He claims these six variables are the key ones constituting a person's experience, contingent upon one's time in history and location in space. I see Flanagan's concern for a complete metaphysics that bridges the explanatory gap between the subjective realist and the objective realist as unattainable so long as the Sellarsian bifurcation remains. On my view, metaphysics must remain incomplete so long as there is experience, which is to say as long as Œ continues. For such a transaction involves our doing things with reality, which is itself still in the making.

The Sellarsian residue—the objectivism of realism—in Flanagan's project may be excised through embracing the pragmatist metaphysics of relations and process as articulated in Johnson's Deweyan conception of the multidimensionality of experience. Reject the view that there is one special moral faculty because moral experience is unique unto itself and thus requires its own special form of inquiry. Johnson states the alternative:

> The fact of the matter is that in most cases a developing experience involves multiple dimensions, which are often inextricably intertwined, so that selecting just one descriptive category misses the richness, complexity, and depth of any given experience. (Johnson 2014, 38–39)

He goes on:

> In claiming that decisions about types of experiences are matters of emphasis and selective interest, Dewey meant that we select certain qualities or dimensions of a situation as salient in relation to some goal, value, interest, or purpose that stands out as important for us in that situation. (Johnson 2014, 39)

The difficulty Flanagan faces so long as he retains a Sellarsian framework, including the implicit conception of inquiry as one that gets behind the appearances, is that one member of the sextet risks becoming the one true and complete account of the experience, setting itself in opposition to the rest. Just as Sellars held that both the scientific image and the manifest image each claim to be the whole story, each member of the sextet risks its own hegemony over the rest.

The way to get over this risk is to reject a number of related commitments Flanagan makes, beginning with the clash between subjective and objective realisms. On that account, the mind or experience is presumed to be a neural event. Flanagan presumes that either science can fully account for the mind by understanding the brain, or it cannot and so there must be something extra-physical that by necessity science cannot explain. There are two problems with this account. The first is the presumption that the only natural account of the mind is that it is just a function of the brain. But as growing research shows, the mind is better conceived as a dynamic pattern of transactions between brain, body, and world—or as I have symbolized it as Œ. On this view, neuroscience is a necessary but not sufficient account of experience. More is required from ecology and anthropology. The second problem is the view that science can be complete and, once complete, we would have nothing further to know. Science is complete only when experience is complete.

To put it into the terms of valuation and allostasis, we inquire when Œ is out of equilibrium. We may be tenacious and let homeostatic mechanisms return the equilibrium to a previously established state. This may work for mundane problems. But to be experimental is not to rely on tradition. The old ways must be questioned, criticized, and modified if not outright rejected. This is the reflective transcendence Putnam describes as one aspiration of philosophy. This aspiration is shared by the scientific enterprise. Through critical inquiry we do not seek a return to homeostatic equilibrium, to the old. Rather, we seek innovation and creative adjustment of ourselves and worlds. Through valuation, we can imagine and deliberately work toward such change. This effort is thoroughly allostatic.

As such it is never complete. There are phases of rest, when our scientific activity has produced reliable enough data to afford new and reliable ways of acting. Yet the dynamics of experience are never such that this rest is final—save when all life function stops. This point is as true of basic biological mechanisms as it is of philosophical practice. Along this line of thought is what Mark Tschaepe has called *philosophical allostasis*. In discussing the virtue of ignorance in scientific inquiry, he writes, "Ignorance is not a given, just as problems are not 'self-set'... Through establishing a specified problem, we in turn establish that which we do not know *but desire to know* in order to solve the problem(s) that causes what we might refer to as *philosophical allostasis*" (Tschaepe 2013a, 609, italics in original). Any form of problem solving—be it artistic, scientific, technological, ethical, political, or spiritual, for instance—is inherently value-laden. It is subjective—not in the Cartesian sense of an isolated atomistic self—but in the sense that the inquiring being (be it an organism, a team, or an institution) is subject to the world around it—a world, Dewey's pragmatic realism reminds us, that often objects to our best-planned actions (see Hickman 2001, 52, and 191, n. 12).

Tschaepe elaborates on the difference between allostasis and homeostasis philosophically:

> *Philosophical allostasis* is here used as a general state of inquisitiveness combined with creative activity aimed at ameliorating exigency, which is brought about by recognized ignorance concerning questions about life, its meaning, actions, and values. This state is distinct from existential crisis, which can be conceived as the individual or collective *feeling* of ignorance with regard to life, its meaning, actions, and values, which results in coping rather than amelioration or solution. Philosophical allostasis is also the antithesis of what might be deemed *philosophical homeostasis*, which would be a general state of stability without inquisitiveness, having designated philosophical questions as meaningless or solved. Those in a state of philosophical homeostasis would likely be dismissive of philosophical questions, denying or ignoring ignorance. (Tschaepe 2013a, 609, n. 3, italics in original)

Flanagan's Sellarsian inheritance keeps his eudaimonic project stuck in philosophical homeostasis. To reach allostasis is not only to subject the philosophical terms of the project to valuation but also the philosophical presumptions of the science. Fortunately, Johnson has done much of the groundwork, clearing away much of the Kantian inheritance in moral psychology.

Such clearing away is more than the job of a handmaiden to science. Philosophy as criticism of criticism has two aspirations. One is to critical thinking—the valorization of reason, the cultivation of intelligence, the allostatic imagination—which has been my predominant focus. The other aspiration is toward justice. As with the first two enlightenments, through critical thinking, actions are changed, generally, but not always, rarely with ease, and not immediately, toward the better, toward justice. From Euthyphro's walking away from the courthouse and Plato's guardian women exercising naked with the guardian men to the rejection of Ptolemy's astronomy and the Divine Right of Kings in favor of modern physics and democracy, philosophical activity has, at times, met its aspirations, if briefly and only partially.

# 5 Cultural politics and cognitive science

Richard Rorty described philosophy's role as cultural politics, concerned over "arguments about what words to use" (2007, 3). I make no pretense to remain true to what Rorty meant by cultural politics. However, I want to close by proposing that those of us interested in radical 4E cognitive science not only draw from the experimental insights of the pragmatists with regard to method or naturalism but also go all the way. By "all the way," I mean not only aspire to critical inquiry and to justice; I also mean to look forward. We must not only look backward to Peirce, James, Dewey, and Mead. We must look forward *from* our current perspective and situation and look forward *to* our imaginative possibilities about how to transform our current situation into a better one. For we are living in very different times than Dewey or Rorty. The advances of our technoscience are astounding and show no sign of slowing down. But how we do technoscience—from our motivations to take it up to our integrations of it into our lives—is radically different than it was not only 100 years ago but even 30. Today it is big business, in which political and economic power has a great stake. We must engage in valuation over not only what words we use but also to what ends.

Consider the tension between the fast and slow processes or systems 1 and 2. At a talk I gave to a psychology association (Solymosi 2015), a retired psychology professor in the audience half-jokingly observed that advertising is the dark side of psychology, where psychology students go to get rich. It's no wonder, then, at least in the United States, that consumers suffer from a deluge of confusing financing deals from cars and computers to mortgages and student loans that are designed to make the consumer feel as though he has gotten a great deal, when, in all likelihood, he has likely been gotten the better of because the cultural milieu encourages the tension between systems 1 and 2 and actively works against the cultivation of the moral imaginative deliberation necessary to see through the deceit. This possibility of deception is a wonderful tool for politicians. Such practices are what led to Plato's concerns about the difference between reality and appearance.

Plato's concerns, of course, are inextricably intertwined with his concerns about justice and the good life. We too should be so concerned, especially with the funding of cognitive science research and its application. As Jonathan Moreno shows (2012), much of what goes on in cognitive science is behind closed doors. Such "black ops," we are told, are necessary for national security. But the compromise we make—we are, to paraphrase Rorty, taking care of security via truth instead of taking care of freedom—is not easily undone. As for the application of the scientific data, there are the concerns over super-soldiers, drones, and better weaponry generally. But we should not forget the role intellectuals have willingly played in the rationalization of torture as seen in the American Psychological Association's direct involvement in torture. As the *New York Times* reported, "The American Psychological Association secretly collaborated with the administration of President George W. Bush to bolster a legal and ethical justification for the torture of prisoners swept up in the post-Sept. 11 war on terror" (Risen 2015). The news article reports on the April 2015 publication of *All the President's Psychologists: The American Psychological Association's Secret Complicity with the White House and US Intelligence Community in Support of the CIA's "Enhanced" Interrogation Program* (Soldz et al. 2015).

With moral mistakes such as this, it is clear to me is that the philosophical vocation of speaking truth to power via critical and creative intelligence has gone into twilight. We must recover philosophy from its isolation from the lived experience and everyday lives of human beings. Cognitive science is providing us with myriad tools for aiding humans. But only through the historical aspirations of philosophy—now set in the evolutionary pragmatist enlightenment, in terms of philosophical allostasis and creative imaginative deliberation—can there be hope for justice.

To that end, consider system 3 as a cultural matrix of affordances (Solymosi 2013, and Johnson 2014, 95).[7] Recall the example of the dieter. His system 1 gets the better of him when there is immediate temptation. His system 2 is able to intervene in specific ways, by making small changes in the hope of altering the habits of system 1. What needs emphasizing here is the role of system 3. For a

---

7 Following Johnson's (2014) discussion of apprenticeship, Bill Bywater (2015) proposes that we would benefit from reconstructing apprenticeship and affordances in what he calls an *apprenticeship of affordances*.

dieter in a typical American city, it is not easy. American culture provides a deluge of affordances to eat with minimal caloric expenditure. The economics of foodstuffs do not afford Americans the healthiest of choices. If one is fortunate enough to have the money to live far from a food desert and close to a fresh produce market, one may have an easier time achieving healthy eating and activity goals. But, if one is not so fortunate, one has a much harder problem.

On this proposed view, Dewey's thoughts on habit are useful. Unlike most people, who think habits are something about the brain or the organism—consider the view that Hebbian learning is habit formation—Dewey argued that habits are as much about the environment as about the organism. Dewey observes that

> Breathing is an affair of the air as truly as of the lungs; digesting an affair of food as truly as of tissues of a stomach. Seeing involves light just as certainly as it does the eye and optic nerve. Walking implicates the ground as well as the legs; speech demands physical air and human companionship and audience as well as vocal organs (Dewey 1922, 14).

Habits "are things done *by* the environment by means of organic structures" (Dewey 1922, 14). Notice the dynamics at play in terms of Œ, especially as Dewey moves from the basic metabolic functions of breathing and eating to the more cultural—but nevertheless regulatory—functions of speech. The more complex the activity—the habit—the greater the dimensionality of the experience. It does not take much imagination for us to see ourselves having a conversation with a friend while sharing a snack as we walk down the street.

Keep this image in mind as well as this view of habit as we consider an important passage from Dewey's *The Public and Its Problems:*

> The influence of habit is decisive because all distinctively human action has to be learned, and the very heart, blood and sinews of learning is creation of habitudes. Habits bind us to orderly and established ways of action because they generate ease, skill and interest in things to which we have grown used and because they instigate fear to walk in different ways, and because they leave us incapacitated for the trial of them. Habit does not preclude the use of thought, but it determines the channels within which it operates. Thinking is secreted in the interstices of habits. The sailor, miner, fisherman and farmer think, but their thoughts fall within the framework of accustomed occupations and relationships... Thinking itself becomes habitual along certain lines; a specialized occupation. Scientific men, philosophers, literary persons, are not men and women who have so broken the bonds of habits that pure reason and emotion undefiled by use and wont speak through them. They are persons of a specialized infrequent habit. Hence the idea that men are moved by an intelligent and calculated regard for their own good is pure mythology. Even if the principle of self-love actuated behavior, it would still be true that the *objects* in which men find their love manifested, the objects which they take as constituting their peculiar interests, are set by habits reflecting social customs. (Dewey 1926, 335–336)

The specializations of our experiences constitute how we think and what we think. Be it in terms of the scientific image or the manifest or Flanagan's sextet, so long as we permit narrow specialization among most human beings, our critical thinking can never aspire to the social changes justice demands.

If we take Dewey's ideas about democracy seriously, we must consider how we structure our environments to promote at least three things. First, we need cultural affordances for the sort of interactions that promote democracy at its day-to-day level. People of different walks of life need to interact with each other regularly. Unfortunately, our current cultural affordances, especially among the tech-savvy youth, prevent such meaningful actions. Instead, we face what Tschaepe has called *dopamine democracy* (2013b). The multitasking behaviors of young people who are regularly distracted by their devices to check myriad forms of social media have been shown to hinder not only the ability to pay focused attention to a task but also to hinder their development of the habits necessary for learning and problem-solving. As Tschaepe sees it, our social media and electronic devices are bringing about the sort of democracy Plato feared: one in which humans pursue their baser pleasures—always looking for that next squirt of dopamine, so to speak. This disharmony between the dopaminergic system of anticipation and the opioid system of consummation is a direct threat to the efficacy of regulatory mechanisms.

The second thing we need to promote for creating democratic cultural affordances goes hand-in-hand with the third: the virtues, or ideals, or ends-in-view of two habits of character. Putnam named one early on, which Johnson echoed in introducing the second: fallibilism and conscientiousness. These intellectual virtues are observed but perhaps not always recognized in many walks of life. The farmer will tell you that he isn't perfect, that if he isn't mindful, he could have a bad harvest. Similarly, the neurosurgeon will tell you that she makes mistakes and needs to ensure she doesn't make them again. But we need more than these virtues within a person's job. We need to find ways of making such virtues a core part of education for *enlightened and creative democracy.* To paraphrase Dewey's democratic faith, it's one thing to be conscientious and fallible in your job but quite another to bring those virtues to public affairs, where absolutes and dogma tend to reign supreme.

Dewey called for the recovery of philosophy nearly a century ago. He was concerned the modern conception of experience led philosophers to ignore the problems of people. In so doing, the epistemology industry started and, despite Dewey's efforts to bring biology into philosophy, the historical relationship between democracy and philosophy was largely ignored. Today, the relationship between philosophy and science is improving, especially among those interested in life sciences. However, what recent events and technoscientific developments of the last few decades indicate is that the democratic concern has lagged. The aspiration of philosophy toward critical thinking is unparalleled. Yet philosophy's aspiration toward justice remains unfulfilled. The problems of everyday people do not undergo critical valuation. Rather, we allow powerful institutions interested in financial profit to dictate the situation in which the difficulties are developed, discerned, and assuaged. As things stand, we are coping because we continue to explore questions formed by another time instead of imagining new questions and redescribing our situations in order to create and discover better solutions for our exigencies. Amelioration is a live option, but what it will take is further criticism, informed by cognitive science, of standing traditions like big science and big business.

Putnam's three main characteristics of the pragmatist enlightenment are its fallibilism and anti-skepticism; its demand that social-scientific research be informed by the problems of people, not by experts, politicians, or the rich; and its recognition that communal life precedes ethics. In recovering philosophy from cognitive science, we must imagine new possibilities for aspiring toward and achieving justice. A key part of doing so is to recognize the transactional nature of human experience so that we can experiment with new ways of interacting with our world through the creation and deployment of new affordances. Given the myriad problems today, from political corruption and war to ecological concerns with food, climate and disease, it is imperative that we begin directing our scientific activity not toward the goal of financial profit but toward the genuine problems of human beings. Our science is crucial as a means for achieving democracy as a moral ideal. So long as scientific activity operates, however, in a legal and ethical justificatory scheme that puts absolutes, principles and fears before and above the fact that communal life precedes ethics, our science will not only be incomplete but also far from ameliorative.[8]

# References

Allen, D. S. (2012). *Why Plato Wrote.* Malden, MA: Wiley-Blackwell.

Bywater, B. (2015). "The Pragmatic and Embodied Origins of Apprenticeship." Pragmatism and Embodied Cognitive Science Conference, Koblenz-Landau University, Koblenz, Germany, May 26 – 28.

8 I wish to thank Bill Bywater and Jim Snyder for their helpful remarks on previous drafts of this chapter. I also wish to thank Roman Madzia and Matthias Jung for their invitation to contribute.

Dewey, J. (1910/2007). *The Influence of Darwin on Philosophy and Other Essays in Contemporary Thought.* ed. by L. A. Hickman. Carbondale: Southern Illinois University Press.

Dewey, J. (1917/1998). "The Need for a Recovery of Philosophy." In: *The Essential Dewey, volume 1*, ed. by L. A. Hickman and T. M. Alexander. Bloomington & Indianapolis: Indiana University Press.

Dewey, J. (1922). *Human Nature and Conduct.* New York: Henry Holt & Company.

Dewey, J. (1925/1989). *Experience and Nature.* In: *The Later Works of John Dewey,* vol. 1. ed. by J. A. Boydston. Carbondale: Southern Illinois University Press.

Dewey, J. (1926/1989). *The Public and Its Problems.* In: *The Later Works of John Dewey,* vol. 2. ed. by Jo Ann Boydston. Carbondale: Southern Illinois University Press.

Dewey, J. (1934/1987). *Art as Experience.* In: *The Later Works of John Dewey,* vol. 10. ed. by J. A. Boydston. Carbondale: Southern Illinois University Press.

Dewey, J. (1939/1989). "Creative Democracy—The Task Before Us." In: *The Later Works of John Dewey,* vol. 14. ed. by J. A. Boydston. Carbondale: Southern Illinois University Press, 224–230.

Flanagan, O. (2007). *The Really Hard Problem: Meaning in a Material World.* Cambridge, MA: MIT Press.

Hickman, L. (2001). *Philosophical Tools for Technological Culture: Putting Pragmatism to Work.* Bloomington and Indianapolis: Indiana University Press.

Hickman, L. (2007). "A Neo-Heideggerian Critique of Technology: Albert Borgmann." In: *Pragmatism as Post-Postmodernism: Lessons from John Dewey.* New York: Fordham University Press, 92–111.

Johnson, M. (2014). *Morality for Humans: Ethical Understanding from the Perspective of Cognitive Science.* Chicago: University of Chicago Press.

Lakoff, G. & Johnson M. (1999). *Philosophy in the Flesh: The Embodied Mind and Its Challenge to Western Thought.* New York: Basic Books.

Moreno, J. (2012). *Mind Wars: Brain Science and the Military in the 21st Century.* New York: Bellevue Literary Press.

Putnam, H. (2002). *The Collapse of the Fact/Value Dichotomy and Other Essays.* Cambridge, MA: Harvard University Press.

Putnam, H. (2004). *Ethics without Ontology.* Cambridge, Mass.: Harvard University Press.

Rawls, J. (1971/1999). *A Theory of Justice.* 2nd Edition. Cambridge, MA: Harvard University Press.

Risen, James. "American Psychological Association Bolstered C.I.A. Torture Program, Report Says." In: *New York Times*, 1 May 2015, A1. Available at: http://www.nytimes.com/2015/05/01/us/report-says-american-psychological-association-collaborated-on-torture-justification.html?_r=0.

Rorty, R. (2007). *Philosophy as Cultural Politics: Philosophical Papers, Volume 4.* New York: Cambridge University Press.

Sellars, W. (1963). *Science, Perception and Reality.* Atascadero, Calif: Ridgeview Publishing Company.

Soldz, S. et al. (2015). *All the President's Psychologists: The American Psychological Association's Secret Complicity with the White House and US Intelligence Community in Support of the CIA's "Enhanced" Interrogation Program.* April 2015. Available at: https://s3.amazonaws.com/s3.documentcloud.org/documents/2069718/report.pdf.

Solymosi, T. (2011). "A Reconstruction of Freedom in the Age of Neuroscience: A View from Neuropragmatism." In: *Contemporary Pragmatism* 8 (1), 153–171.
Solymosi, T. (2013). "Against Representations: A Brief Introduction to Cultural Affordances." In: *Human Affairs* 23, 594–605.
Solymosi, T. (2014). "Descendants of Pragmatism: Reconciliation and Reconstruction in Neopragmatism, Neurophilosophy, and Neuropragmatism." In: J. Shook & T. Solymosi (Eds.) *Pragmatist Neurophilosophy: American Philosophy and the Brain.* London and New York: Bloomsbury Press, 83–110.
Solymosi, T. (2015). "Habits, Brains, and Cultures." Northwestern Pennsylvania Psychological Association Annual Academic Research Panel, 24 February 2015.
Solymosi, T. & Shook J. (2013). "Neuropragmatism and the Culture of Inquiry: Moving Beyond Creeping Cartesianism." In: *Intellectica* 60, 137–159.
Tschaepe, M. (2013a). "Reconsidering Philosophical Questions and Neuroscientific Answers: Two Pillars of Inquiry." In: *Human Affairs* 23, 606–615.
Tschaepe, M. (2013b). "The Dopamine Democracy: Leaving Leadership to Tyrants." Philosophy Club Leadership Summit, Prairie View A&M University, 20 November.

Teed Rockwell

# The Embodied "We": The Extended Mind as Cognitive Sociology

**Abstract:** This paper explores and questions the philosophical assumptions of Game Theory—the branch of computer science that assumes that society can only be understood as the interaction of isolated rational autonomous agents. If the Game Theory of the future were to follow the lead of cutting edge cognitive science, it would replace computational models with dynamical ones. Just as Extended Cognition theories recognize that the line between mind and world is a flexible one, Dynamic social theories would recognize that the line between mind and mind is equally flexible—that we must be understood not as autonomous individuals with selfish interests, but rather as fluctuating tribes or families dynamically bonded, and motivated not only by selfishness, but by trust, loyalty and love.

## 1 Individualist or embedded or extended?

As a defender of the Hypothesis of Extended Cognition (HEC), I am often asked the following question: If the mind extends into the environment, and the environment includes a society with other minds in it, doesn't that imply that our minds are embodied by other minds? Doesn't this imply, in other words, that our minds are part of a larger group mind, and not isolated Cartesian subjects? This possibility was dealt with at great length in Wilson 2004, especially on p. 265–309, using a deflationary position Wilson called *the social manifestation thesis.* He argued that the most sensible and ontologically parsimonious way of describing so-called group minds was to acknowledge that individual autonomous subjects had certain qualities which only manifested when those individuals socially interacted with each other.

> In contrast to individualism, the externalist psychology demarcated by the social manifestation thesis views psychological states as both taxonomically and locationally *embedded* in broader social systems. In contrast to the group mind hypothesis, it does not ascribe psychological states themselves to entities, such as the group, the community, or the nation, larger than the individual and to which the individual belongs. (Wilson 2004, 301, italics added)

The key word in this passage is "embedded", which was also embraced by Rob Rupert in his criticisms of HEC. Rupert attacked HEC by contrasting it with what he called HEMC, or the Hypothesis of Embedded Cognition (Rupert 2009). HEMC theory acknowledges the ontological dependence of the mind on the world. It thus rejects what Wilson calls "individualism" in the quote above i.e. the belief that the mind and the world are merely "present at hand" for each other, and have no ontological interdependence at all. However, it also rejects the idea that the mind itself extends into the world, which also implies that, although the mind can be part of some larger group, that group cannot have its own mind. The mind is what it is because of its relationship with its world, but relationships can only exist between things that are distinct from each other. As far as I can see, Wilson and Rupert are essentially in agreement on this point. The above passage emphasizes Wilson's rejection of an extended group mind, but it also relies on the same principles that prompted Wilson to distinguish between "core realizations" and "background conditions" when discussing the physical embodiment of a single mind (Wilson 2001, 11). The background conditions are causally coupled to the core realizations, which are the only things that constitute/embody mental states. Similarly, the appearance of something like a group mind is nothing but a pattern that manifests when individual minds are causally coupled with each other. Each mind is ontologically embedded in a society and a world, but is also ontologically distinct from its society and world.

This point of view is plausible, but I do not believe it is the only one that fits the data. Consider this passage from George Herbert Mead.

> What I want particularly to emphasize is the temporal and logical preexistence of the social process to the self-conscious individual that arises in it. [...] It cannot be said that the individuals come first and the community later, for the individuals arise in the very process itself, just as much as the human body or any multi-cellular form is one in which differentiated cells arise. There has to be a life-process going on in order to have the differentiated cells; in the same way there has to be a social process going on in order that there may be individuals. It is just as true in society as it is in the physiological situation that there could not be the individual if there was not the process of which he is a part. (Mead 1934, 85–86)

Wilson's social manifestation thesis says that the individual is ontologically fundamental, and that society emerges from the interactions of individuals. Mead reverses this ontological relationship by saying that the individual emerges from, and presupposes the existence of, society. In much the same way, James' radical empiricism denies that experience is assembled from discrete sense data. James claims instead that experience is fundamentally unified, and that it is analysis, not synthesis, which requires a distinct mental act. Dewey made similar claims about the metaphysics of all part/whole relationships, at one point evoking the same cellular analogy that Mead uses in the above quote.

> Metaphysically, the case was begged as soon as an affair[,] no matter how elaborate in structure, is regarded as *being* composite. To *be* a composite is one thing; to be capable of reduction to a composite by certain measures, is another thing. (Dewey 1929/1958, 143, italics in original)
>
> Unless macroscopic things are recognized, cells, electrons, logical elements, become meaningless. The latter have meanings only as elements *of*. Dewey 1929/1958, 144, italics in original)

So who is right about this, Mead and the other pragmatists, or Wilson and Rupert? This question is, I think best phrased as "is mind extended, or only embedded?" For our present purposes, I will define the distinction between embeddedness and extension thusly. If the mind is an object with rigid borders, it makes more sense to say the mind is embedded. Language is the classic case of ontological embeddedness. The word "dog" is decisively ontologically dependent on its relationship to dogs. Without that relationship, it would only be a noise, or marks on paper, or bits stored in a computer. Yet there is no blurry line between dogs and the word "dog". Everyone agrees that there is no overlap of where the word begins and ends, and where the dog begins and ends. However, if the mind is a field with fluctuating borders, it makes more to sense to say the mind is extended. The metaphor of embeddedness presupposes something like stones in cement, or separate words on a page, and is meant to stress that there is a clear line that uncontroversially determines which is which. That is why no one would say that a field is embedded in its medium. Fields can stabilize in recurring patterns sometimes called attractor basins, which create an illusion of interacting individual substances. Nevertheless, ultimately and fundamentally, these "individuals" are as unified as the "distinct" waves in the same ocean. If the relationship between mind and world, and mind and mind, were significantly like this, it would be more accurate to describe the mind as extended, rather than embedded.

The key weasel word, of course, is "significantly". Everyone acknowledges that there is some truth to Heraclitus' claim that you can't step into the same river twice. But does this metaphysical claim possess for cognitive science what James called *cash value?* We can only decide this by asking more specific questions like "are there any concrete factual differences between these views of the mind/mind relationship? What do we lose and/or gain if we accept the social manifestation thesis and deny the existence of genuine group minds? Conversely, what truths about human nature are we currently missing because our best cognitive and social sciences assume that minds are fundamentally isolated and autonomous individuals?" To answer these questions, we need to find a science which relies on the traditional individualistic view of the mind, and see whether it has problems which could be solved by taking Mead's holistic view. The most dramatic way of resolving this controversy—one that would take it out of the realm of philosophy and into the realm of science—would be to discover mathematically expressible laws using dynamic and holistic principles which could account for scientific facts that were inexplicable when using the assumption that society is nothing but an aggregate of autonomous individuals.

I don't have the mathematical skills, or the space in this paper, to pull off a scientific revolution of this magnitude. I think, however, that I have discovered a science, which arguably could benefit from such a revolution, and I will try to tell a Jules Verne-like tale of what such a revolution might look like. This tale will rely not on math, but on the sort of stories and anecdotes that help us comprehend common sense experience, seen through speculations about possible implications of dynamic systems theory. Such stories are not science, of course, but they can indicate crisis points that might show where a scientific revolution is needed.

## 2 Game theory as GOFAI

HEC emerged as a reaction against what John Haugeland called Good Old Fashioned Artificial Intelligence or GOFAI. This paradigm for cognitive science assumed that minds are best understood as computers which contained symbolic representations, and which understood the world by manipulating those representations using the laws of logic. This view of the mind is atomistic in two senses. First of all, the mind itself is an item distinct from the world in which it dwells, including the other minds that dwell in that world. The mind is thus seen as fundamentally isolated, and its goal as an epistemic engine is to bridge that isolation and connect itself with the world and with other minds. Secondly, this kind of mind deals with information as discrete bits, which need to be assembled and manipulated. This is an inevitable consequence of the first assumption. If the mind starts as an "empty box" distinct from the world, the only way it can comprehend the world is to analyze it into parts that are small enough to fit inside the box. (The other alternative, positing that the knowledge is already in the box *a priori*, forces us to conclude that we never get to see the outside world at all.)

Most other GOFAI theorizing is concerned only with a single isolated computer system processing information input from the external word. Game Theory also starts with the presupposition that each human being is a GOFAI computer, but then throws them together into a socialized world where each competes with the rest to serve its various interests, using rational deliberation to figure out the best way of optimally maximizing those interests. Game Theory defines rational deliberation as 1) deciding the possible courses of action a person can take in a situation. 2) Deciding which outcome results from each course of action. 3) Deciding which outcome is best (given the person's value system). If the person in question is rational, she will then 4) take the action, which produces the best outcome (given the person's value system).

This use of the word "rational" is both similar and importantly different from the way it is ordinarily used in cognitive science. Game theory computer models are systems that not only cogitate, but also make choices and have preferences. That is why game theory should be seen as modeling the *self*, not just the *mind*. In most cognitive science literature, the object of investigation is usually referred to as a mind most of the time, but every so often it becomes necessary to shift over to the word "self" without explaining why. The books both criticizing and praising HEC mostly use the words "mind" and "cognition", but often seamlessly substitute conjunctions like "Self, mind, and agenthood" or "The mind as the seat of reason and of the self" (Clark 1998, 217, 69) or "Human cognition, and with it the human mind and self" or "The self, the mind, and cognition" (Rupert 2009, 4, 12).

Because Rupert conflates the mind and the self, his challenges to HEC are aimed at thinking machines that are solving purely epistemic problems, or biological systems conceived as such. Machines of this sort don't ask questions like "who am I?" or "what do I want?" The machines we build have only derived intentionality, because they serve our purposes, and do not choose purposes for themselves. No one wants a smart bomb that is so smart it turns pacifist and refuses to go off. There are admittedly robots that can make "choices" long after their programmers have abandoned them to their own devices. (An especially appropriate metaphor here.) The robots on Mars make their "own" decisions about what are the exact best moves for fulfilling their programmer's purposes. But these robots do not choose purposes for themselves. They do not, for example, decide to give up data collecting and take up interpretive dance. So how can game theory uses computer modeling to model the rational choice of those creatures from whom derived intentionality is derived? How, in other words, can we model the behavior of a creature which has not just a mind, but also a self?

Game Theory has gone with the obvious answer to this question: you can give a machine a self by making it selfish. In other words, it tries to explain the behavior of social groups by analyzing those groups into autonomous rational agents that only look out for their own posited preferences, then calculating what would manifest when these agents interact with each other. It thus presupposes Wilson's social manifestation thesis, and may arguably in some forms presuppose the reductionist individualism that Wilson criticizes. There is certainly no talk of group minds permitted in orthodox game theory. Behavior that benefits a community must be interpreted as the individual personally placing value on the group, not as the community having values of its own.

There is no denying that this kind of individualistic Game Theory can effectively predict social interactions in a variety of contexts. Like every theory, however, it has situations where it doesn't work very well, and these places are similar to other situations where GOFAI systems break down. Conversely, the dynamic theories that have offered potentially new solutions to these GOFAI problems in other areas have relied heavily on some form of HEC, recognizing that in Tim Van Gelder's words "the cognitive system cannot be simply the encapsulated brain; rather it is a single unified system embracing the [...] nervous system, body and environment" (Port & Van Gelder 1995, 13). A dynamic HEC approach to social theory would study the implications of the fact that the environment also includes other minds, and therefore my mind is always in some sense embodied by other minds. In other words, it would embrace something like Mead's idea that the group mind is more fundamental than the individual mind.

In the rest of this paper, I will be outlining parallels between the weaknesses of GOFAI, especially those outlined in books like Dreyfus' *What Computers still can't Do*, and the weaknesses of Game Theory. The existence of these parallels leads me to believe that parallel solutions are called for. If the way out for GOFAI problems in other branches of Cognitive Science is dynamic, embodied, and extended, it seems likely that a post-GOFAI game theory would also be dynamic, embodied and extended.

## 3 What game theory can't do

Historically, Game Theory appears to be an offshoot of economics. Categorically, however, economics is a subset of Game Theory. Its first principles began to emerge when economists recognized that their discipline is not the study of money but the study of incentives. Many people respond to incentives other than money, and economists now acknowledge that people will often trade money for those other incentives. People give money away to charities to receive status or high self esteem, and politicians spend campaign contributions to acquire votes. Furthermore, a great deal of rational economic behavior consists of co-operation, not competition. The whole point of "prisoner's dilemmas" is that it is often to the benefit of the selfish individual to look out for the interests of others. Consequently, economics does not require or claim that rational people must value money over everything else, nor accept that the market place is necessarily "Red in tooth and claw". These concessions are important to combat the dangerously simplistic belief that being economically rational is merely equivalent to being skillfully greedy, but they do not go far enough.

Game Theory has a view of human nature which needs to be questioned: the belief that we are fundamentally individual autonomous agents competing for resources, and that cooperation is rational only when it results in more resources for each individual that cooperates. This is an inevitable implication of the inference that because societies are *divisible* into isolated autonomous individuals they must therefore be *assembled from* autonomous individuals. If you start with that assumption, you create the following puzzle: Why do those individuals stay together? What's in it for them? This has been a very fruitful puzzle, resulting in such discoveries as the Nash equilibrium, which explains why a social interaction endures when no single individual can benefit herself by changing it. Nevertheless, this puzzle takes a lot for granted that we have every right, perhaps every obligation, to question.

Suppose that our first experience of the world into which we are thrown is of being, not an isolated individual, but rather a self which is seamlessly bonded to a family or community? In that case, there would still be a need to explain how social bonds are broken and reformed. But there would be certain fundamental bonds that we start off with as conscious beings. Having those bonds would be part of the default mode of consciousness, and consequently we would not need to assume that they would endure only when they serve the selfish interests of each individual agent. I believe that jettisoning this assumption could be the first step in solving many of Game theory's problems. In the next few sections of this paper I will discuss some of those problems, and try to show how positing an extended social mind might help to solve them.

## 4 Happiness and the hedonic calculus

When Game theory is dealing with money as the motivating incentive, it is easy to quantifiably measure which incentive is stronger. All other things being equal, we would all prefer to have \$100 rather than \$50. The problem is, all other things frequently aren't equal, and game theory has a lot of trouble when they aren't. It's only strategy is to treat all other incentives as if they could be quantifiably ranked the same way that money is. This strategy is inspired by Bentham's concept of the *Hedonic calculus.* All pleasure is assumed to be measurable in fundamental units called "utils". Pleasurable activities and objects are assigned numbers of utils in proportion to the amount of pleasure they give. Game theorists realize that there is lot of arbitrariness as to how these numbers are assigned. In most cases, the only function of the utils is to identify the fact that factor X is preferable to factor Y. In such a case, the calculations usually work as long as the number of utils assigned to X is greater than the number assigned to Y, and it doesn't matter mathematically if the difference is one or a million.

Game theorists also recognize that no activity or object gives exactly the same amount of pleasure all the time. There is a game theory formula that deals with one aspect of the dynamics of pleasure, but it is misleadingly incomplete. The concept of *Marginal Utility* is borrowed from economics as an extrapolation from the concept of *Marginal Productivity. (*Khan Academy) Marginal productivity measures the fact that each new worker added to a company has a decreasing positive effect on that company's productivity. To some degree, this is uncontroversial. A company with three workers will have a very large boost in productivity if it hires a fourth. A company with a thousand workers will have a much smaller boost if it hires a thousand and first. The concept of Marginal Utility, however, posits that pleasure follows a similar trajectory, only measurable in utils instead of dollars or goods. If the first candy bar you eat gives you 120 Utils of pleasure, then the next one will give you 100 Utils, the one after that will give you 80 and so on. With this formula, one can allegedly maximize one's pleasure by buying the products that deliver the greatest amount of utils. Let us accept the above formula for chocolate utils, and posit that the first apple you eat gives you only 80 utils of pleasure. The rational action would be to buy two chocolate bars before you buy your first apple, because the first two chocolate bars give you more pleasure.

There are a lot of obvious objections to this example, and it may seem at first that the util theory has answers to them. What about the fact that the person who buys the two candy bars has a greater chance of getting diabetes? Assign a negative util value to the diabetes, and use the hedonic calculus to prove it is more rational to eat three apples. Aren't there situations where the pleasure increases the more times you use something? There are certain movies I didn't like the first time I saw them, and a new style of music needs to be heard a few times before it can be fully appreciated. All we have to do is plot the fluctuations of utils as starting slow, then rising to a peak. In fact, we could probably use the principle of marginal utility again once that peak is reached. Don't we start to get sick of a song after it's been playing on the radio a lot? At least if we hear the song a lot in a short period of time. We could probably hear the same song (or eat the same kind of chocolate bar) once a year for twenty years and never get tired of it, even though eating the same twenty chocolate bars within the space of an hour would probably destroy the possibility of getting pleasure out of the twenty first. Do we need to add a temporal variable to the Hedonic Calculus to account for this?

Some people may consider these problems to be difficult, but not insurmountable, and not significantly different from the kinds of problems that every science has when trying to subsume a particular case under universal laws. I submit, however, that these “solutions” face an insurmountable obstacle: they are only descriptive, not explanatory. After discovering these facts about human experience, you can use util measurements to give *post hoc* descriptions of differences in human pleasure, but this theory is of no help in telling you why certain things are pleasurable and others aren't. In a way the util system is a lot like behaviorist psychology, which collected reams of data about human behavior, and refused to provide any theory of mind that could make any sense out of it (see Rockwell 1996). There are lots of very different reasons why people's pleasure fluctuates while doing essentially the same thing, or stays the same while doing different things. Artificially quantifying all of these experiences with the same util measurement obfuscates these crucial differences, and thus produces more confusion than knowledge.

Game theorist Scott Stevens tacitly admits that util measurement can never be an exact science when he warns, “it often doesn't make sense to compare the payoffs of one person to the payoffs of another” (Stevens 2008, lecture 9). The idea of a science, which only applies to one person, is clearly a contradiction in terms. We all agree that money is valuable, and what it is worth, more or less. But my utils are not your utils, and any attempt to find equivalencies between them will be successful only through coincidence or luck, not through scientific laws. I think however, that Stevens' example implies even more than this. He discusses a hypothetical example of two families who both regularly visit the same campsite.

> We can see that the Joneses are evidently bigger nature lovers then the Smiths. The payoff for the Joneses in a given visitation pattern is always twice as great as the pay off for an equivalent visitation pattern for the Smiths. This brings up an interesting question, for those interested in equity, in fairness. If it were your decision who would you give more time at the campsite? The Jones get more pleasure per visit to the site, does this mean they should get more time? Or that the Smiths should, since it takes them more time to get the same degree of pleasure? (Stevens 2008, lecture 9)

This suggestion is clearly ridiculous, and yet it follows inevitably from the util system for measuring happiness. Stevens recognizes that this example is a problem for game theory, but he underestimates the size. This problem does not arise only from comparing the payoffs to different people. There are equally preposterous absurdities that arise when we quantify utils for the same individual. Let us consider an individual who gets ten utils from eating a hamburger and ten thousand from visiting Paris. Let us further suppose that it is much cheaper to buy a thousand hamburgers than to visit Paris. Would we consider it irrational to spend the extra money to go to Paris, rather than stay home and eat a thousand hamburgers? In this age of universal tolerance, we may be inclined to refrain from judging the person who stayed home and ate the hamburgers, but the util theory requires us to actually condemn as irrational the person who goes to Paris.

I am tempted to finish that argument with the above rhetorical question, but another fact about this argument points towards the solution to (or at least an escape from) this paradox. If there is a *Homo Economicus* in the audience who believes the rational thing to do is to stay home and eat the hamburgers, there is nothing I can say to talk her out of this position. Our experience tells us that a thousand hamburger experiences is not always preferable to a trip to Paris, and we cannot explain this any more than we can explain the experience of color to a blind person. Why do you and I know that the contrary claim is crazy, even though a *Homo Economicus* computer does not? Because unlike that computer, you and I possess "the background of common sense that adult human beings have by virtue of having bodies, interacting skillfully with the material world, and being trained into a culture." (Dreyfus 1972, 3) Making wise judgments about the nature of the good life requires such a background, and I would add that part of being trained into a culture involves being seamlessly interwoven into a family and a community with real human beings you empathize with and care about.

An epistemic engine that can only attempt to understand human happiness by breaking human experience up into discrete bits cannot be part of such a community. A util is the emotional equivalent of a sense datum, with all of the problems and paradoxes of that rightly discarded theory. We can of course break our experiences up into discrete moments of Me-Happy-Now, but to do so makes it impossible to understand why I am happy, who I am that is being happy, and why those two factors are necessarily connected. We do not become happier by acquiring more and more happiness, the way we become richer by acquiring more and more money. One problem with this view of happiness is that it ignores John Stuart Mill's reminder that a worthwhile human existence requires a higher class of utils than those enjoyed by swine. For the util system to work, all utils have to be equally fungible (for a single person), or the mathematical calculations cannot go through. But a bigger problem, which Mill's qualifier could not solve anyway, is that certain kinds of happiness are clearly ontologically dependent on particular kinds of suffering. We can't have the experience of triumph without first having the experience of adversity, any more than we can be parents without having children. The util theory makes it impossible to account for this, because it necessarily ignores the fact that happiness consists of a balance of a variety of factors in a system. Some of these factors produce happiness, but some of them also produce unavoidable suffering. Happiness is thus a not a single commodity that exists as a separate substance, which can be stored up independently from the rest of our experience. To understand why I appear to be "piling up the utils" while doing X, and losing them while doing Y, I need to understand my own value system, which in turn requires me to understand my relationship to my work, my family, and my society.

All the major religions teach that an agent who spends her entire life focused on maximum acquisition of utils will end up with a very small supply of them. This is often called gaining the world and losing your soul, but it would be more accurately described as losing both the world and your soul, by trading both for a random collection of glittery things. A person who has a world has responsibilities within that world, and to the other people who share that world. If you find a place for yourself in that world, you will be happy some of the time, and consumed by drudgery and frustration some of the time. If you never commit to any sort of responsibility, and instead focus entirely on grasping at the happiness and running from the drudgery, you will have no happiness at all. I probably can't convince you of this if it hasn't been confirmed by your experience. But I hope it is clear that an epistemic engine that does nothing but hunt and store utils would be incapable of discovering this fact, if it is true.

## 5 Love and desire

Once we acknowledge that we are ontologically constituted by the network of social relationships that define our loves and responsibilities, we can account for a fact which is largely a mystery to Game theory. Why is it that we feed and clothe our spouses and children, but do not feed and clothe other people's spouses and children? Game theory could describe this fact by saying I acquire more utils from feeding my own children, but that is no more explanatory than saying opium makes us sleep because it has dormitive powers. The only explanations that current game theory can possibly offer are clearly inadequate. We are forced to say things like "we are bartering food for cuteness" when we feed our own children. This, however, does not account for the fact that we do not automatically abandon our children when we find cuter ones. Admittedly, behavior similar to this is not uncommon in some people's interactions with spouses. Sexual pleasure, unlike filial affection, has often been treated as an economic commodity or service, hence the odious term “oldest profession”. Even in relationships ostensibly built on love, there have always been affluent people who “trade up” to “trophy spouses”, and trophy spouses who accept such trades as purely economic transactions. Nevertheless, there is a general consensus that there is something wrong with this behavior, and that long-term suffering is the inevitable consequence of it. We are not only wired up to pair up, we are wired up to create far more complex emotional bonds, and those bonds fundamentally determine who we are.

This is, once again, not easy to see if we see ourselves as cognitive engines who process information in discrete bits. Such an engine comprehends the world by analyzing it into components, which requires us to see each individual the way Locke and Hume did i.e. as a cluster of perceptible qualities that can never be more than the sum of its parts. In the economic sphere, these qualities are seen as commodities that can be bought and sold, and the value of a person to me is determined by whether or not she is able to enter into a mutually beneficial exchange of those commodities. In relationships based on love, however, it is the person we love, not their attributes. That is why we take care of them when we are sick, and do not abandon them when they are no longer useful to us. It is why people want to be loved “for themselves”, not for their looks, or fame, or money. It is why we are horrified when we hear the story of the psychopath who responds calmly to the death of her child by saying “I can always have another one”.

Yet, as John Sullins has pointed out, (personal communication) the behavior of rational agents as described by game theory is essentially indistinguishable from psychopathy. We think it is OK to treat clients and employees as nothing more than means to our ends, and game theory uses this as the definition of rationality. The thing that makes psychopaths horrifying is that they also treat all of their personal relationships this way. But why is it morally acceptable to say that employees are replaceable and interchangeable, but not acceptable to say the same thing about children and spouses? In America, this distinction is widely accepted. There is, however, nothing inherently rational or necessary about drawing a sharp line between personal and business relationships entirely by blood relation. In Japan, employees are reportedly seen as more like a family, and firing a loyal long-term worker is seen as being almost as bad (or at least the same kind of bad) as selling one of your own children. My guess is that feelings of benevolence taper off for a Tokyo CEO as we move from his daughter to his butler to the janitor of his factory in Hokkaido. But regardless of whether he has even met most of his employees, the CEO can still define himself as someone who has responsibility for them. When he does this, his self is no longer selfish in the ordinary sense of the term. He is no longer an “I”, but a “we”, and this kind of "we" can function as a unit in game theory equations: For example, as a corporation with "selfish" interests of its own.

Economics recognizes that these social selves are dynamic and unstable, which creates what is called the principle/agent problem. A decision that will create a large profit this year, and drive the company out of business in five years, might benefit the CEO financially if she is planning on leaving the company before then. Corporations, created by agreements, which presuppose that all parties involved are autonomous agents seeking to maximize self-interest, are especially vulnerable to this kind of conflict. However, the vast majority of social units are bonded by love and affection (or at worst, only history and habit), and are not defined by any sort of specific contract. For this reason, it is probably more accurate to think of families, tribes, gangs etc. as more like attractor basins in a dynamic system, rather than agreements between discrete interacting agents. Many experts believe that these kinds of dynamic explanations are the most effective path to understanding biological phenomena. This could be equally true of social phenomena, if the real causal power in the universe resides at the level of attractors rather than at the level of discrete substances. In that case, game theory's paradigm of rational autonomous agents would be as inadequate as trying to explain a rainstorm by studying the molecular structure of water.

## 6 Love and money: a dynamic communitarian re-interpretation of game theory

Game theory starts with the assumption that a rational individual will terminate altruistic behavior whenever that behavior stops benefiting that individual. And yet, people do not always do this, and this fact is one of many that a dynamic and communitarian game theory might be able to explain more effectively. Such a theory would take as fundamental not that individuals compete with each other, but rather that communities do.

Hobbes was wrong when he famously described life in the state of nature as solitary, poor, nasty, brutish and short. All of these adjectives apply except "solitary". What we see instead is various kinds of tribal warfare—groups competing against groups, not individuals competing against individuals. This is what we find in places where government breaks down, such as pre-Islamic Arabia, Somalia before its recent formation of a central government, or Mafia-dominated Sicily. Colin Turnbull describes one real society, which arguably resembled Hobbes' state of nature: The Ik of Uganda. (Turnbull 1972) They were forbidden to hunt by Idi Amin, given seeds, and told to plant crops. Because they knew nothing about farming, they were forced to eat the seeds, and the resulting competition for very limited resources, turned them into a selfish society with essentially no sense of community at any level. Turnbull claimed that Ik society was not just a competition between families or clans, but a zero sum game in which parents forcefully took food from their own children, and vice versa. However, Ik society as Turnbull describes it was not a state of nature, because it was socially created by the brutal whims of a tyrant. People can be broken by extreme circumstances, and such broken people can become selfish autonomous agents in this Hobbesian manner, just as psychopaths can become similarly selfish through neural malfunctions. But that does not mean that all communities are contracts created by selfish autonomous agents. The community is ontologically more fundamental than the individual. The fact that some communities degenerate to having only single members does not change the fact we are fundamentally members of communities, not lone individuals.

All societies contain groups that interact with each other, but Game Theory ignores that there are two importantly different kinds of interactions between groups. It acknowledges that there are groups that compete *against* each other, but it ignores the fact that there also groups that compete *for* each other. Game Theory tries to explain human behavior by saying that each individual is either competing or cooperating with every other individual, and that each individual cooperates only when it is in her best interests to do so. This ignores the fact that many, perhaps even most, economic decisions are made on the basis of love, not on the basis of individual incentives. This is not true of everyone. Some people are clinical psychopaths, and if business and economics degrees do not include effective courses in ethics, it is likely that we will create more socially constructed psychopaths even among those who are not biologically damaged. But throughout history, such people have been in the minority, even among the most powerful members of the capitalist class. As ruthless as a businessperson might be while competing against other businesses, the main motivation for this competition is to enable that person to bring home the bacon to protect and nourish those people she is competing for.

## 7 Who's my tribe?

At first, the above points might seem to imply a more positive view of human nature than the rational selfishness presupposed by traditional economics. But the benevolence and loyalty that inspires us to take care of our parents, children, and pets also has its dark side. It is responsible for nepotism and corruption, racism and jingoism. It explains why a boss is more likely to fire the workers on the factory floor than the middle manager he has lunch with. Heidegger called this phenomenon being-with, and considered it a fundamental ontological structure of Dasein. But once we start saying that the group is more fundamental that its members, and that loyalty to the group is more important than justice and individual rights, aren't we in danger of trading democracy for fascism? Could this have been the factor that attracted Heidegger to the Nazi rhetoric about blood and soil? Cartoonist Ruben Bolling artfully satirized this principle by encoding it in the rule "The more a living being is like you, the nicer you must be to it" (see figure 1). The genius of this cartoon is that it shows the same principle as both acceptable and horrifying. We all agree that we should make great sacrifices for our children, and feel no twinge of guilt when destroying a radish. So how can we avoid making that same principle lead us to things like "If you view your racial group as your community, drop members of other races to 'outsider' status"?

This is a common criticism of Communitarianism, but I believe Markate Daly has found the solution for it. When static objective features like blood and soil define the boundaries of your community, it is very difficult to avoid creating a kind of fortress that keeps us in and them out. The boundaries that define us must therefore be kept fluid and dynamic by means of the principles she calls *reciprocity, care, and openness*. (Daly 2014, 271–279). Summarized and greatly simplified: 1) Reciprocity is the social foundation of responsibility. It implies recognition of the fact that if someone does something for me, I have an obligation to do something for them. 2) Care is a recognition of the fact that we can have obligations to people in our community precisely because they are helpless and cannot reciprocate the help given to them. This includes, but is not limited to, the very old and the very young. 3) Openness means that we must grant the consideration described in 1) and 2) to whomever we encounter in our daily lives.

Openness is importantly different from the abstract principle of universal justice for all. It necessarily implies the act of encountering a particular person, which creates a concrete relationship that must change you as a person. This means that every "you" becomes a "we" to some degree, and thus the line between self and others becomes blurred. Heidegger was wrong to think that Nazism rejected Cartesian Individualism. Instead it expands the Cartesian Individual to the size of a *Volk*, and replaces the centralized GOFAI brain with the central control of a *fuehrer*. Daly's dynamic version of communitarianism is not vulnerable to this kind of abuse, because it follows the inclusive principle usually expressed as "love thy neighbor". If your neighbor is Jewish or Gay, then the principle of openness requires you to include him or her in your "we".

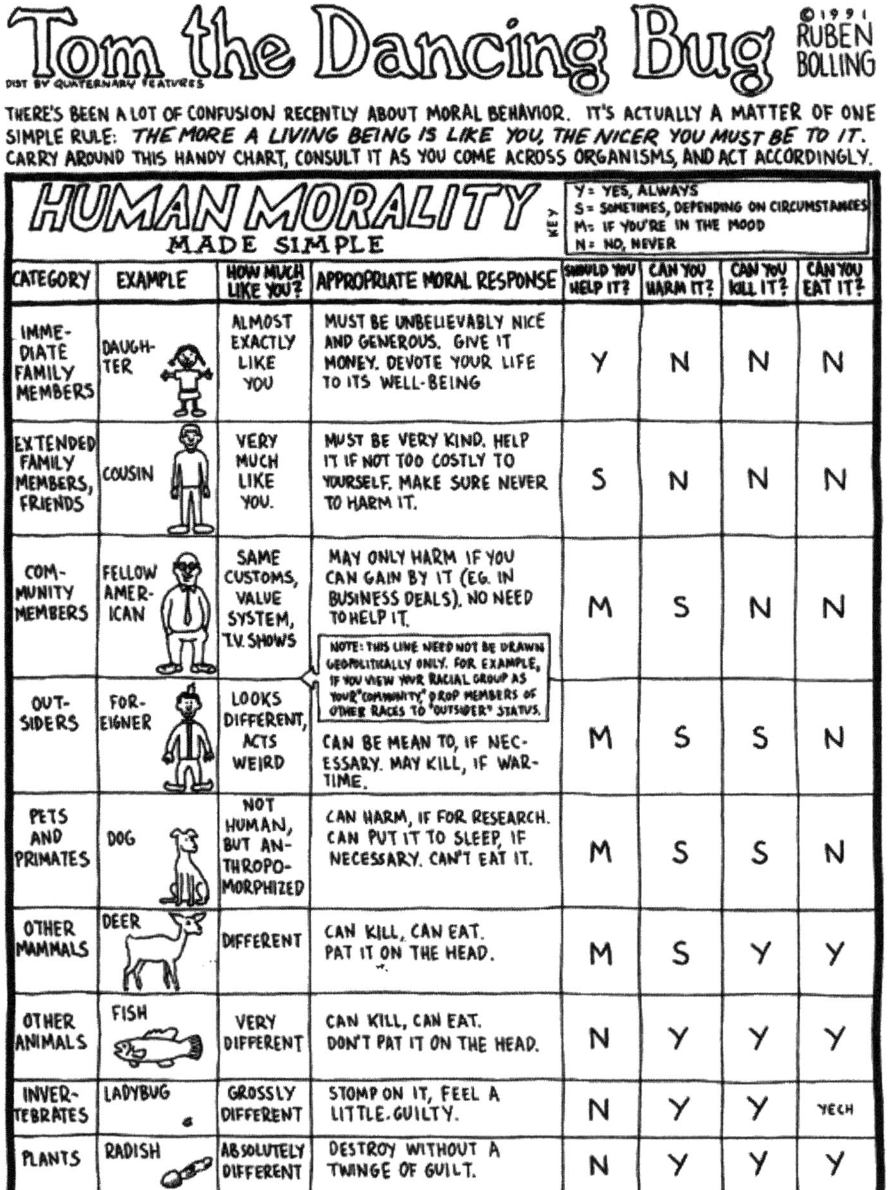


**Figure 1:** Ruben Bolling, "Tom the dancing bug"

Unless that individual or her group does something that threatens us or our principles, (or we are manipulated into thinking she has done so) our natural empathetic faculties perform their proper function, and we spontaneously include her as one of us.

Unlike GOFAI versions of Game Theory, a dynamic version of game theory would use some form of calculus to measure the degrees of this blurriness, rather than logical calculations which assume that the interests of each self is distinct from the others. Biological individuals would be only one example of the attractor basins that included families, tribes, and nations, and it would be the shifting borders of these attractor basins that would be the fundamental explanatory factors in what I will call *Dynamic Game Theory* (DGT). If Rupert and Wilson balk at the idea of granting shifting borders to the biological individual, I would refer them to Dennett's multiple drafts theory and Dawkins' Meme theory, which both deny the primacy of what Dennett called the Central Meaner. Dennett and Dawkins already acknowledge that the mind is fundamentally a cultural artifact, not just a biological organ. (Dennett 1991, 199). DGT just extends this principle one step further by saying that we are constituted not only by other people's ideas and constructions, but also by those other people as well. Because DGT recognizes that the same principles govern both the inside and the outside of the biological individual, and that these principles are dynamic rather than static, DGT thus accepts that the mind should be seen is extended rather than imbedded. As I mentioned earlier, if the mind is a field with fluctuating borders, it makes more to sense to say the mind is extended. If we could explain social phenomena by means of GOFAI game theory, then the borders between self and world would be clear and unchanging, Wilson's social manifestation theory would be correct, and a mind would be only embedded in its environment and society. If game theory progresses by transforming itself into DGT, then we would be implicitly accepting that mind is extended, not just embedded. The question is still undecided, but my money is on DGT.

## References

Clark, A. (1998). *Being There: Putting Brain, Body, and World Together Again*. Cambridge, MA: Bradford Books.

Daly, M. (2014). "Pragmatist Ethics: a Dynamical Theory Based on Active Responsibility." In: T. Solymosi & J. R. Shook (Eds.). *Neuroscience, Neurophilosophy, and Pragmatism: Brains at Work in the World*. New York: Palgrave Macmillan, 264–290.

Dennett, D. (1991). *Consciousness Explained*. Boston: Little, Brown and Co.

Dewey, J. (1925/1958). *Experience and Nature*. New York: Dover publications.

Dreyfus, H.L. (1994). *What Computers still can't do*. Cambridge, Mass: MIT Press

Khan academy, Marginal Utility. https://www.khanacademy.org/economics-finance-domain/microeconomics/choices-opp-cost-tutorial/marginal-utility-tutorial/v/marginal-utility.
Mead, G. H. (1934). *Mind, Self, and Society.* Chicago: University of Chicago Press.
Port, R. F. & Van Gelder, T. (Eds.) (1995). *Mind as Motion: Explorations in the Dynamics of Cognition.* Cambridge, MA: MIT Press.
Rockwell, T. (1996). *The Effects of Atomistic Ontology on the History of Psychology.* https://www.academia.edu/1133065/The_Effects_of_Atomistic_Ontology_on_the_History_of_Psychology.
Rupert, R. (2009). *Cognitive Systems and the Extended Mind.* Oxford, UK: Oxford University Press.
Solymosi, T. & Shook, J. R. (Eds.) (2014). *Neuroscience, Neurophilosophy, and Pragmatism: Brains at Work in the World.* New York: Palgrave Macmillan.
Stevens, S. (2008). *Games People Play: Game Theory in Life, Business and Beyond.* Chantilly, VA: The Great Courses.
Turnbull, C. (1972). *The Mountain People New York:* Simon and Schuster.
Wilson, R. E. (2004). *Boundaries of the Mind. The Individual in the Fragile Sciences.* Cambridge, UK: Cambridge University Press.

Guido Baggio

# Sympathy and Empathy: G. H. Mead and the Pragmatist Basis of (Neuro)economics

**Abstract:** The paper sketches a parallel between Mead's bio-social theory of mind and the self and cognitive sciences that would contribute to shed light on some topical economic issues concerning people's social attitudes. In particular, it would contribute to elucidate two main issues current neuroeconomics is concerned with. The first issue is epistemological: namely, how understanding the process by which people take the role of the others can help us clarify what knowledge people can reasonably have about other people's beliefs, intentions and motives. The second issue is motivational: if empathetic identification leads not only to a better understanding of another person's intentions and behavioral attitudes, this may well undermine purely self-interested choices, and instead promote other-regarding behavior. If this were the case, it would provide a sound reason for rejecting the paradigm of the economic agent as a self-interested individual. An in depth examination of the bio-social basis of empathy would permit us to link the positive, or descriptive, approach to human economic behavior to the ethical approach to it.

## 1 Behavioral economics and neuroeconomics

Neuroeconomics being a recent branch of Behavioral Economics (BE) is a significant field of economic research that focuses on the emotional and non-rational human attitudes in economic subject's behavior (Mullainathan & Thaler 2000). Intersecting economics and psychology, the BE contributes to highlight and overcomes theoretical as well as empirical limits of the neoclassic economics paradigm, which presupposes a *rational* subject that is able to act and make economic choices on the basis of the criterion of rational utility.[1] Such a new theoretical paradigm analyzes the economic behavior of individuals in situations of uncertainty, focusing their investigations on the role that the emotional tension plays in the decisions of individual economic actors. It highlights the anomaly of the prospect of reducing human behavior to the elements of coherence and ration-

1 Kahneman and Tversky (1979; 1981) were first to show empirically that emotional dimension and non-rational context play a central role in economic decisions as well as in individuals' assessment regarding their perceptions of satisfaction and well-being.

ality, rehabilitating Bentham's experienced utility theory, and supporting the idea that human preferences are not stable and people very often do not know act to maximize them.[2]

More recently, certain empirical studies have shown that economists have to regard people's social preferences to understand adequately, amongst other things, the effects of competition on market outcomes, the rules governing cooperation and collective action, and important forces shaping social rules and market failures (Camerer & Fehr 2006; Fehr & Fischbacker 2002; Van Lange 2000). More generally, economic science opted to pay attention to behavioral, social- and neuro-sciences to investigate mind-reading and empathy in pro-social economic behavior (Singer and Fehr 2005) as well as the ways in which individuals share social emotions and acts of reasoning of others (Artinger et al. 2014; Fehr & Fischbacher 2006; Gaertner, Schokkaert 2010; Charness & Rabin 2002). Numerous experiments, in particular of Public Goods Game, have demonstrated that, as opposed to the predictions of the standard economic model, participants are cooperative. Nevertheless, the cooperation is fragile due to the free-riding behavior of selfish participants (Camerer 2003).[3]

Neuroeconomics, which studies the biological basis of human economic conduct, is taking the research in social economic behavior further, by studying the biological basis of the economic behavior of human beings. The basic idea behind this research program is that "Knowing how the brain solves problems, and what specialized systems it has at its disposal to do so, challenges some of

2 One of the main limits of such a paradigm up to the recent past lied in still leaving unresolved the issue of the way the psychological significance of the preferences of the subjects implies not only the immediate pleasure (or pain relief). Kahneman's reductionist tendency identifies pleasure and pain with psychophysical reactions, bringing back the process of choosing a person to the body's reaction to sensitive stimuli. In other words, it confounds pursued pleasures with pleasures of achievement (Kahneman 2011). As William James had already pointed out against Bain's philosophy of pleasure, a pleasant act is in itself a perfectly distinct concept of the act of pursuing pleasure, "though they coalesce in one concrete phenomenon whenever pleasure is deliberately pursued" (James 1890, 812). Minimizing the overall experience of unpleasantness as a result of impulse is different from pursuing pleasure, the latter being a cause for human acts. And, even if the pleasure of achievement can, on special occasions, become a pursued pleasure, "it does not follow that everywhere and always that pleasure must be what is pursued" (James 1890, 813). For a more extended exposition of this point see Baggio (2016a).

3 A free-rider is a person who enjoys a benefit accruing from a collective effort, but contributes little or nothing to the effort (Baumol 1952). According to these studies, people can be divided in two preference types: selfish and cooperative, namely other-regarding. A vast array of studies has examined behavior in Game Theory experiments (Dictator Game, Prisoner's Dilemma Game, Public Goods Game, and Ultimatum Game).

our fundamental assumptions about how people differ from one-another when it comes to economic behavior" (Camerer, Loewenstein & Prelec 2005, 32).[4]

Despite the useful intersecting of economics, psychology and neurosciences, there seem to be at least two main limitations in the recent neuroeconomical research program. The first limit is the lack of a unitary framework that can shed light on the meanings and limits of its research (Politser 2008, 9), the second is the risk of its reductionist tendency to identify neural circuits activations with empathy (which is the basic mechanism in the creation of social preferences), or with morality (Decety & Cowell 2014). By means of interpreting behavior as an emergent event of an interplay between neural processes, neuroecomists identify empathy with social preferences, and in this way they avoid certain problematic questions concerning the mechanisms through which empathy may produce social preferences. In other words, neuroeconomists identify social preferences as the economic language translation of what empathy would be in neurosciences, so reducing empathy to brain functioning (Thoron, to appear).[5]

## 2 Mead's bio-social theory of the self and human conduct

Now, in this field of research a Mead-informed bio-social approach to human conduct would offer some interesting insights to adequately highlight the interdependency between the Self and others generally attributed to empathic attitude. Such an approach to empathic attitude would, in fact, take into account the hypothesis of Kirman and Teschl's (Kirman & Teschl 2010, 314) according to which empathy can be modulated by a number of social-cognitive factors such as emotion regulation and selective attention, as well as the kind of experiences individuals have lived through and their closeness to certain people.

4 For an overall view see Politser (2008); Ross, Sharp, Vucinich, Spurrett (2008); Ross (2005); Glimcher (2003). Recently, Emonds et al. (2011) have found differences in brain activation between preference types in the Public Good Game, showing that cooperation level between individuals is influenced by preference type and inferences about group members' likely actions. While pro-selves seem to be very sensitive to the presence of external incentives, pro-socials approach social exchange situations regarding at a mutually beneficial outcome as their most desirable option.

5 Scanning methods are impressive but not perfect. For instance, fMRI directly measures the oxigen level in different parts of the brain; however, it is an indirect measure of neuronal activity and the measurement of single neurons will be necessary before our knowledge about mirror neurons in humans will be secured.

Stressing the *innate* social dimension of human beings, Mead's social theory of mind and Self (Mead 1934; 1964; 2001) proposes a complex idea of the Self as the result of mental and non-mental behavioral attitudes, the latter being the expression of human abilities to act in a social-natural environment. According to Mead, individuals acquire their identity through communication processes in which the organized community or social group gives the individuals their unity as Selves. The sum of the vocal and gestural replies of others to an individual's stimuli constitute the attitude of the whole community towards the individual's behavior. Mead called such a social attitude "generalized other", depicting it as the crystallizing of all particular attitudes into a single attitude or standpoint (Mead 1934, 90). Individuals take the general attitude towards the various phases or aspects of the common social activity or set of social undertakings in which, as members of an organized society or social group, they are engaged.[6]

Behind the common social activity there is a set of social primitive instincts by the nature of which the distinction between the 'me' and the other is given. The instincts are social in the sense that the individual's gestural and vocal responses arise in answer to indications of various movements by other individuals of the group. This means that the beginning of human communication and thus, of the genesis of the Self, is not in imitation, but rather in *cooperation* where conduct differs and the act of the one answers to and calls out the act of the other. As Mead argues:

> Imitation becomes comprehensible when there is a consciousness of other selves, and not before. However, an organization of social instincts gives rise to many situations which have the outward appearance of imitation, but these situations—those in which under the influence of social stimulation, one form does what others are doing—are no more responsible for the appearance in consciousness of other selves that answer to our own than are the situations which call out different and even opposed reactions. Social consciousness is the presupposition of imitation [...]. (Mead 1964, 100)

6 "Only in so far as [the individual] takes the attitudes of the organized social group to which he belongs toward the organized, cooperative social activity or set of such activities in which that group as such is engaged, does he develop a complete self or possess the sort of complete self he has developed. And on the other hand, the complex cooperative processes and activities and institutional functioning of organized human society are also possible only in so far as every individual involved in them or belonging to that society can take the general attitudes of all other such individuals with reference to these processes and activities and institutional functionings, and to the organized social whole of experiential relations and interactions thereby constituted—and can direct his own behavior accordingly." (Mead 1934, 155)

## Pro-social emotions and sympathy

Behind the above-mentioned idea is Mead's and Dewey's *Theory of emotion* (Mead 1895; Dewey 1894; 1895), according to which the expression of certain emotional attitudes are the marks of the passing over of emotional attitude into *communicative gesture*. According to such a theory, there is a distinction between movements that are preparatory to a set of acts and movements that are useful in themselves as accomplished ends, the first being a *mode of behavior*, i.e., the culmination of which is the performance of the act.[7]

The early stages of these activities give rise in the individual, in parallel to antisocial socio-physiological impulses (Mead 1934, 303–304), to what is currently called human *pro-social emotions*, namely psycho-physiological reactions to movements of other group members that induce agents to engage in cooperative behavior. The bio-social mechanism is, in fact, at the basis of the feeling of *sympathy*, which consists, according to Mead's definition, in taking:

> a distinctively human, that is, self-conscious, social attitude toward another individual, or to become aware of him as such, is to identify yourself sympathetically with him, by taking his attitude toward, and his role in, the given social situation, and by thus responding to that situation implicitly as he does or is about to do explicitly. (Mead 1934, 300)

Sympathetic identification is here depicted as a natural process presenting both cognitive and affective aspects, since it refers to both the capacity to represent the other's intentions and beliefs (attitudes) and to share his/her feelings.

---

7 As Dewey argues: "The emotion is, psychologically, the adjustment or tension of habit and ideal, and the organic changes in the body are the literal working out, in concrete terms, of the struggle of adjustment" (Dewey 1895, 185). The explanation of the rhythmical character of action represents the most important contribution of Mead's physiological perspective to Dewey's theory of emotions, according to which: "all well-arranged or economical action is rhythmical". Mead argues that the arousal of emotional tones of consciousness that lie in the physiological reply to symbolic stimuli connected to the rhythmical repetition of physiological stimulation. In particular, the rhythmical aspect, which is at the basis of human actions, represents the link between symbolic stimuli (which have an aesthetic value expressed at the beginning of war and love dances in primitive populations) and the instinctive acts at their basis. It is noteworthy that more recently Gregory (1999) pointed to rhythmic aspect in dance and music as aspects of the role bodies have in unconscious mutual adaptations and that Panksepp (2001; 1998) has advanced a theory of emotions similar to that of Dewey and Mead. In particular, Panksepp addresses the need to complement neurophysiological terms with behavioral, psychological, and cerebral explanations of emotions, in a triangulation which he defines "affective neuroscience".

Mead's notion of sympathy has its philosophical basis in David Hume's *Treatise of human nature* and in Adam Smith's *Theory of moral sentiments*, in which sympathy is referred to as a quality of human nature at the basis of communication and social rules. According to Hume, in fact, (Hume 1740, Bk II, Part II, sec. V) sympathy is a disposition to share sentiments, to participate in the emotional life of others and to form general rules to pursue. Smith, who further elaborated Hume's use of sympathy to explain economic phenomena,[8] considers sympathy as individual's "fellow-feeling with any passion whatever" which may arise from "the view of a certain emotion in another person", but not "from the view of the passion, as from that of the situation which excites it" through the imagination (Smith 1759, I, I, 1, §10). This is due to the fact that people have *learned from experience* the ways to approve or disapprove of the opinions of others (I, I, 3, §§ 3 and 4), as well as their own (III, 1, §§ 4 and 5). According to Smith, sympathy is a complex imaginative process that presents emotive content and that helps the individual to evaluate the complex situation to be approved or disapproved. Sympathy is, therefore, strictly intertwined with moral evaluations which interest also self-evaluations.[9]

---

**8** On Smith's further elaboration of Hume's sympathy see Lecaldano (2014, 17–19; 24–30). An interesting insight of the cognitive and affective dimensions of Smith's notion of sympathy is elaborated in Bréban (2015).

**9** It would be interesting to develop a comparison between Mead's "generalized other" and Smith's "impartial spectator", aiming at highlighting the crucial role imagination assumes in both authors' moral theories, and specifically in their ideas concerning individual's learning of social rules through the taking "the role of the other" as well as the perspective of the "interior man" (as Smith called the impartial spectator in his Theory of Moral Sentiments). Moreover, it would be interesting to compare the passage of Mind, Self and Society where the distinction between "me" and "I" is exposed with the paragraph 6 of the third part of the Theory of Moral Sentiments regarding the distinction between two "I", "the examiner and judge" and "the person whose conduct is examined into and judged of". As Smith argues, (III, 1, § 6) the first I "is the spectator, whose sentiments with regard to my own conduct I endeavour to enter into, by placing myself in his situation, and by considering how it would appear to me, when seen from that particular point of view. The second is the agent, the person whom I properly call myself, and of whose conduct, under the character of a spectator, I was endeavouring to form some opinion". Similarly, Mead speaks of the me as both the historical Self as well as the Social perspective about rules opposed to the I as the active and impulsive Self. However, this is not the subject of this article so we cannot delve further with it here.

# 3 Sympathy and empathy. Meadian pragmatist basis for (neuro-)economics

Although Mead's notion of sympathy is rooted in Hume's *Treatise* and Smith's *Theory*, in my opinion a more fecund parallel with Mead's theory of sympathy can be nowadays found in the current neuroscientific and neurophenomenological hypothesis on "empathy" related to the recent studies on *mirror neurons*. A parallel between Mead and neurosciences would, in fact, contribute to shedding light on some topical economic issues concerning people's social attitudes. In particular, it would contribute to elucidate two main issues which current neuroeconomics is concerned with.[10] The first issue is epistemological: that is, understanding the process by which people take the role of the other will help to clarify what knowledge people can reasonably have about other people's beliefs, intentions and motives. The second issue is motivational: that is, if empathetic identification leads not only to a better understanding of the other's intentions and behavioral attitudes, this may well undermine purely self-interested choices, and instead promote other-regarding behavior. If this were the case, it would again provide a good reason for rejecting the paradigm of the economic agent as a self-interested individual (Kirman & Teschl 2010, 305). Examining in depth the bio-social basis of empathy would permit us to link the physio-psychological approach to human economic behavior to the ethical approach to it.

Now, generally speaking economics is based on the assumption that people are capable of predicting the actions of others, that is, they are able to put themselves in other people's shoes and understand their motives and beliefs (this is particularly true in Game Theory, in which the outcome is related to successful prediction of the actions of others). However, as the range of economic studies expanded, the results have not become much clearer. Economists still know little about the reasons that enable people to put themselves into the roles of others, and how this ability interacts and conditions their own beliefs and preferences.

More recently some authors argued that refererring to Theory of Mind (ToM)[11] and empathy is necessary to explain an individual's decision-making process. According to Leipold et al. (Leipold et al. 2013, 246–247), for instance, referring to ToM would help to better understand how individuals take the intentions of others into account when making decisions. Along the same lines, Artinger, Ex-

10 For more general topics of urgent interest across neuroscience and philosophy from the perspective of pragmatism see Shook and Solymosi (2014).

11 Theory of Mind is the ability to attribute mental states such as intentions and emotions to oneself or to others (Premack & Woodruff 1978).

adaktylos, Koppel, and Sääksvuori have put together empathy and Theory of Mind to develop an experimental model for investigating the underlying psychological mechanism of social preferences in a neutrally framed context, arguing as conclusions of the tests that "fair individuals exhibit more accurate beliefs about the behavior of others but that Tom can be used to pursue both self-interest and pro-social goals depending on the principle objectives of the person" (Artinger et al. 2014, 2). In these studies, "empathy" is interpreted as a human capacity to share social emotions, whereas ToM as the capacity to construct the mental states of others making inferences about beliefs, intentions and emotions.

Nonetheless, referring to ToM as an integration of a "vague" human capacity to share social emotions, economists—and this is also particularly true for neuroeconomists (see Politser 2008, 7)—do not have a clear theoretical framework yet that can enable them to interpret notions that do not have an origin in the economic science, for example, the notion of empathy. Moreover, they tend to distinguish between affective and cognitive dimension in social interactions and then in empathetic identification, revealing a latent cognitive dualism. Finally, it also has to be noted that in modern economics notions of empathy and sympathy have been used synonymously, either to render the idea of interdependent utility functions, or to convey the process of imagining oneself in someone else's place.[12] However, psychologists distinguish clearly between the two concepts.[13] Sympathy refers to a process of our affection for the *same* feeling of another, namely an identification with others, while 'empathy' is conceived as a capacity to take the role of the other and understand what is inherently *foreign* in others with regard to our own self. If empathy tends to expand our relational field, on the contrary, sympathy could be considered as an egoistic natural mechanism of identification with those who are *like* us. From this perspective, sympathizing with others' feelings and beliefs refers to a total identification of her feelings and beliefs with ours. On the contrary, to empathize with other's intentions and beliefs is to put ourselves in his/her situation, that is, to take his/her attitude. Like Mead's reference to sympathetic identification, the concept of

---

**12** According to Fontaine (Fontaine 2001, 388), a distinction should be made between a "partial empathetic identification" which refers to an imagined change of objective circumstances with another, and a "complete empathetic identification" which consists in the imagined change of objective circumstances and subjective features with another. For sympathy/empathy distinction on social psychology see Wispé (1986; 1991).

**13** I refer here in particular to Corradi-Fiumara (2009). A similar elaboration of the notion of "empathy" in psychology can be found in Rogers (1975). For the elaborations of the notion of "empathy" see Gieser (Gieser 2008, 307–313).

empathetic identification depicted here refers to "an individual who perfectly naturally arouses a certain response in himself because his gesture operates on himself as it does on the other" (Mead 1934, 300). Thanks to empathetic identification, an individual can comprehend the other's behavioral and expressive attitudes and interact with him through a so-to-speak *natural social-enlargement* in which the affective and cognitive dimensions are closely related, without ceasing to be able to clearly separate which beliefs and feelings belong to which self.[14]

This process has its *conditio sine qua non* in the social interaction through which individuals assimilate and introject others' behavioral attitudes, and its biological support in the simulation function of mirror neurons. Empathy is generally depicted in neurosciences as a pre-reflexive embodied mechanism that plays a role in understanding the intentional actions of others, by ascribing meaning to their unreflective facial expressions and being able to "read" their emotionally-laden bodily attitudes.[15] It is worth noting that nowadays Mead's theory of gestural conversation as the expression of an innate social dimension of human beings is undergoing a new revival among neuroscientists. Rizzolatti and Sinigaglia (Rizzolatti & Sinigaglia 2006, 48–49; 148–149), for instance, explicitly refer to the work of Mead as supporting the hypothesis according to which the arising of human language would have had at its basis a gestural attitude linked to a pre-conscious mechanism for understanding the gestural attitudes of others. And McNeill (2005, 30–31) refers to what he has called "Mead's loop theory" as "a capacity, not present in other primate brains, for the mirror neuron circuit to respond to one's own gestures as if they belonged to someone else".

Mead himself argued that all that is innate

> in connection with minds and selves is the physiological mechanism of the human central nervous system, by means of which the genesis of minds and selves out of the human social process of experience and behavior—out of the human matrix of social relations and interactions—is made biologically possible in human individuals (Mead 1934, 237n).

14 Cf. Bower (1983). Although recently both cognitive and affective components have been seen as crucial factors of empathy (see Vreeke & van der Mark 2003; Preston & de Waal 2002), Giusti and Locatelli (Giusti & Locatelli 2007, 11) have pointed out that Mead has been among the first authors to introduce the cognitive element in the affective meaning of "empathy".

15 See among others Borghi, Cimatti (2009); Gallese (2006; 2009a; 2009b); Gallese, Goldman (1998); Rizzolatti, Sinigaglia (2006).

I would like to make it clear, however, that Mead did not intend to give a reductionist interpretation of the human mind and I do not either. It is difficult to trace the origin of human understanding of the minds of others in mere motion of electrochemical synaptic impulses between neurons. To *understand* preferences and attitudes of others means to refer to higher cognitive abilities in which mirror neurons have a functional role. People's ability to understand the intentions and preferences of others is not identical to people's brain activation of the neural mimetic mechanism. There would be, otherwise, a specular identification, a mere imitative mirroring. And, as I mentioned before, according to Mead, imitation becomes comprehensible when there is a consciousness of other selves, and not before. As he writes:

> as the problem of communication is the problem of gestural understanding, also the sensible stimulus cannot be considered as a stimulus to reproduce what the organism has seen or heard, rather it is a stimulus contributing to the social act [...] And, to be saying that the ideas which we have of the actions of others are ideo-motor in their character [...] does not make out of imitation the means of their becoming ideo-motor (Mead 1964, 100).[16]

After all, the notion of "simulation" has not been given a uniform, invariable definition in neuroscience. Some neuroscientists use the concept with reference to the neural mechanism through which we understand the minds of others,[17] whereas others consider simulation as a conscious process depending on a deliberate re-activating of some beforehand accomplished actions.[18]

Thus, a distinction should be made between a *simulation mechanism* referred to as an embedded "proto-intentional" mechanism (similar to what Madzia and Jung called "bodily-interactive" intentionality)[19] of recognition of the attitudes of others (namely, a neural recognition of a co-specific's intentions),[20] and a *simulation process* which refers to interpersonal interactions in which

---

**16** More recently Hickok (Hickok 2014, 214) has argued against the "myth" of neuron mirrors mechanism of imitation that we mimic because we are social, and that social situation guide imitation, rather than the opposite.

**17** Gallese, Goldman (1998).

**18** Decety, Ingvar (1990).

**19** Madzia and Jung (2016) argue that intentionality can be divided into "bodily-interactive" and "social-symbolic" intentionality. The former shows that intentionality is not primarily a property of the mind but of the active body and may can serve as a conceptual ground for social (second-order) intentionality.

**20** Cf. Gallese (2009b). The concept of "proto-intention" refers to Damasio's concept of "proto-self" (Damasio 1999, 170 ss.), which is the biological preconscious function, which pre-exist to the self-feeling. Like the proto-self, "proto-intention" is a function arising through the interaction between neural signals and social and natural environment stimuli.

the individual's higher cognitive and affective abilities, that have simulation mechanism as their material causal condition but are not reducible to it,[21] are involved in a process of identification with the attitudes, desires, preferences, and intentions of others.[22]

From this perspective, a pragmatist pluralistic approach to empathetic identification would provide a sketch to adequately highlight the interdependent and reciprocally conditioning, organic relation between the self and others generally attributed to the empathetic attitude. The simulation process, in fact, mediates between our ability to share meanings about actions, intentions, feelings and emotions, and the empathetic identification that allows us to put ourselves in the shoes of others and attempt to interpret, more or less correctly, the beliefs, intentions and motives of others.

## 3.1 Pragmatist pluralistic approach to empathetic identification

The basic idea behind such an approach originates from Gallagher's suggestion (Gallagher 2015) that a pluralist approach to social cognition would treat simulation theory, theory-theory, and interaction theory as different perspectives on conscious, non-conscious, personal and subpersonal dimension, as well as brain activation, social and cultural environments, individual processes and intersubjective interactions implied in situated contexts (Gallagher 2015, 461).[23]

It is worth noting that Gallagher (2014) has recently referred explicitly to pragmatist conception of cognition as supporting an integrative view different from enactive or extended conceptions. He points out, in particular, Mead's characterization of the reachable peripersonal space as the "manipulatory area" which refers to a readiness to grasp (Mead 1938, 103) instead of a representation of the perceived object. And he referred to Dewey's notion of "situation" as the coupling of the organism-environment and its characterization (taken from

21 There is a distinction between what is a "causal condition" and what is a "cause". It is obvious that without the brain there is no mind, the brain is the material side of thought, but alone does nothing without some external "stimulation". The brain is an active cognitive apparatus that it does not mean a consciousness apparatus but the causal condition of it.

22 On this point, it would be possible to find out some potential connections to the use of neuron mirrors and the paradox Khalil (2011) highlights about the rising of divergent processes from the same neuron mirrors activation.

23 See also Fiebich, Gallagher, and Hutto (2015).

Mead, and not taken over and made clear by Mead) as always social (Gallagher 2014, 120).

Therefore, a pragmatist pluralist approach seems to be particularly useful for developing an ontogeny of empathetic identity processes which include embodied, minimal experiential, affective, intersubjective, and psychological aspects. From such a perspective we will refer here to Goldman's (1995) and Gordon's (1995) simulation theory (ST) as well as to Martin et al.'s (2008; 2013) neo-Meadian interactionist theory and Gallagher's notion of enactive social perception (Gallagher 2015; 2005; Gallagher & Hutto 2008).

According to Goldman everyone has a set of mental mechanisms that allow one to simulate and ascribe mental states to oneself and others. As he argues, to use "the simulation heuristic, one would feed pretend initial states into such a mechanism, namely, the initial states of the targeted agent, and see what further states the mechanism produces" (Goldman 1995, 718). In other words, Goldman considers the argument from analogy according to which one first recognizes his/her own mental states under hypothetical conditions and then infers similar states in others. Goldman's theory, however, seems to overestimate the cognitive mechanism at the basis of speculative knowledge of the mind of others without proportionately considering the affective dimension that empathy manifests. To balance out Goldman's perspective, Gordon emphasizes the emotional side of the simulation process, referring to a process of "imaginative shift" through which we transform ourselves and become the other. Gordon points out that the imaginative shift reflects "a much deeper, more important shift" (Gordon 1995, 734), similar to an actor's re-centering their egocentric map in order to become the character they play. Although unlike actors we don't transform ourselves audibly or visibly, both the actor's knowledge of fictional roles and our knowledge of actual minds belong in the category of practical knowing how to transform ourselves into others. As Gordon argues, the "imaginative shift in the reference of indexicals reflects a much deeper, more important shift in the emotional empathetic simulation, a kind of 'contained' emotional contagion that we recognize the emotional expressions of the others" (Gordon 1995, 738).

Even if Gordon's notion of "imaginative shift" balance Goldman's theory, the main limit of their approach is that it is too self-referential and does not offer a theoretical explanation of the situated, embodied, dynamic, and coordinated way of being and acting of selves. To put it differently, it regards a first person perspective only, based on the analogical argument of "as if", a first-person model used to infer the beliefs and desires of others (Gallagher 2007a; 2007b; 2008). Moreover, it presupposes the capacity to simulate others basing it on a neural activation, that is neutral, coupled with a social intentional capacity (or as Madzia and Jung put it in the introduction to this volume, a "social-symbolic" intentionality), that, on the contrary, has to be implemented. This does not mean that ST is not relevant to social cognition, in particular in its being a higher order cognitive and affective processing. However, it cannot be considered as a default way individuals understand others.

To counterbalance the limit of the pervasive way of ST and to offer an explanation of the understanding of the actions of others, Martin, Sokol and Elfers, neo-Meadian interactionists, proposed a neo-Meadian approach to human agency. The neo-Meadian interactionist theory is in favor of a situated, embodied, dynamic, and coordinated way of being and acting of selves, highlighting the biosocial basis of human beings (Martin et al. 2008). Martin and Gillespie showed how human agency should be conceptualized in terms of distancing from immediate experience, highlighting the process of how social interactions, institutions, and symbolic resources foster the development of agency in increasingly complex ways (Martin and Gillespie 2010). Moreover, Martin elaborated a new method for conducting studies of life narratives, what he called Life Positioning Analysis (LPA), consistent with a theory of how personhood develops through embodied, situated participation in social interactions (Martin 2013). Related to neo-meadian social interactionism is also the idea that we understand other's actions, intentions, and emotions in their embodied comportments in contexts of dynamic interactions in everyday social situations, avoiding the need to engage in mind-reading or simulating (see Madzia 2013a).

Gallagher's notion of enactive social perception (Gallagher 2015; Gallager & Hutto 2008) integrates the theoretical framework of the genesis of understanding other's intentions, emotions and desires, through a genesis based on the idea that gestures implicitly contribute to the formation of our cognition. In particular, Gallagher (Gallagher 2005, 121) argues that even if we are not explicitly aware of our gestures, and also in circumstances where these offer no contribution to the communication process, they implicitly contribute to the formation of our cognition. In other words, gestures emerge as interacting parts of a cognitive ability that involves the entire organic relationship. Gallagher believes that the body and the "body-oriented" assume a "*pre-noetic*" role previous to conscious acts of perception, memory and judgment: body activities contribute to structure consciousness though they are not explicit in its content (Gallagher 2015, 32). As it is well known, Gallagher's approach takes root in the broader theory of embodied mind, according to which body influences cognition and affects the action of the individual in the same way the environment in which the individual is located does, conditioning the ability to select the stimuli.[24] The agent perceives the

---

**24** See also Gallagher and Zahavi (2008). Between latest versions of embodied theories, two are

world from a defined spatial perspective (see Mead 1926) which shapes his experiences. In this context, gesture assumes the task of structuring cognition in the fulfillment of thought. Some aspects of the mind are the result of gestures, expressive movements, and communicative practices (Gallagher 2005, 121).[25]

Now, what is important in these different approaches is that they present some elements useful for sketching a pragmatist pluralistic approach to empathetic identification. Such an approach would intertwine the simulation process with social interaction and enactive social cognition in order to meet the need for a social ascribing practice as the condition for the individual's choice processes linked to other-regarding preferences.

The basic idea of this theory is that everyone has embodied interactive experiences in a socio-cultural context since birth thanks to his/her ideo-senso-motor capacities of interacting with and reacting emotively to physical and social objects surrounding him/her (see Gibson 1988; Gallagher & Hutto 2008; Trevarthen 2011). At the beginning, through gestural and vocal mimetic interactions referring in the first phases to emotive and affective expressions, the individual affectively reacts to their co-specifics' non-mental behavioral attitudes (see Bowlby 1979; Panksepp 1998; Liotti 2005). Through the mimetic act of role-taking in social situations (see Cozolino 2006; Franks 2010),[26] an individual experiences a set of affective and cognitive mechanisms which stimulate him to generate more complex mental and non-mental behavioral attitudes from the initial ones. Such processes, which compose the affective and cognitive experiential background,[27] are based on the *apperception* (Mead 1903; Smith 2010), namely

the more interesting for our purpose. One highlights action as crucial to the embodied cognition (Glenberg 1997), the other sees sensori-motor system as crucial and in particular the simulation mechanism referred to the context in which action takes place, and to body states (cf. B.C. Smith 1999; Barsalou 2008).

**25** Recent research (Borghi & Cimatti 2009; Borghi et al. 2011) on the link between knowing and acting in reference to the connection between the subject's perception and the perceived object has shown that the enacted contact between subject and physical world is characterized by the reciprocal conditioning between visual perception and action as well as between action, perception and semantic dimension.

**26** According to Franks (Franks 2010, 55–59), the human brain is organized for sociality but "What is needed is an interactional environment of mutual responsiveness which involves the active participation of both the baby and the caregiver. Without any environmental supports for this sociality (most of which is emotional), we do not develop the foundations for anything else" (Franks 2010, 55). Zlatev (2008) speaks about proto-mimesis indicating an embodied, emotional, sensori-motor and non-conceptual practice.

**27** Emotive expressions can be seen as the product of an instinctual mark attributable to affective dimensions. From this perspective, human language is the extension of some pre-linguistic the immediate process of attention, of consciously directed conduct which can be seen as a non-derived intentionality, or as an embodied form of transaction non reducible to physical processes in the brain.[28] They are generally employed to understand others' movements, facial expressions and more general embodied comportments, facilitating sympathetic identification with others. This does not involve a cognitive simulation yet, but an affectively perception-based understanding in face-to-face intersubjective experiences (see also Gallagher & Hutto 2008).[29]

This helps in a later stage to increase the human ability to predict others' behaviors. In fact, as soon as the primary stage the perceptual and interactional capacities are reinforced and the infants jointly attend with others to the surrounding environment, engaging in various practices, learning to distinguish the others as social objects (Mead 1964, 135), their attitudes and actions in social contexts. We can suppose that these first stages of intersubjectivity are related to what Mead called *play*, namely the first stages of the becoming of the self (Mead 1934, 153). We can, then, refer to what Zlatev (2008) called the *diadic mimesis*, namely the subject's capacity to place himself in the shoes of the other and understand what he is attending to.

As the child passes from the stage of *play* to that of *game*, he learns to assume the attitudes of other players which he presupposed to be organized into a sort of unit, and it is that organization which controls his responses, and the responses of others to his responses. (Mead 1934, 154) In this process, the child gradually becomes a social being in his own experience, acting toward himself in a manner analogous to that in which he acts towards others. The meanings Trevarthen (2011) speaks about arise in the process through which, as Mead argues, "the individual takes the attitude of another toward himself, and in some sense arouses in himself the tendency to the action, which his conduct calls out in the other individual" (Mead 1964, 244). The *meaning* is therefore the readiness to respond in certain ways to certain stimuli, and "this indication

---

communicative behaviors at the base of our empathetic identification. As we have seen, this hypothesis has been anticipated by Dewey and Mead's theory of emotion.

**28** We have to distinguish between the notion of apperception as referred to by Mead to that of Smith (2010). The latter consider the apperception in line with Husserl's phenomenology. However, as I pointed out in Baggio (2015), according to Mead in the organic interaction, the mind is seen as a property of a particular field of events that work together. Now, Mead does not refer to it in terms of phenomenological intentionality, rather in terms of nature, function, and the role that the apperception plays in the human psychic process.

**29** As Trevarthen (2011) suggests, the development of cultural intelligent in an infant is motivated by the innate intersubjective sympathy that an alert infant can show shortly after birth in sharing meanings with adults. See also Reddy (2008) and Rochat (2009).

must be made in the attitude or role of the other individual to whom it is pointed out or to whom it may be pointed out" (Mead 1964, 244). At this point, we can see the first steps of "children extending and expanding on their embodied practical abilities and skills" (Hutto 2008, 131). We can also refer to what Zlatev called *triadic mimesis* (Zlatev 2008), in which there is the understanding of the communicative use of one's bodily motion and object, and "*the same meaning for* the addressee as for the sender". As Zlatev argues, having the same meaning implies a development of third order mentality, putting the focus on cooperation, as Mead (2001) argued. As we have already seen, according to Mead, through the sympathetic placing of themselves in each other's roles, "and finding thus in their own experiences the response of the others, what would otherwise be an unintelligent gesture, acquires just the value which is connoted by signification, both in its specific applications and in its universality" (Mead 1964, 246).[30]

In this case, the passage to empathetic identification seems to be crucial in the passage from *triadic mimesis* to the process of language (in the sense of conventional signs)[31] as well as in learning more and more about the complexity of the experienced relations. We can suppose that the development of language puts the capacities to interact in and contribute to social acts in more complex and articulated ways. Since at the basis of mimesis there is a process of *cooperation*, it seems plausible to suppose that the process of expanding and sophisticating abilities to understand the emotions, desires, intentions of others, is related to the active role the individual assumes in the *construction of the common relational world*. In other words, it is by *experiencing the relations* that we learn to understand others, it is then the result of our becoming accustomed to rules of acting and expressing linguistically and behaviorally in social contexts. As Gallagher and Hutto (2008) point out, we learn how we ought to behave in various circumstances through shared training about roles and rules of the common world, and at the same time we learn how the other ought to behave

---

**30** We can also suppose that in the secondary stage of interaction an infant would apprehend the cooperative actions in the way that they are able to re-enact the goal-directed behavior that someone else fails to complete (Meltzoff & Brooks 2001).

**31** Senghas, Kita and Özürek (2004) showed that the spontaneous emergence of Nicaraguan Sign Language has its origin in triadic mimesis but has quickly become conventional and systematic. We can also refer to Quine (Quine 1992, 37–38), who argued that "Each of us learns his language by observing other people's verbal behavior and having his own faltering verbal behavior observed and reinforced or corrected by others". And at the basis of this process there is "an uncanny knack for empathizing another's perceptual situation, however ignorant of the physiological or optical mechanism of his perception. The knack is comparable, almost, to our ability to recognize faces while unable to sketch or describe them" (Quine 1992, 42–43). For a comparison between Mead and Quine see Baggio (2016b).

as well. Such a process of learning does not depend on our internalizing explicit rules to behave, it rather depends on our comprehension of regular attitudes to behave in a contextualized relational world.[32] To learn the regularity of certain behavioral attitudes of others goes beyond the observation of the behavior and involves, in particular, the dynamics implied in *interactions* and *relationships* related to our experience with others. This is possible on the basis of empathetic identification, namely the capacity we have to comprehend others' expression and general body movements, as well as to take their roles, understanding their behavioral and expressive attitudes. Thanks to empathetic identification, we share affective and cognitive experiences through a "natural extension" of relational field. We are then able to attribute a certain psychical state to another individual because of a previous process of learning his regular tendencies to act in a shared world of common sense. By experiencing our relational field we learn through interactive stimulations our understanding of mental states of others, by observing behavioral regularities in different situations that present certain common features.

The pragmatist approach to empathetic identification implies the possibility to *apprehend* at interpreting the dispositions of others to behave thanks to *habits* —that are social rules—apprehended with the development of our ability to take the role of others. This allows us to consider empathetic identification as a function of anticipation of the behaviors of others, and so place it at the core of social interaction in the same social environment.[33] The theoretical framework sketched above lets us suppose a partial understanding of the desires and preferences of others on the basis of a shared relational experience. A feeling can be attributed to a subject observed on the basis of the consequences it has on the behavior that it expresses. This possibility is rooted in an instinctive "residual" in the form of natural human relationship with respect to which our language is the extension and sophistication of certain pre-linguistic communicative behaviors. One might say that in the process of understanding the other's mind a mechanism similar to what Stern (1987) has indicated as the "tuning of affection" is in action, namely the working of complex a-verbal processes expressed either in consequential behavior, single acts, or vocal gestures through which the social

32 For more on the problem of normativity and rule-following in Mead, see Madzia (2013b).
33 Cf. Turner (2007). We refer here to regularities as not opposed to normativity, rather as part of what Frega (2015) calls a normative structure of the ordinary (Frega 2015).

act is realized (Stern refers, in particular, to the affective participation between a mother and an infant). These tuning processes operate since birth.[34]

Such an approach is particularly concerned with subjective experiences, namely the "inner phases" of social acts, that "work from the outside to the inside instead of from the inside to the outside" (Mead 1934, 8). It is based on the idea that our feelings and thoughts are largely constructed from metaphors taken from our community's behavioral dispositions, in which mimicry based on cooperative acts is the "social glue" (Lakin et al. 2003) that allows people to feel a sense of psychological connection with each other and to maintain harmonious relationships with fellow group members binding them together. It would, therefore, concur with Kirman and Teschl's suggestions to reply to the two issues regarding empathy. In appealing to the results of some experiments they suggest that empathy "can be modulated by a number of social-cognitive factors such as emotion regulation, selective attention but also by the kind of experiences individuals have lived through and their closeness to certain people. This would suggest that empathy is situation-specific or at least context dependent" (Kirman & Teschl 2010, 314). Mimicry, in fact, increases perceived similarity which increases helping behavior (Yabar et al. 2006). Moreover, it is worth noting that, according to Decety and Cowell, the moral rules of a community is a way to take account of the emotional empathy. They argue, "empathy has a vital role in social interaction, from bonding between mother and child, to understanding others' feelings and subjective psychological states" (Decety & Cowell 2014, 337).

All this justified the idea that individuals attribute and interpret other-regarding preferences in a context of economic choices.[35] This does not mean that another's states of mind are totally accessible without reference to anything other than behavior. There are some subjective experiences which are necessarily confined to a particular individual, and which cannot be shared by others. This is especially true for the motivations at the basis of personal choices.[36] However, the subjective states either have, or are assumed to have, objective reference, since generally anyone lives in a situated socio-cultural context that contributed

---

**34** In economics, this has been confirmed by some experiments that showed that in order to explain an individual's behavior, it is necessary to take into account the details of the so-called "network architecture" (Choi, Gale & Shachar Kariv 2005).

**35** In a more philosophical way, Rorty (Rorty 1979, 189) argued that this "is to be explained on the basis of that sort of community feeling which unites us with anything humanoid. To be humanoid is to have a human face, and the most important part of that face is a mouth which we can imagine uttering sentences in synchrony with appropriate expressions of the face as a whole".

**36** For more on this point see Sen (1973).

to the constitution of his rule-governed behaviors. Here, we are assuming, with Mead, that in spite of the "privacy" of some inner states there is also "an implication of something that has objective reality" in them, something that may be discovered through an act of "reconstruction and interpretation", and this is possible because privacy is not something exclusively "inner". Instead it "does not imply necessarily anything more than difference of access or of perspective on the part of the different organisms" (Mead 1964, 271–272).

It seems, therefore, that people's capacity to interpret others' intentions and behavioral dispositions in psychology can be referred to, in economics, as the elicitation of beliefs and expectations. As some recent experiments have also shown, the cooperative acts influence one's own decisions in reference to other persons' decisions, that is by preference type and inferences about group members' likely actions (Leipold et al. 2013; Fischbacher & Gächter 2010; Croson 2007). From this perspective, the pragmatist pluralistic approach to empathetic identification would contribute to sketch an integrated account of identity and agency which would overcome the tension between self-preference and other-regarding preferences.

The experiences that individuals live through and their closeness to other people occur in relation to what we saw Mead calling the "generalized other", namely the "organized community or social group which gives to the individual his unity of self" (Mead 1934, 154). The attitude of the "generalized other" is the attitude of the whole community, namely the common *ethos* which gives unity and consistency to the person's *identity*. The community's identity is, therefore, embedded in the self as a co-determining factor of his valuating processes of preferences and choices. People's social attitudes, which in part contribute and in part are subject to a community's *values* (often incommensurable values) offer the basis for empathetic identification through which private subjective preferences are potentially sharable. Behavioral dispositions seem thence strictly related to behavioral mimicry as adaptive mechanism, and, as some studies demonstrate, prosocial behaviors increase after being mimicked (Müller et al. 2012; Fischer-Lokou et al. 2011; Guègin et al. 2011; van Baaren et al. 2004; Lakin et al. 2003). This would offer a theoretical explanation of the fact that if a person lives in a community which promotes a culture of egoistic rule-governed behaviors, he/she will quite likely act referring mainly to self-regarding preferences.

## 4 Concluding remarks

Referring to pragmatism (and Mead's theory of the self in particular), as a possible theoretical framework through which one might clarify meanings and limits of new economic models, would contribute to a more realistic model of human economic behaviors. Such a model would refer to both moral dimensions as well as to those normally covered exclusively by economic theory. Mead's pragmatist approach to explaining human action should inspire neuroeconomics not to reduce the explanation of economic human behavior (and in particular the mechanism of empathy) to brain functioning.[37] To consider neuroeconomics as a new approach that might help us to understand different components of human decision-making capacities, clarifying the meaning and limitations of neuroeconomic research would contribute to interpreting the role habits with a functional value assume in the decision-making processes. This means to refer adequately to the imagined events or states that represent the outcomes in the economic models, contributing also to set up a reflection on the "evaluation" process involved. Mead's pragmatism would help us overcome the obstacle of the lack of a systematic framework for clarifying the structure of existing studies and the meaning of the evidence they provide. It would be useful to interpret studies developed in neuroeconomics and in the more general behavioral economics, and so contributing to select and compared new studies subjects. It would, therefore, be possible to develop an alternative theory of interpretation and prediction of the intentions of others in cooperative actions by using a framework based on Mead's bio-social theory of mind as sketched above. Such a sketch would offer the basis for the elaboration of a theoretical framework useful for furthering certain experimental hypothesis on prosocial human behaviors. Moreover, it would offer a theoretical map that can be implemented for interpreting real-situated contexts, which would test the idea that progress involves expanding our concern from the family to humanity as a whole (Rockwell 2016). This would be, in my opinion, possible through an approach to apprenticeship learning based on *practical mimesis.*[38]

---

37 See on this point also Tschaepe (2014).

38 This last point has to be better developed. Gieser (2008) moves a first step on the elaboration of a theoretical framework of the learning process of embodied cultural abilities (which, however, I would prefer to indicate as behavioral dispositions). In particular, he proposes a phenomenologically elaboretad concept of empathy as an "ideal-type" that would provide a theoretical notion useful to describe "a hitherto 'silent' process of learning through practical engagement or apprenticeship" (Gieser 2008, 314). Even if his contribution wants to be directed to antropologists and ethnographers, I think it may also be used for our purposes.

# References

Artinger, F., Exadaktylos, F., Koppel, H. & Sääksvuori, L. (2014). "In Others' Shoes: Do Individual Differences in Empathy and Theory of Mind Shape Social Preferences?" In: *Plose One* 9 (4), 1–9.

Baggio, G. (2015). *La mente bio-sociale. Filosofia e psicologia in G. H. Mead.* Pisa: ETS.

Baggio, G. (2016a). "Mead's Bio-social Theory of the Self and the Economics of Happiness. For a Pragmatist Philosophy of Economics." In: K. Stikkers & K. Skowronski (Eds.). *Philosophy in the Time of Crisis.* New York: Routledge [forthcoming].

Baggio, B. (2016b). "Il behaviorismo epistemologico di G. H. Mead." In: A. M. Nieddu (Ed.). *La filosofia sociale di George H. Mead. Analisi, interpretazioni, prospettive.* Milano: Mimesis, 49–62.

Barsalou, L. W. (2008). "Grounded Cognition." In: *Annual Review of Psychology* 59, 617–645.

Baumol, W. (1952). *Welfare Economics and The Theory of the State.* Cambridge, MA: Harvard University Press.

Borghi, A. & Cimatti, F. (2009). "Embodied cognition and beyond: acting and sensing the body." In: *Neuropsychologia* 5, 763–773.

Borghi, A. et al. (2011). "Manipulating objects and telling words: a study on concrete and abstract words acquisition." In: *Frontiers in Psychology* 2 (2011), 1–14.

Bower, G. H. (1983). "Affect and Cognition." In: *Philosophical Transaction of the Royal Society of London*, Series B, 302, 387–402.

Bowlby, J. (1979). *The Making and Breaking of Affectional Bonds.* London: Tavistock.

Bréban, L. (2015). "An Investigation into the Smithian System of Sympathy: from Cognition to Emotion." In: *The Adam Smith Review*, Vol. 9.

Camerer, C. & Fehr, E. (2006). "When Does 'Economic Man' Dominate Social Behavior?" In: *Science* 311, 47–52.

Camerer, C., Loewenstein, G. & Prelec, D. (2005). "Neuroeconomics: How Neuroscience Can Inform Economics." In: *Journal of Economic Literature* XLIII, 9–64.

Camerer C. (2003). *Behavioral game theory: Experiments in strategic interaction.* Princeton, NJ: Russell Sage Foundation.

Charness, G. & Rabin, M. (2002). "Understanding Social Preferences with Simple Tests." In: *Quarterly Journal of Economics* 117 (3), 817–869.

Choi, S., Gale, D. & Kariv, S. (2005). "Behavioral aspects of learning in social networks: an experimental study." In: J. Morgan. *Advances in Applied Microeconomics, volume 13: Experimental and Behavioral Economics.* Oxford: Elsevier.

Corradi-Fiumara, G. (2009). *Spontaneity. A Psychoanalytic Inquiry.* London and New York: Routledge.

Cozolino, L. J. (2006). *The neuroscience of human relationship: Attachment and the developing social brain.* New York: Norton.

Croson, R. (2007). "Theories of commitment, altruism and reciprocity: Evidence from linear public goods games." In: *Economic Inquiry* 45, 199–216.

Damasio, A. R. (1999). *The Feeling of What Happens. Body and Emotion in the Making of Consciousness.* Boston, MA: Harcourt.

Decety, J. & Cowell J. M. (2014). "The complex relation between morality and empathy." In: *Trends in Cognitive Sciences* 18 (7), 337–339.

Decety, J. & Ingvar, D. H. (1990). "Brain structures participating in mental simulation of motor behavior: a neuropsychological interpretation." In: *Acta Psychologica* 73, 13–24.
Dewey, J. (1894). "The Theory of Emotion. I" In: *John Dewey: The Early Works*. Vol. 4. ed. by J. A. Boydston. Carbondale, IL: Southern Illinois University Press 1971.
Dewey, J. (1895). "The Theory of Emotion. II" In: *John Dewey: The Early Works*. Vol. 4. ed. by J. A. Boydston. Carbondale, IL: Southern Illinois University Press 1971.
Emonds, G. et al. (2011). "Comparing the Neural Basis of Decision Making in Social Dilemmas of People With Different Social Value Orientations, a fMRI Study." In: *Journal of Neuroscience, Psychology, and Economics* 4 (1), 11–24.
Fehr, E. & Fischbacher, U. (2002). "Why Social Preferences Matter—The Impact of Non-Selfish Motives on Competition, Cooperation and Incentives." In: *The Economic Journal* 112 (478), C1–C33.
Fiebich, A., Gallagher, S. & Hutto, D. D. (2015). "Pluralism, interaction, and the ontogeny of social cognition." In: J. Kiverstein (Ed.). *Routledge Handbook of "The Philosophy of the Social Mind"*. https://www.academia.edu/15413357/Pluralism_Interaction_and_the_Ontogeny_of_Social_Cognition.
Fischbacher, U. & Gächter S. (2010). "Social preferences, beliefs, and the dynamics of free riding in public goods experiments." In: *American Economic Review* 100, 541–556.
Fischer-Lokou, J. et al. (2011). "Mimicry and propagation of prosocial behavior in a natural setting." In: *Psychological Reports* 108, 599–605.
Fontaine, P. (1997). "Identification and Economic Behavior: Sympathy and Empathy in Historical Perspective." In: *Economics and Philosophy* 13 (2), 261–280.
Fontaine, P. (2001). "The Changing Place of Empathy in Welfare Economics." In: *History of Political Economy* 33 (3), 387–409.
Franks. D. D. (2010). *Neurosociology. The Nexus between Neuroscience and Social Psychology*. New York: Springer.
Frega, R. (2015). "The Normative Structure of the Ordinary." In: *European Journal of Pragmatism and American Philosophy* 7 (1), 54–76. http://lnx.journalofpragmatism.eu/wp-content/uploads/2015/07/Frega-EJPAP-Symposia-2015–1.pdf.
Gaertner, W. & Schokkaert, E. (2010). *Empirical Social Choice: Questionnaire-Experimental Studies on Distributive Justice*. http://darp.lse.ac.uk/Resources/Books/Gaertner-Schokkaert_%28CUP12%29.pdf.
Gallagher, S. (2005). "Phenomenological Contributions to a Theory of Social Cognition." In: *Husserl Studies* 21, 95–110.
Gallagher, S. (2007a). "Simulation Trouble." In: *Social Neuroscience* 2 (3–4), 353–365
Gallagher, S. (2007b). "Logical and phenomenological arguments against simulation theory." In: D. D. Hutto & M. Ratcliffe (Eds.). *Folk Psychology Re-Assessed*. Dordrecht: Springer, 63–78.
Gallagher, S. (2014). "Pragmatic Interventions into Enactive and Extended Conceptions of Cognition." In: *Philosophical Issues* 24, 110–126.
Gallagher, S. (2015), "The new hybrid: continuing debates on social perception." In: *Consciousness and Cognition* 36, 452–465.
Gallagher, S. & Hutto, D. (2008). "Understanding others through primary interaction and narrative practice." In: Zlatev, J. et al. (Eds.). *The Shared Mind. Perspectives on intersubjectivity*. Amsterdam: John Benjamins, 17–38.

Gallagher, S. & Zahavi, D. (2008). *The Phenomenological Mind. An Introduction to Philosophy of Mind and Cognitive Science.* London/New York: Routledge.
Gallese, V. (2009a). "Motor abstraction: a neuroscientific account of how action goals and intentions are mapped and understood." In: *Psychological Research* 73 (2009), 486–498.
Gallese, V. (2009b). "Mirror Neurons, Embodied Simulation, and the Neural Basis of Social Identification." In: *Psychoanalytic Dialogues* 19, 519–536.
Gallese, V. & Goldman, A. I. (1998). "Mirror neurons and the simulation theory." In: *Trends in Cognitive Science* 2, 493–501.
Gibson, E. (1988). "Explanatory Behavior in the Development of Perceiving, Acting, and the Acquiring of Knowledge." In: *Annual Review of Psychology* 39, 1–41.
Gieser, T. (2008). "Embodiment, emotion and empathy: A phenomenological approach to apprenticeship learning." In: *Anthropological Theory* 8, 299–318.
Giusti, E. & Locatelli, M. (2007). *L'empatia integrata. Analisi umanistica del comportamento motivazionale nella clinica e nella formazione.* Roma: Soveria Multimedia.
Glenberg, A. M. (1997). "What memory is for." In: *Behavioral and Brain Sciences* 20, 1–55.
Glimcher, P. W. (2003). *Decision, Uncertainty, and the Brain. The Science of Neuroeconomics.* Cambridge, MA: MIT.
Goldman, A. I. (1995). "Simulation and Interpersonal Utility." In: *Ethics* 105 (4), 709–726.
Gordon, R. M. (1995). "Sympathy, Simulation, and the Impartial Spectator." In: *Ethics* 105 (4), 727–742.
Gregory, S. (1999). "Navigating the sound stream of the human voice." In: D. Franks & T. Smith (Eds.). *Mind, Brain, and Society: Toward a Neurosociology of Emotion. Social Perspectives on Emotion*, vol. 5. Stanford, CT: Jai Press.
Guègen, N., Martin, A. & Meineri, S. (2011). "Mimicry and helping behavior: an evalutation of mimicry on explicit helping request." In: *Journal of Social Psychology* 151, 1–4.
Hickok, G. (2014). *The Myth of Mirror Neurons. The Real Neuroscience of Communication and Cognition.* New York/London: W. W. Norton & Company.
Hume, D. (1739). *A Treatise of Humane Nature.* ed. by L. A. Selby-Bigge. Oxford, MA: Clarendon Press, 1896.
Hutto, D. D. (2008). *Folk Psychological Narratives: The Sociocultural Basis of Understanding Reasons.* Cambridge, MA: MIT Press.
Kahneman, D. E. (2011). *Thinking, Fast and Slow.* New York: Farrar, Straus and Giroux.
Kahneman, D. E & Tversky, A. (1979). "Prospect Theory: An Analysis of Decision under Risk." In: *Econometrica* 47 (2), 263–291.
Khalil, E. A. (2011). "The mirror neuron paradox: How far is *understanding* from *mimicking?*" In: *Journal of Economic Behavior and Organization* 77 (1), 86–96.
Kirman, A. & Teschl, M. (2010). "See Selfish or selfless? The role of empathy in economics." In: *Philosophical Transactions of The Royal Society.* B (2010) 365, 303–317.
Lakin, J. L. et al. (2003). "The Cameleon Effect as Social Glue: Evidence for the Evolutionary Significance of nonconscious Mimicry." In: *Journal of Nonverbal Behavior* 27 (3), 145–162.
Lecaldano, E. (2014). "Introduzione." In: A. Smith. *La teoria dei sentimenti morali*, trad. it. di Stefania di Pietro. Milano: Rizzoli.

Leipold, K. et al. (2013). "Individual and Developmental Differences in the Relationship Between Preferences and Theory of Mind." In: *Journal of Neuroscience, Psychology, and Economics* 6 (4), 236 – 251.
Liotti, G. (2005). *La dimensione interpersonale della coscienza.* Roma: Carocci.
Madzia, R. (2013a). "Mead and self-embodiment: imatation, simulation, and the problem of taking the attitude of the other." In: *Österreichische Zeitschrift für Soziologie* 38, 195 – 213.
Madzia, R. (2013b). "The Concept of Rule-following in the Philosophy of George Herbert Mead." In: T. Burke & K. Skowroński (Eds.). *George Herbert Mead in the Twenty-first Century.* Lanham, MD: Lexington Press, 61 – 70.
Martin, J., Sokol, B. W. & Elfers, T. (2008). "Taking and Coordinating Perspectives: From Prereflective Interactivity, through Reflective Intersubjectivity, to Metareflective Sociality." In: *Human Development* 51, 294 – 317.
Martin, J. (2013). "Life positioning analysis: An analytic framework for the study of lives and life narratives." In: *Journal of Theoretical and Philosophical Psychology* 33 (1), Feb 2013, 1 – 17. http://psycnet.apa.org/psycinfo/2011 – 24367 – 001/.
Martin, J. & Gillespie, A. (2010). "A Neo-Meadian approach to human agency: relating the social and the psychological in the ontogenesis of perspective-coordinating persons." In: *Integrative psychological and behavioral science* 44 (3), 252 – 272.
Mead, G. H. (1895). "A Theory of Emotions from the Physiological Standpoint (Abstract of a paper read to the third annual meeting of the American Psychological Association, 1894)." In: *Psychological Review* 2, 162 – 164.
Mead, G. H. (1934). *Mind, Self, and Society.* ed. by Charles W. Morris. Chicago: University of Chicago Press.
Mead, G. H. (1938). *The Philosophy of the Act.* Chicago: Chicago University Press.
Mead, G. H. (1964). *Selected Writings.* Chicago: Chicago University Press.
Mead, G. H. (2001). *Essays on Social Psychology.* Edited by Mary Jo Deegan. New Brunswick, NJ: Transaction Publishers.
Meltzoff, A. N. & Brooks, R. (2001). "'Like Me' as a building block for understanding other minds: Bodily acts, attention, and intention." In: B. Malle, L. J. Moses & D. A. Baldwin (Eds.). *Intentions and Intentionality: Foundations of Social Cognition.* Cambridge, MA: MIT Press, 171 – 191.
Müller, B. C. N. et al. (2012). "Prosocial Consequences of Imitation." In: *Psychological Reports* 110 (3), 891 – 898.
Mullainathan, S. & Thaler, R. (2000). "Behavioral economics", NBER working paper 7948.
Panksepp, J. (2001). "The Neuro-evolutionary Cusp between Emotions and Cognition." In: *Evolution and Cognition* 7, 141 – 163.
Panksepp, J. (1998). *Affective Neuroscience: The Foundation of Human and Animal Emotions.* Oxford: Oxford University Press.
Politser, P. (2008). *Neuroeconomics. A guide to the new science of making choices.* New York: Oxford.
Premack, D. & Woodruff, G. (1978). "Does the chimpanzee have a theory of mind?" In: *Behavioral and Brain Sciences* 1, 515 – 526.
Quine, W. V. O (1992). *Pursuit of Truth.* Cambridge, MA: Harvard University Press.
Reddy, V. (2008). *How infants know minds.* Cambridge, MA: Harvard University Press.
Rizzolatti, G. & Sinigaglia, C. (2006). *So quel che fai.* Milano: Raffaello Cortina.

Rochat, P. (2009). *The infant's world.* Cambridge, MA: Harvard University Press.
Rockwell, T. (2016). "The Embodied 'We': The Extended Mind as Cognitive Sociology." In: R. Madzia & M. Jung (Eds.). *Pragmatism and Embodied Cognitive Science.* Berlin/New York: De Gruyter.
Rogers, C. (1975) "Empathic: An Unappreciated Way of Being." In: *The Counseling Psychologist* 5 (2), 2–10.
Rorty, R. (1979). *Philosophy and the Mirror of Nature.* Princeton NJ: Princeton University Press.
Ross, D. (2005). *Economic Theory and Cognitive Science. Microexplanation.* Cambridge, MA: MIT Press.
Ross, D. et al. (2008). *Midbrain Mutiny: The Picoeconomics and Neuroeconomics of Disordered Gambling. Economic Theory and Cognitive Science.* Cambridge, MA: MIT Press.
Sen, A. (1973). "Behaviour and the Concept of Preference." In: *Economica* 40 (159), 241–259.
Senghas, A., Kita, S. & Özyürek, A. (2004). "Children creating core properties of language: Evidence from an emerging sign language in Nicaragua." In: *Science* 305, 1779–1782.
Singer, T. & Fehr E. (2005). "The Neuroeconomics of Mind Reading and Empathy" In: *The American Economic Review* 95 (2), 340–345.
Shook, J. R. & Solymosi, T. (Eds.) (2014). *Pragmatist Neurophilosophy. American Philosophy and the Brain.* London/New York: Bloomsbury.
Smith, A. (1759). *The Theory of Moral Sentiments*, ed. by D.D. Raphael and A.L. Macfie. Oxford: Oxford University Press, 1976.
Smith, B. C. (1999). "Situatedness/Embeddedness." In: R.A. Wilson & F.C. Keil (Eds.). *The MIT Encyclopedia of the Cognitive Sciences.* Cambridge, MA: MIT Press.
Smith, J. (2010). "Seeing other people." In: *Philosophy and Phenomenological Research* 81 (3), 731–748.
Stern, D. (1985). *The Interpersonal World of the Infant.* New York: Basic Books.
Thoron, S. (to appear). "Des sentiments moraux d'Adam Smith à la neuro-économie. Vers une théorie positive de la morale?" In: *Oeconomia. History, Methodology, Philosophy.*
Trevarthen, C. (2011). "What Is It Like to Be a Person Who Knows Nothing? Defining the Active Intersubjective Mind of a Newborn Human Being." In: *Infant and Child Development* 20, 119–135.
Tschaepe, M. (2014). "Neuropragmatic Reconstruction. A Case from Neuroeconomics." In: J. R. Shook & T. Solymosi (Eds.). *Pragmatist Neurophilosophy. American Philosophy and the Brain.* London/New York: Bloomsbury, 111–126.
Turner, S. (2007). *Human emotions: A sociological theory.* New York: Routledge.
Tversky, A. & Kahneman, D. (1981). "The Framing of Decision and the Psychology of Choice." In: *Science* 211 (4481), 453–458.
van Baaren, R. B. et al. (2004). "Mimicry and prosocial behavior." In: *Psychological Science* 15, 71–74.
Van Lange, P. A. M. (2000). "Beyond Self-interest: A set of Propositions Relevant to Interpersonal Orientations." In: *European Review of Social Psychology* 11 (1), 297–331.
Vreeke, G. & Van der Mark, I. (2003). "Empathy, an Integrative Model." In: *New Ideas in Psychology* 21 (3), 177–207.
Wispé, L. (1986). "The Distinction Between Sympathy and Empathy: To Call Forth a Concept, A Word is Needed." In: *Journal of Personality and Social Psychology* 50 (2), 314–321.

Wispé, L. (1991). *The Psychology of Sympathy.* New York: Plenum Press.
Yabar, Y. et al. (2006). "Implicit behavioral mimicry: investigating the impact of group membership." In: *Journal of Nonverbal Behavior* 30, 97 – 113.
Zlatev, J. (2008). "The co-evolution of intersubjectivity and bodily mimesis." In: J. Zlatev et al. (Eds.). *The Shared Mind. Perspectives on intersubjectivity.* Amsterdam: John Benjamins, 214 – 244.

# Part III: The Pragmatists as Pioneers of Situated Cognition: Embodied Language, Reasoning, and Feeling

Roman Madzia

# Mind, Symbol and Action-Prediction: George H. Mead and the Embodied Roots of Language

**Abstract:** The essay "Mind, Symbol and Action-Prediction: George H. Mead and the 'Embodied' Roots of Language" presents Mead's radically embodied and anti-intellectualist view of language and symbolic behavior. This paper develops a Meadian perspective on the phenomenon of the participatory, embodied sense-making called language and will connect it with the theory of 'prediction error minimization' in order to elucidate the role which symbolic thinking plays in the process of action-understanding. It also demonstrates the way in which Mead understood significant symbols as encoding environmental and social affordances. Further, this paper re-defines some key concepts of philosophy of language like that of meaning, reference, intension and extension, etc. Finally, this paper also demonstrates that the classical pragmatist understanding of concepts such as goal-directedness of action, resolution of action-problems, etc., is indispensable if we want to come up with a persuasive account of embodied roots of linguistic behavior.

In one of his most illuminating articles called 'A Nice Derangement of Epitaphs', Donald Davidson wrote: "there is no such thing as a language, not if a language is anything like what many philosophers and linguists have supposed. There is therefore no such thing to be learned, mastered, or born with. We must give up the idea of a clearly defined shared structure which language-users acquire and then apply to cases" (Davidson 1986/2001, 446). Although there are considerable differences between Davidson and George Herbert Mead—the main philosophical figure of this paper—they both, each in his own right, deeply problematized the picture of language as has been usually understood. In conventional rendition, language is depicted a set of mental labels (names), referring to things in the world, usually in a one-to-one fashion, and is mastered by our grasping of abstract rules which, in turn, make our utterances about the world conform to this abstract system and, therefore, intelligible to others. If our normative idea of language[1] really is such a transcendental set of labels and rules which is,

1 We should note that in his article, Davidson talks about ordinary languages (as opposed to artifical languages used, e.g., in computer science, etc.).

in the course of human life, acquired by individuals, then, as Davidson maintains, languages not only do not, but in fact, never even have existed. If we look at the phenomenon of language *sub specie aeternitatis*, in other words—if we analyze it in the classical philosophical sense as a phenomenon without a history and development, we indeed most likely arrive at a definition which resembles the one given earlier. According to the pragmatists, exactly this is the error commited all too often by philosophers who still today (almost exactly like those from a hundred years ago) tend to disregard the fact that we no longer live in the static Aristotelian world of substances but in a Darwinian cosmos where change and chance are the crucial factors in the development of events.

George H. Mead was one of the first philosophers who unambiguously articulated the need of looking at the phenomenon of language from a Darwinian perspective as a means of social adaptation. Surely, from the strictly conceptual point of view, understanding language (or symbolism, to speak more broadly) as a means of adaptation between social organisms does not, by itself, prove its embodied roots. In order to provide such proof, it will be necessary to demonstrate that the *means* and *adaptation* are identical—in other words—that language *is* a form of adaptive behavior between human bodies. That is exactly what this paper focuses on. Thus, language here will be presented as a name for a specific embodied activity in which one body attunes its action to that of another body in the course of their dynamic dealings with environmental structures and events. Humberto Maturana has pertinently called this phenomenon 'languaging': "language is a manner of living together in a flow of coordinations of coordinations of [...] behaviors or doings that arises in a history of living in the collaboration of doing things together" (Maturana 2002, 27). Understanding language as a phenomenon which *grows* out of bodily interactions, rather than as something which 'merely' *governs* them (as if it were a mere software uploaded to our heads) might, in the most desirable case, help set us free from conceiving of it as a set of names for objects which is used according to explicit rules.

The so-called '4E' theories of mind and cognition have recently made impressive progress with regard to proposing new, interactive accounts of cognition: "embodied cognition is sweeping the planet" wrote Fred Adams not so many years ago (Adams 2010, 619). However, as some of their critics (de Bruin and Kästner 2012; de Bruin and de Haan 2012; Cowley and Nash 2013), contend (not totally unjustifiably)—for the most part, these theories deal predominantly with lower-level, on-line cognitive capacities, and are often limited to a perspective of an individual cognitive agent. Consequently, according to the critical voices in the debate on situated cognition, the 4E approaches still face considerable difficulties when they try to propose accounts of the alleged 'higher-level' mental phenomena, such as the origins and use of language. Although there have been significant endeavors to fill in this gap in the recent literature, these have been effectuated almost exclusively from the phenomenological-enactivist perspective (Fuchs and De Jaegher 2009; Cuffari, Di Paolo and De Jaegher 2014, etc.).

The following paper is going to present a view on the relation between embodiment and language stemming from the pragmatist tradition. Long before contemporary enactivists, Mead came up with a naturalist explanation of mental phenomena which puts aside the traditional dualisms of the body and mind, subjectivity and intersubjectivity, individual and society, etc. Besides this fact, which is rather of historical than systematic importance, there is another reason why looking into Mead might be worth our efforts. Namely, if we intend to analyze the phenomenon of language from the point of view of situated cognition, it should be noted that for Mead any explanation of the mind was unthinkable aside from the explanation of linguistic (symbolic) behavior. I entirely agree with Erkki Kilpinen's contention that rather than as a sociologist or a social psychologist we should view Mead as a "psychologically oriented philosopher of mind and theorist of action" (Kilpinen 2011, 139). In Mead's philosophy of action, neither mind, nor language is treated as independent of, or prior to the other. Moreover, from his perspective, conceptual thinking as well as understanding the action of others is not an intellectual but primarily a motor, bodily process. Upon these assumptions, Mead managed to propose an account of symbolic behavior in (what we today would call) fully enactivist and 'embodied' terms. This paper is going to present a Meadian perspective on the phenomenon of the participatory, embodied sense-making called language and will connect it with Jakob Hohwy's theory of 'prediction error minimization' in order to elucidate the role which symbolic thinking plays in the process of action-understanding. It will also demonstrate the way in which Mead understood significant symbols as encoding environmental and social affordances. Further, this paper will re-define some key concepts of philosophy of language like that of meaning, reference, intension and extension, etc. Finally, it will demonstrate that the classical pragmatist understanding of concepts such as goal-directedness of action, resolution of action-problems, etc., is indispensable if we want to come up with a persuasive account of embodied roots of linguistic behavior.

## 1 Enkinaesthesia and the bodily origins of meaning

The beginnings of Mead's autonomous work on the theory of language could be found in his 1904 paper "The Relations of Psychology and Philology", dedicated in the first place to a critical examination of Wilhelm Wundt's *Völkerpsychologie*. Being fully aware of the shortcomings of the theories of language stemming both from the rationalist as well as empiricist tradition, Mead praises Wundt's suggestion that the beginnings of symbolic articulation are to be found in primitive bodily expressions of emotion, i. e., in gestures and recognizes a close interrelatedness between language and motor activities of the body:

> Articulation, as a muscular process, is explained in the same way that movements of the face, of the hands, of the whole body are accounted for under the influence of emotional tension. Instead, therefore, of having to assume unknown or exceptional conditions as the antecedents of the origin of speech, we can find the conditions, present in our own movements, in the first activities of children, in the gesture languages of primitive peoples or the deaf-mutes. (Mead 1904a, 380)

Although the years following the publication of the aforementioned article are characteristic by ever more critical stance *vis-à-vis* his former German mentor,[2] Mead never abandoned his basic conviction of the necessity of viewing the phenomenon of language as a series of sensori-motor events by means of which social relationships between human bodies are established and gain their specifical dynamics. There are two crucial concepts of embodiment present in Mead's philosophy of language, namely: i) attitude and ii) gesture. In Mead's understanding, it is the concept of an 'attitude' that needs to suplement Wundt's basically correct but incomplete treatment of expressivity in order to provide for a satisfactory account of how animal (and consequently also human) social interaction takes place.[3] In the most basic sense, Mead defines 'an attitude' is an incipient movement of the body, or as a beginning of an action. On the basis of the organism's active behavior and problem-solving activities, attitudes come into existence as neural pathways encoding goal-directed bodily habits which respond to specific kinds of environmental stimulation.[4] From today's point of view, attitudes could be characterized as neurologically instantiated bodily dispositions encoding perceived objects in terms of possible reactions toward them. Mead talks about these incipient movements as essentially being encoded in the central nervous system, but since they anticipate later phases of an overt act, and occur simultaneously with present overt movement we could just as well interpret them as initial movements of the body as such. Mead holds that attitudes are goal-directed cognitive items detecting environmental affordances and con-

2 For a more detailed account of the relation between Mead's and Wundt's theory of language, see Joas (Joas 1985, 90–98).
3 The final constructive depart from Wundt's theory of language can be found in Mead (1910).
4 See Mead (Mead 1934/1967, 5–13).

stantly controlling the course of an act from the very beginning until its successful completion (cf. Mead 1934/1967, 11). This is also why Mead characterizes attitudes as encoding objects as "invitations to action" (Mead 1938, 12).[5] Attitudes are behavioral items encoding the world in instrumental terms as a space of *praxis*; as such they manage the behavior of an individual cognizer and can, on the more primitive levels, be analyzed without any regard to the social environment of the cognitive agent.

By gestures, Mead means attitudes which are being responded to by others. Gestures are, therefore, attitudes seen and analyzed from a social perspective—just like attitudes 'gestures' are incipient movements of one organism to which (an) other organism(s) motorically react: "Back of all gestures, then, lie tendencies to act. Gestures may be regarded as the beginnings of acts" (Mead 1914/1982, 36). Certain attitudes therefore become gestures just by being detected and responded to by others in a social act. That means, in principle, that any attitude (incipient movement) can become a gesture if it is responded to by another organism: "The term 'gesture' may be identified with these beginnings or early stages of social acts which serve as stimuli for the response of other forms" (Mead 1934/1967, 43). What is crucial for Mead's treatment of gestures (as well as attitudes) is that although these neurological/behavioral items are means of organization of the social act, they are fundamentally non-intentional. More precisely, there does not have to be a communicative intention in the back of gestures to be transmitted from one organism to another in order for an attitude to become a gesture. Being a) a goal-directed bodily movement and being b) responded to by others in a social act are both a necessary and sufficient conditions in order to become a gesture. Consequently, gestures can be something as simple as:

5 Mead often illustrated the functioning of attitudes with an example of a person reaching for a hammer: "If one approaches a distant object he approaches it with reference to what he is going to do when he arrives there. If one is approaching a hammer he is muscularly all ready to seize the handle of the hammer. The later stages in the act are present in the early stages—not simply in the sense that they are all ready to go off, but in the sense they serve to control the process itself. They determine how we are going to approach the object, and the steps in our early manipulation of it. We can recognize, then, that the innervation of certain groups of cells in the central nervous system can already initiate in advance the later stages of the act" (Mead 1934/1967, 11). Recent research into mirror neurons has proven the validity of this insight made by Mead. For instance, according to Rizzolatti and Sinigaglia, there, indeed, exists a class of neurons (which has been called canonical neurons) responsible for such a process of transforming visual information regarding an object into appropriate motor acts. In this respect, both these authors explicitly acknowledge the validity Mead's theories with regard to what he called 'attitudes.' See Rizzolatti & Sinigaglia 2008, 79–115.

> the glance of the eye that is the beginning of the spring or the flight, the attitude of body with which the spring or flight commences, the growl, or cry, or snarl with which the respiration adjusts itself to oncoming struggle, and they all change with the answering attitudes, glances of the eye, growls and snarls which are the beginnings of the actions which they themselves arouse. (Mead 1910/1964, 124)

For Mead it is the dynamic 'dance' of mutual motor adjustments between bodies where the roots of meaning and symbolic communication are to be found. The most fundamental embodied social event in which this occurs was called a 'conversation of gestures' (Mead 1910/1964, 124) by Mead.[6] A conversation of gestures is a process of ongoing sensori-motor adjustments between two or more participants of the social act where an execution of a gesture by one individual is an impulse for an adjustive response by another. Since each gesture (being an instance of an attitude) contains in its earliest phases information about its final, consummatory phase, Mead sometimes described gestures as "truncated acts" (Mead 1909/1964, 102) informing other participants of the social act about what will be the outcome of the action undertaken by the individual executing that gesture. Mead liked to present this seemingly very theoretical situation with the example of a dog-fight. Let us, therefore, imagine a situation in which there are two dogs approaching each other in a hostile manner. One dog walks around the other while the other one observes it and starts to crouch low to the ground; in response, the first dog starts to growl, etc.[7] Mead points to the fact that in this kind of affectively charged situations the bodily action of one dog becomes a stimulus for the response of the other, which, in turn, becomes a stimulus for the first, etc., in a possibly endless reciprocal shifting of positions and movements. Regardless whether intentionally or unintentionally, in a social act, one individual reveals his attitude to another by means of non-reflective bodily movements which become an impulse for an adjustive response by others. Gestures, being the most primitive element of communication, are also non-reflective articulations of emotions and inform other participants of the social act about what the individual performing them is up to next. If in the process of a conversation of gestures, gestures of one body provoke a motor response of other bodies, then these gestures bear meaning. In other words, the meaning of a gesture is the bodily response it provokes in the other: "The gesture stands for a certain resultant of the social act, a resultant to which there is a definite response on the part of the individuals involved therein; so that meaning is

6 Sometimes Mead also used the term 'conversation of attitudes.' See, e.g., (Mead 1934/1967, 358).
7 See further Mead (Mead 1934/1967, 14, 42–43).

given or stated in terms of response" (Mead 1934/1967, 76). In the transactional process of conversation of bodily gestures, the action of one body spills into the experience of the other in that it provokes his active response and thus establishes a dynamic reciprocity of affective co-engagement. Mead's anti-intellectualist, enactivist position with regard to meaning therefore rejects the idea that meaning should be studied by examining 'mental content'. On the contrary, meaning is the way others *enact* our bodily movement into their own by means of responding to it in an adjustive manner. Meaning is, therefore, objective and external—it exists in the structure of the social act which transcends the possible intentions of individual bodies.

In what sense is meaning further embodied? I believe we can elucidate the embodied element of Mead's theory of gestural communication very pertinently by means of Susan Stuart's concept of *enkinaesthesia* (Stuart 2011; 2012; 2013). As opposed to classical, individual-centered theories of the mind, Stuart's phenomenology emphasizes the experiential entanglement of agent and agent, and agent and object (see below). Borrowing from Stuart's beautiful prose and projecting it onto Mead's theory of the origins of embodied meaning we can say that meaning is born in the affective reciprocity between bodies in concrete pragmatic situations, in their:

> [F]olding, enfolding and unfolding, which co-constitutes the conscious relation and the experientially recursive dynamics that lead to the formation and maintenance of the deep integral enkinaesthetic structures and melodies which bind us together, even when they pull us apart. Such deeply felt enkinaesthetic melodies emphasize the dialogical nature of the backgrounded feeling of being. (Stuart 2012, 167)

One could contend that meaning is an emergent event which results from the enkinaesthetic process of mutual adjustments between bodies. It does not belong to any individual 'mind' but it is an emergent trait of the entire enkinaesthetic system. In such a kind of social transaction its partakers enter into what Andy Clark (Clark 1997a, 476) has coined *continuous reciprocal causation* which "occurs when some system is both continuously affecting and simultaneously being affected by activity of some other system" (Clark 2008, 24). According to Clark, appropriately coupled systems capable of entering into the process of continuous reciprocal causation can create unique dynamical systems which may give rise to emergent properties belonging to neither of its components exclusively but only to the system as a whole.[8] Taken from the perspective of Mead's philosophy

8 See Clark (1997b, 164).

of language, such an emergent property could be the meaning[9] of gestures which represents its objective characteristic, even though—at the level of a conversation of gestures—neither of its partakers is *aware* of this meaning: "Awareness or consciousness is not necessary to the presence of meaning in the process of social experience [...] The mechanism of meaning is thus present in the social act before the emergence of consciousness or awareness of meaning occurs" (Mead 1934/1967, 76). Mead defines meaning in terms of a "threefold relationship of gesture to adjustive response and to the resultant of the given social act" (Mead 1934/1967, 80). In contrast to vast majority of modern (and contemporary) epistemologists, he was convinced that objectivity of meaning of the social act is a precondition for the emergence of consciousness and not the other way around.

## 2 Mindedness, symbol, and enkinaesthetic action-prediction

The situation of a 'conversation of gestures' is, in Mead's understanding, a prelinguistic sense-making situation. It is so because, although meaning originates already at this stage of embodied social transaction,[10] the participants of the conversation of gestures are not aware of it; in other words, they are not aware of the sort of response their own comportment is going to provoke in other participants of the social act. In a conversation of gestures, bodies actively make sense of each other but are not aware of this fact, i.e., they are not yet able to regard their own comportment 'from the perspective of the other body' and are not able to anticipate its bodily response. In moving to another level, Mead's enactivist theory of language still neatly embraces enkinaesthesia because, just like

9 "Meaning arises and lies within the field of the relation between the gesture of a given human organism and the subsequent behavior of this organism as indicated to another human organism by that gesture. If that gesture does so indicate to another organism the subsequent (or resultant) behavior of the given organism, then it has meaning. In other words, the relationship between a given stimulus—as a gesture—and the later phases of the social act of which it is an early (if not the initial) phase constitutes the field within which meaning originates and exists. Meaning is thus a development of something objectively there as a relation between certain phases of the social act; it is not a psychical addition to that act and it is not an 'idea' as traditionally conceived" (Mead 1934/1967, 75–76).

10 In the present context I will be using John Dewey's term 'transaction' in order to refer to social activities that various organisms (and notably human beings) perform together. As opposed to the term 'interaction'—transaction underlines the fundamental entanglement of social actors and their irreducible mutual co-constitution. See Dewey & Bentley (1949).

in Stuart's approach, "the focal point moves from the agent and [...] individual agency [conversation of gestures—R.M.] to the necessity of our being co-agential in a co-dynamically continuous, affectively laden enkinaesthetic processual horizon of experience" (Stuart 2012, 170). The other level of social transaction we are talking about here Mead called 'taking the attitude of the other.' Mead held that the capacity of taking the attitude of the other gives rise to mindedness (or consciousness of meaning) and consists basically in the capacity of calling out in oneself the same sort of motor response one's movement (gesture) would call out in the other participant of the social act. "Gestures become significant symbols when they implicitly arouse in an individual making them the same responses which they explicitly arouse, or are supposed to arouse, in other individuals" (Mead 1934/1967, 47).

Given this capacity, gestures can become 'significant' for they bear the same meaning for both participants of the social act. Phenomenologically speaking, the capacity of taking the attitude of the other could also be understood as enkinaesthetic anticipation or enkinaesthetic imagination by means of which we dive into the world of meanings of the other by predicting his motor response to our own movements. In this sense "we might say that we inhabit the other's activity, for that's how we learn, how we become enculturated, and how we develop our sensory and kinaesthetic and enkinaesthetic imagination that enables us to anticipate what the other might do" (Stuart 2012, 170–171). Stuart's enkinaesthetic terminology is, in the context of Mead's philosophy of mind and language, particularly pertinent for it captures what he was up to even better than the terminology of 'embodiment.' Not only do meanings reside in bodies and their behavioral dispositions but, first and foremost, they originate by means of coupling more motorically inter-acting and, thus, trans-acting bodies together. The mutual kinaesthesis in which meanings leak from one body to another must therefore be regarded as a condition of possibility for the appearance of higher mental functions like language and self-consciousness (mindedness). To sum up, for Mead, gestures which arouse the same sort of motor response in the body that produces them as in the body which receives them become 'significant symbols': "where the gesture reaches that situation, it has become what we call 'language'" (Mead 1934/1967, 46). Although such a definition omits a handful of important traits of languages (e.g. syntax), it, nevertheless, captures their most important property, i.e., the mechanism of origination and conveyance of meanings. Whereas traditional, individualist approaches consider language to be a means of expressing pre-existing ideas, Mead sees the phenomenon of language as a culturally and historically conditioned sediment of enkinaesthetic meanings, originating and stabilizing themselves into regularities over long periods of human embodied transactions: "Language, then, has to be studied from the point of view of the gestural type of conduct within which it existed without being as such a definite language. And we have to see how the communicative function could have arisen out of that prior sort of conduct" (Mead 1934/1967, 17). At this point we can recall Davidson's words from the first paragraph, according to which there is nothing like a language in the classical, transcendental understanding of that term. Per Linell, a Swedish linguist, expressed aptly a similar idea when he wrote:

> we do not first have a grammar a priori, and then start building utterances in accordance with it. Instead, we are always immersed in languaging, in and through which we develop habits of using and reusing elements of language and routines of enacting communicative projects through linguistics and other semiotic means. As a result of this continuous process, regularities emerge and rules of linguistic behavior get established and conventionalized. (Linell 2009, 56–57)

Let us, however, examine the capacity of 'taking the role of the other' a bit more closely for it could be argued that if we adopt the enkinaesthetic approach to Mead's understanding of the origins of language, some important lessons may be learnt from Mead with regard to the recently burgeoning approach to cognition called 'prediction error minimization' (Hohwy 2013). We have already made clear that certain bodily movements can become gestures if the body executing them is aware of the sort of motor response it is going to provoke in others. From Mead's point of view, it seems, therefore, that the origins of common meanings are directly dependent on the capability to correctly anticipate the movement of the other. It may be that what Mead calls 'taking the attitude of the other' is a pragmatic matter of minimizing the uncertainty of sensory input in the process of social transactions, or of minimizing the factor of error about what the other is going to do next. The evolutionary process might have pushed human beings into such a form of cooperation in which they started to construct hypotheses about the motor responses of others towards their own movements, which capacity, in turn, enabled them to better manage the enkinaesthetic social process as a whole and push it into qualitatively new levels of interaction. It should be noted that what we call 'hypotheses' need not be understood in strongly representational (contentful) terms at all. What might serve as a hypothesis about the bodily comportment of others *vis-à-vis* our own movement might well be, as Stuart writes, a matter of "kinaesthetic memory and kinaesthetic imagination" (Stuart 2012, 167). If there is more or less regular perceptual consequence to my movement (a certain sort of action of the other), than this consequence gets triggered whenever I perform this sort of action in particular circumstances.[11] In this respect, Stuart refers to Bernhard Hommel's 'theory of event coding'[12] according to which perception, attention, intention and action share a common 'representational' domain (where 'representation', again, refers to a common code for perceptual and motor action) within which perceiving an event will trigger the associated action, and *vice versa*. According to Hommel:

> According to TEC [theory of event coding—RM], the core structure of the functional architecture supporting perception and action planning is formed by a common representational domain for perceived events (perception) and intended or to-be-generated events (action) [...] we believe that it makes sense to assume that the stimulus representations underlying perception and the action representations underlying action planning are coded and stored not separately, but together in a common representational medium. This implies that stimulus and response codes are not entities of a completely different kind, but only refer to, and thus represent, different events in a particular task and context. (Hommel et al. 2001, 849)[13]

The individuals who are able to generate the most correct anticipations (i. e. predictions) about the motor responses of others have, at the same time, the best chance to succeed in the process of social communication. Let us now elucidate the levels of embodiment as far as significant symbols are concerned. It has already been made clear that significant symbols are gestures which call out the same motor response by both participants of the social act. However, how does this work in the case of the relations between speakers and the world of objects? We can say that in Mead's understanding this threefold relationship must be analyzed on two levels, namely that individual/object and individual/individual. The key link which connects both these levels is that of an attitude. As described earlier, attitudes are, for Mead, neurologically instantiated bodily dispositions encoding perceptual objects in terms of possible pragmatic reactions towards them.[14] Already in one of his earliest papers "Suggestions Towards a Theory of the Philosophical Disciplines" Mead talks about the 'telological nature

**11** What these circumstances are can not be entirely determined upfront. As the situation repeats itself and as the cognizer (often after numerous repetitions) minimizes the perceptual surprise, he will finally 'get it right.'

**12** Hummel's theory is, after all, not dissimilar from William James' *ideomotor theory* according to which every representation of movement awakes in ourselves to some degree the movement it represents. See James (James 1890, 522–527).

**13** For a pragmatist argumentation in favor of the idea of such a common coding of stimuli and responses, see Dewey (1896).

**14** In describing objects in this particular manner, Mead clearly anticipated James Gibson's concept of 'affordances'. See Gibson (Gibson 1979, 127). For a closer examination of a pragmatist dimension of Gibson's work, see Heft (2001).

of the concept' according to which "the meaning of the object is derived entirely from our reaction upon it, or [...] our use of it" (Mead 1900/1964, 8).[15] Therefore, on the inferior 'individual/object' level—attitudes encode concepts, i.e., they encode perceptual objects in terms of their use for the individual. An object has as many meanings for the individual as many ways there is for that individual to utilise them in order to meet his pragmatic goals. In short, attitudes encode what one could call 'individual affordances.' Hence, if attitudes encode individual affordances and if gestures are attitudes to which others respond, then significant gestures (or significant symbols) encode what one might call 'social affordances', because such significant symbols call out the same sort of response towards environmental structures:

> What language seems to carry is a set of symbols answering to certain content which is measurably identical in the experience of the different individuals. If there is to be communication as such, the symbol has to mean the same thing to all individuals involved. If a number of individuals respond in different ways to the stimulus, the stimulus means different things to them (Mead 1934/1967, 54).

However, what does it mean that the response in case of significant symbols is supposed to be the same? Obviously, in case of any social act, numerically different bodies must respond by means of numerically different responses; in what sense are then these responses expected to be identical? This question brings us fully to the individual / individual level of enkinaesthetic transaction. If I, for instance, tell a friend of mine to fetch a chair because I feel tired, and that person brings over a chair from a nearby classrom, his response is sufficiently identical to what I anticipated (given that I didn't mean a particular chair).[16] In other words, such a response would pass for being, what David L. Miller has coined —'functionally identical'[17]—in that it enables the social act to continue and thus stands the test of practical action. If, however, that person (without further

---

**15** Again, by such an understanding of the term 'concept' Mead anticipates the theoreticians of the embodied mind such as Mark Johnson, for whom concepts are "neural activation patterns [...] recurring structures of experiences actual and possible, retained in the organism as synaptic weights" (Johnson 2007, 159), or Alva Noë who holds that "Concepts [...] are rather skills for taking hold of what there is. To say that perceptual experience is *conceptual* from this standpoint, is to say that perceptual experience is a skillful grappling with what there is [...]. Don't think of a concept as a label you can slap on a thing; think of it as a pair of calipers with which you can pick the thing up" (Noë 2012, 35–36).
**16** The determination of the question whether that person has correctly understood my order will necessarily be strongly context-dependent. More on the general question of rule-following in G. H. Mead in Madzia (2013).
**17** See Miller (Miller 1973, 12–13).

explanation) brings over a bottle of water, the gesture 'chair' means very different things to both of us. The condition of identity of response can, therefore, be adjudicated on the background of the 'pragmatic test', i. e., on whether our usage of words enables our practical action to continue in a relatively straightforward manner.

## 3 The hand-made mind: language between the body and the world of things

Mead believed that the human capacity of 'enkinaesthetic anticipation' or embodied action-prediction was, in the human evolutionary history, a result of a growing need of organizing our social relations with others: "The mechanism of human society is that of bodily selves who assist or hinder each other in their cooperative acts by the manipulation of physical things" (Mead 1932, 169). That is why Mead identified the roots of language in the manipulatory phase of the act where the acting body, confronted with a problem, inhibits its conduct, suspends the consummatory phase of the act, 'takes a step back' and thinks. For Mead, conceptually speaking, the initial phases of symbolic interaction can be identified with problematic situations, where the gesturing organism is forced to reflect on the sort of action which is going to bring about a desired motor response of the other. In other words, since we always tend to minimize the perceptual uncertainty in the course of social action, the process of social learning consists in the effort of always getting better and better at predicting the responses of others. This is why Mead believed that until our gestures become strongly habitualized, or 'absorbed' into our social conduct, the moments of social communication bear the traits of 'problematic situations' in which it is vital to reflect on what we are going to do or say next.[18] From Mead's perspective, language is not only closely connected but, in fact, originates in the need to solve action-problems. As David L. Miller puts it: "Mead makes it abundantly clear that if men could live by habit alone there would be no occasion for thinking" (Miller 1973, 148).

We have already elucidated the levels of embodiment of significant symbols. For Mead, however, significant symbols are only one half of the story. The second half, which plays a crucial role in reinforcing the process of symbolic exchange,

**18** Every learner of a foreign language, who finds himself for the first time in a country where that language is spoken, knows these situations more than well.

is provided by the physiological construction of the human hand. Cornelis de Waal even goes as far as to contend that for Mead

> the hand with the juxtaposition of the thumb, is in many respects even more characteristic of human intelligence than the brain [...] it would be impossible for entirely disembodied spirits, say, for angels or extraterrestrial intelligent vapors, to develop the concept of a physical object as we know it. For that, one needs hands (de Waal 2002, 26).

From Mead's naturalistic point of view, since it is the hand which is the main organ of manipulation by human beings, and since the hand finds itself 'in the middle of the road' between objects of visual perception and the mouth (the primary organ of consummation) human bodies have the natural tendency to inhibit, or at least delay, the consummatory phase of the act in favor of manipulation, disassembling, and/or creative rearranging of physical things, which capacity, in turn, gives them additional time to reflect upon possible affordances of the physical object:

> There is [...] another very important phase in the development of the human animal which is perhaps quite as essential as speech for the development of man's peculiar intelligence, and that is the use of the hand for the isolation of physical things. Speech and the hand go along together in the development of the social human being. There has to arise self-consciousness for the whole flowering-out of intelligence. But there has to be some phase of the act which stops short of consummation if that act is to develop intelligently, and language and the hand provide the necessary mechanisms. (Mead 1934/1967, 237)

In Mead's philosophy of embodiment, the human hand and significant symbols mutually reinforce each other in the process of inhibition and enable the human body to analyze problematic objects in terms of new possible responses toward them. Mead held that significant symbols refer to objects in two complementary ways—denotation and connotation, intension and extension.[19] On the one hand, significant symbols always 'denote' (or name) objects or things in the world. On the other, significant symbols as names, always also 'connote' sets of possible responses toward those objects. Mead held (pace most of analytic philosophers of language) that extension and intension are not mutually exclusive, i.e., if a significant symbol is to mean anything, it has to be composed of both these elements. Let us take a closer look at both these embodied properties of significant symbols: i) Denotation has to do with what one could call 'perceptual content' of significant symbols, which is essentially twofold a) the *acoustic* (or written, etc.) and b) *gesture* (e.g. the English 'dog') which calls out a certain *mental image*. The

19 See Mead (Mead 1922/1964, 246).

denotative element of significant symbols enables us to receive, comprehend and identify objects in the world. This idenfitication takes place by ii) connoting the denotative element of significant symbols with sets of different responses which we can carry out toward the object, referred to by the mental image: "Meanings, universals, therefore come into existence (as relations between the form and its environment) by virtue of an awareness of the responses, implicit or explicit, which they arouse" (Miller 1973, 80). While the mental image which gets called out by a vocal (acoustic) gesture serves as sort of a 'mental prototype' that represents the successful achievement of action, it is the connotation, or practical bodily response towards objects in the world, which is, for Mead, the decisive factor for the objectivity of meaning. Why is this so? David L. Miller pertinently points out to the fact that if individuals (perceptual stimuli, mental images, etc.) gain meaning, it is because the response towards them is universal (in principle shareable with others). Therefore we can not say anything about individuals as individuals. This is not so because individuals are not perceived but because in order for the human experience (permeated by symbols) to be meaningful, that experience must be shareable with others and since meaningful experience is only achievable by means of significant symbols, this meaning must consist in shareability of responses.[20]

Probably this is the reason why Ludwig Wittgenstein (in the §§ 1–34 of the *Philosophical Investigations*) calls into question the possibility of learning the meaning of words by means of a simple ostensive definition. The meaning of a word (say a 'chair') cannot be determined by however long a chain of particular objects or images but only by making clear what all these particulars have in common; and the thing they all have in common is a socially shareable manner of their practical usage. As opposed to the 'picture theory of meaning' (Locke, etc.), neither ostensions, nor any sort of mental images can possibly represent a sufficient condition for understanding words. Although Mead was convinced that mental images accompany all significant symbols, however abstract they may be (Mead 1904b, 606), these images exclusively cannot determine their meaning for the simple reason that images are *signs* and, therefore, have the same logical status as words.[21] If, however that were the case, in other words,

**20** It should be noted that as opposed to some analytic theories of meaning which, very counter-intuitively after all, understand 'meaning' as existing on the level of propositions—Mead does not have a problem with saying that it is already single words that have meaning since they evoke concrete mental images and sets of bodily responses.

**21** Let us stay with the example of a 'dog'. If one were to say that the meaning of the term 'a dog' is one's mental image, then the image itself does not help one bit as any mental image if the meaning of a sign was just another sign, than this other sign would need to be interpreted by means of yet another sign, etc., *ad infinitum*. The meaning of a sign (in Mead's terminology of significant symbols) must, in the last instance, reside in something more than just signs, namely in a certain sort of socially shareable practice. Consequently, also for Mead, meaning and understanding are not things which we achieve in our heads but things we do in the world.

With Andy Clark, we might call significant symbols "cognitive shortcuts" (Clark 2003, 70), because a) being intersubjectively shareable items (denotation) and b) they subsume sets of possible motor responses towards objects of the common practical world (connotation). In the problematic situation, it is usually the element of connotation, i.e., the individual instance of a concept that becomes an object of reflection. The problematic situation emerges when certain parts of the world do not respond to our habitualized ways of comportment the way we predicted, which is to say—the way which permits us to achieve our practical goals. Now, if an object is a set of possible responses which we can carry out towards a particular stimulus and this object becomes problematic, then in the reflective mind, an object becomes a linguistically mediated content of the mind because it is language which enables the reflective mind to single out, decontextualize and analyze particular problematic properties of the object in question. Mead called this stage of action *manipulation*. Thanks to the physiological construction of the hand, human beings are able to creatively enact the wordly structures in an unprecedented number of ways: "Intelligence is essentially the ability to solve the problems of present behavior in terms of its possible future consequences as implicated on the basis of past experience" (Mead 1934/1967, 100).

Since significant symbols enable us to refer to objects and their mutual relations, Mead held that in the case of socialized, reflective beings which operate with significant symbols the action-problems, which normally take place in the world by means of the organism's bodily skills, become *internalized* and scrutinized not necessarily by means of operations of the human hand but often exclusively on the basis of an internal dialog of the social mind with itself: "When you are reasoning you are indicating to yourself the characters that call out certain responses—and that is all you are doing" (Mead 1934/1967, 93). To paraphrase another one of Clark's terms, significant symbols are means of 'off-loading' the energetically demanding process of 'external' trial-and-error to the 'inner' process of internal conversation:

of a dog can, in principle, signify an infinite number of properties (a mammal, a fluffy animal, an animal with a tail, a barking animal, an animal with sharp teeth, etc.).

> Public speech, inner rehearsal [...] are all potent tools that reconfigure the shape of computational space. Again and again we trade culturally achieved representation against individual computation. Again and again we use words to focus, clarify, transform, offload, and control our own thinkings. Thus understood, language is not the mere imperfect mirror of our intuitive knowledge. Rather, it is part and parcel of the mechanism of reason itself. (Clark 1997, 207)

We should, nevertheless, keep in mind that, in Mead's eyes, the process of reflective deliberation and reconstruction of the act by means of significant symbols is derivative of the embodied practice of transforming, manipulating and exploiting the environmental structures by means of the human hand: "Language, Mead holds, would not have emerged had not the hands been freed from supporting the body and capable, through the juxtaposition of the thumb and the fingers, of dissecting and reassembling objects. Without the hand, language would have no function, and by language the increase in the ways of manipulating the physical object is unlimited" (Miller 1973, 62). From Mead's perspective, thinking is manipulation with, and creative rearrangement of, connotations of significant symbols. Thinking is also profoundly future-oriented and has a character of predictive inference: "When we speak of reflective conduct we very definitely refer to the presence of the future in terms of ideas [...] it is the picture [...] of what future is going to be [...] that is the characteristic of human intelligence" (Mead 1934/1967, 119).

Reflective thinking is, however, not to be understood as a sort of pure computation, taking place in our heads. On the contrary, it takes place in the the world and directly enacts its structures: "While the conflict of reactions takes place within the individual, the analysis takes place in the object. Mind is then a field that is not confined to the individual, much less is located in a brain" (Mead 1922/1964, 247). It seems, therefore, that for Mead—the human mind was to be defined as an 'elastic' field which not only originates in the interactions between the body and the environment but always also reaches out to the world and works in a close collaboration with it: "Mind involves [...] a relationship to the characters of things. Those characters are in the things, and while the stimuli call out the response which is in one sense present in the organism, the responses are to things out there. The whole process is not a mental product and you cannot put it inside of the brain" (Mead 1934/1967, 124–125). In this respect, Mead would undoubtedly agree with Andy Clark's statement contending that the "capacity to creatively distribute labor between biology and the designed environment is the very signature of our species" (Clark 2003, 174). Since action-problems appear in real-time, and are intensely felt as pressing difficulties, the task of producing new and efficient predictions should not be understood as an effort of trying to come up with the best possible solution. From the point of view of Mead's pragmatist externalism, embodied agents tend to act along the lines of what Clark has called the 'principle of ecological assembly' according to which "the canny cognizer tends to recruit, on the spot, whatever mix of problem-solving resources will yield an acceptable result with a minimum of effort" (Clark 2008, 13). The task of producing efficient predictions does not consist in creating an 'ideal copy of reality' but to enable the act to continue in an efficient, acceptable manner. In the case of 'minded' beings, like us, this task is fulfilled by means of symbolic thinking, in other words—language.

As indicated at the beginning of this paper, the main value of Mead's philosophy consists in viewing language and mind as two sides of the same coin. Of course, even today there are pragmatistically inclined philosophers, such as Robert Brandom, who view the relation between language and mind in a similar way. As opposed to the 'new pragmatists', however, George H. Mead was aware of the necessity to do justice to the embodied roots of language and mind.[22] Mead was able to conceptually cover the implications of the embodied roots of reflective thinking with unusual depth and intellectual clarity. With respect to the current progress in embodied cognition studies, this makes him a worthwhile source which researchers are well advised to take into account.

## References

Adams, F. (2010). "Embodied Cognition." In: *Phenomenology and the Cognitive Sciences* 9 (4), 619–628.

Cowley, S., & Nash, L. (2013). "Language, Interactivity and Solution Probing: Repetition Without Repetition." In: *Adaptive Behavior* 21 (3), 187–198.

Clark, A. (1997a). "The Dynamical Challenge." In: *Cognitive Science* 21 (4), 461–481.

Clark, A. (1997b). *Being There: Putting Brain, Body, and World Together Again.* Cambridge: MIT Press.

Clark, A. (2003). *Natural-born Cyborgs. Minds, Technologies, and the Future of Human Intelligence.* Oxford: Oxford University Press.

Clark, A. (2008). *Supersizing the Mind. Embodiment, Action and Cognitive Extension.* Oxford: Oxford University Press.

Cuffari, E. C., Di Paolo, E., & De Jaegher, H. (2014). "From Participatory Sense-making to Language: There and Back Again." In: *Phenomenology and the Cognitive Sciences.* DOI 10.1007/s11097-014-9404-9.

Davidson, D. (1986/2006). "A Nice Derangement of Epitaphs." In: *The Essential Davidson.* Oxford: Oxford University Press, 251–265.

22 In this regard, it is stunning how the embodied element is almost entirely absent in Brandom's treatment of these issues. See Brandom (2000).

de Bruin, L. & de Haan, S. (2012). "Enactivism & Social Cognition: In Search of the Whole Story." In: *Journal of Cognitive Semiotics* 4 (1), 225 – 250.
de Bruin, L. C. & Kästner, L. (2012). "Dynamic Embodied Cognition." In: *Phenomenology and the Cognitive Sciences* 11 (4), 541 – 563.
Dewey, J. (1896). "The Reflex Arc Concept in Psychology." In: *Psychological Review* 3 (1), 357 – 370. Retrieved from: http://www.brocku.ca/MeadProject/Dewey/Dewey_1896.html [09.05.2015].
Dewey, J., & Bentley, A. F. (1949). *Knowing and the known.* Boston, MA: Beacon Press.
Fuchs, T. & De Jaegher, H. (2009). "Enactive Intersubjectivity: Participatory Sense-making and Mutual Incorporation." In: *Phenomenology and the Cognitive Sciences* 8 (4), 465 – 486.
Gibson, J. J. (1979). *The Ecological Approach to Visual Perception.* Boston: Houghton-Mifflin.
Heft, H. (2001). *Ecological Psychology in Context: James Gibson, Roger Barker, and the Legacy of William James.* East Sussex: Psychology Press.
Hommel, B. et al. (2001). "The Theory of Event Coding (TEC): A Framework for Perception and Action Planning." In: *Behavioral and Brain Sciences* 24, 849 – 937.
Hohwy, J. (2013). *The Predictive Mind.* Oxford: Oxford University Press.
James, W. (1890). *The Principles of Psychology.* London: Macmillan and co., Ltd.
Joas, H. (1985). *G. H. Mead: A Contemporary Re-examination of His Thought (Studies in Contemporary German Social Thought).* Cambridge: MIT Press.
Johnson, M. (2007). *The Meaning of the Body: Aesthetics of Human Understanding.* Chicago, IL: University of Chicago Press.
Kilpinen, E. (2011). "Social Theory." In: S. Pihlström (Ed.). *Continuum Companion to Pragmatism.* London, UK/New York, NY: Continuum, 139 – 149.
Linell, P. (2009). *Rethinking Language, Mind, and World Dialogically: Interactional and Contextual Theories of Human Sense-making.* Charlotte: Information Age Publishing.
Madzia, R. (2013). "The Concept of Rule-following in the Philosophy of George Herbert Mead." In: T. Burke & K. Skowroński (Eds.). *George Herbert Mead in the Twenty-first Century.* Lanham: Lexington Press, 61 – 70.
Madzia, R. (2015). "Presentation and Re-presentation: Language, Content, and the Reconstruction of Experience." In: *Timeliness of G. H. Mead*, D. L. Huebner, & H. Joas (Eds.). Chicago: Chicago University Press.
Maturana, H. R. (2002). "Autopoiesis, Structural Coupling and Cognition: A History of These and Other Notions in the Biology of Cognition." In: *Cybernetics & Human Knowing* 9 (3 – 4), 5 – 34.
Mead, G. H. (1900/1964). "Suggestions Towards a Theory of Philosophical Disciplines." In: *Selected Writings.* A. J. Reck (Ed.). Chicago: University of Chicago Press, 6 – 24.
Mead, G. H. (1904a). "The Relations of Psychology to Philology." In: *Psychological Bulletin* 1, 375 – 391. Retrieved from: http://www.brocku.ca/MeadProject/Mead/pubs/Mead_1904c.html [03.05.2015].
Mead, G. H. (1904b). "Image or sensation." In: *Journal of Philosophy* 1 (2), 604 – 607. Retrieved from: http://www.brocku.ca/MeadProject/Mead/pubs/Mead_1904b.html [03.05.2015].
Mead, G. H. (1909/1964). "Social Psychology as Counterpart to Physiological Psychology." In: *Selected Writings. George Herbert Mead.* A. J. Reck (Ed.). Chicago: University of Chicago Press, 94 – 104.

Mead, G. H. (1910/1964). "Social Consciousness and the Consciousness of Meaning." In: *Selected Writings. George Herbert Mead.* A. J. Reck (Ed.). Chicago: University of Chicago Press, 123–133.
Mead, G. H. (1922/1964). "A Behavioristic Account of the Significant Symbol." In: *Selected Writings. George Herbert Mead.* A. J. Reck (Ed.). Chicago: University of Chicago Press, 240–247.
Mead, G. H. (1982). "1914 Class Lectures in Social Psychology." In: *The Individual and the Social Self: Unpublished Work.* D. L. Miller (Ed.). Chicago: University of Chicago Press, 27–105.
Mead, G. H. (1934/1967). *Mind, Self, and Society.* Chicago: University of Chicago Press.
Mead, G. H. (1938). *Philosophy of the Act.* Charles W. Morris et al. (Ed.). Chicago: University of Chicago Press.
Miller, D. L. (1973). *George Herbert Mead. Self, Language, and the World.* Chicago, IL: University of Chicago Press.
Noë, A. (2012). *Varieties of Presence.* Cambridge, MA: Harvard University Press.
Stuart, S. A. J. (2011) "Enkinaesthesia: The Fundamental Challenge for Machine Consciousness." In: *International Journal of Machine Consciousness* 3 (1), 145–162.
Stuart, S. A. J. (2012). "Enkinaesthesia: The Essential Sensuous Background for Co-Agency." In: *Knowing Without Thinking. Mind, Action, Cognition, and the Phenomenon of the Background.* Z. Radman (Ed.). New York: Palgrave Macmillan, 167–186.
Stuart, S. A. J. (2013). "The Union of Two Nervous Systems: Neurophenomenology, Enkinaesthesia, and the Alexander Technique." In: *Constructivist Foundations* 8 (3), 314–323.
Wittgenstein, L. (1953/2009). *Philosophical Investigations.* P. M. S. Hacker & J. Schulte (Eds.). West Sussex: Blackwell Publishing.

Matthew Crippen

# Dewey, Enactivism, and Greek Thought[1]

**Abstract:** In this chapter, I examine how Dewey circumnavigated debates between empiricists and *a priorists* by showing that active bodies can perform integrative operations traditionally attributed to "inner" mechanisms, and how he thereby realized developments at which the artificial intelligence, robotics and cognitive science communities only later arrived. Some of his ideas about experience being constituted through skills actively deployed in cultural settings were inspired by ancient Greek sources. Thus in some of his more radical moments, Dewey refined rather than invented the wheel, and I suggest that prominent embodiment figures have done the same, Dewey having anticipated them, particularly Noë and his version of enactivism. I urge that cognitive science may progress into relatively unexplored territory by traveling Dewey's historically sensitive path.

## 1 Introduction

In this paper, I argue that enactivism has antecedents in Deweyan pragmatism partly via Dewey's appropriation of ancient Greek thought. Consequently the radicalness of Dewey's ideas, which are advanced by the measure of contemporary cognitive science, derives partly from his historical sensitivity. This suggests some variants of enactivism and pragmatism are, as James put it in the subtitle of his (1907) book *Pragmatism*, "A New Name for Some Old Ways of Thinking."

I begin by reviewing Dewey's history of concepts of experience, and how he believed biology was resuscitating ancient notions. Next I elaborate on Gallagher's (2009) claim that Dewey anticipated enactivism: the view, in the words of its founders, "that cognition is not the representation of a pregiven world [...] but rather the enactment of a world and a mind on the basis of a history of the variety of actions that a being in the world performs" (Varela, Thompson & Rosch 1991, 9). In broad strokes, I examine how Dewey circumnavigated debates between empiricists and *a priorists* by arguing that active bodies in environments can perform integrative operations traditionally attributed to "inner" mechanisms, and how Dewey foresaw developments only later realized by the AI, ro-

1 I would like to aknowlegde the help of my former research assistant, Salma El-Galley, on this chapter.

botics and cognitive science communities. I focus on striking parallels between Dewey's account and Noë's (2004, 2009; O'Regan & Noë 2001) version of enactivism; and, with other researchers (e.g., Lakoff & Johnson 1999; Rockwell 2005; Menary 2007; Gallagher 2009; Fingerhut 2011; Solymosi 2011, 2013; Skorburg 2013; Madzia 2013), strive to show that Dewey's work is generally in the spirit of current embodied approaches to cognitive science. I conclude by stressing that Dewey knowingly drew on ancient ideas. He refined rather than invented the wheel, and stated so. I suggest that enactivists and most prominently Noë have done the same, albeit less knowingly, and that enactivists might progress their field into relatively unexplored areas by following a historically sensitive path.

## 2 Dewey's history of experience

Dewey observed that empiricists and rationalists "do not thoroughly agree among themselves, but they agree that experience never rises above the level of the particular, the contingent, and the probable" (Dewey 1920, 78). In other words, both have skeptical attitudes towards experience, only the former hold nothing else is available to build knowledge.

Dewey wanted to resuscitate the status of experience, and in doing so, he returned to ancient ideas, observing that "the key to the matter" is curiously found in the Greek notion that "experience [is] itself a product of experience" (Dewey 1920, 79), which is to say, having experience means being experienced or skilled. Among much else, the Greek word *Experiri* "means try, to put to the test" (Lennon 1960, 316), and in Plato and Aristotle, Dewey elaborated, "you find experience conceived as something primarily practical in nature," namely, "the association of experience with the kind of intelligent skills that is formed in the arts by the accumulation of the trial and error method" (Dewey 1910–1911, 133), and therefore accumulated from the past. This ancient account "agrees very closely with what the modern psychologist knows as the method of learning by trial and error as distinct from the method of learning by ideas," and "[i]f another conception of experience is now possible, it is precisely because the quality of experience as it may now be lived has undergone a profound social and intellectual change from that of earlier times" (Dewey 1920, 79).

Elaborating on the radicalness of the shift from pre-Modern to Modern concepts of experience, Dewey explained that in the Modern era, "instead of being the representative factor that binds man to the past," experience "is the liberating factor; it is not the memory of the old, but the perception of the new, which vouchsafes, conveys itself free from the distortions which intellect tends to introduce" (Dewey 1910 – 1911, 133 – 134). Modern philosophers "usually had a critical purpose in mind," and therefore rejected or devalued ancient notions of experience. We see this prominently in Descartes, who expressly blamed "custom" for many of our delusions. We also see it in Hume, who, while equating experience with sense impressions, and while acknowledging a role for custom, also used the latter to explain why knowledge is little more than a conditioned response.

Dewey believed "[t]he effect of the development of biology has been to reverse" the standard stimulus-response picture of human mind that became dominant in the Modern era (Dewey 1920, 84), and to bring about a partial return to ancient concepts of experience. James' psychology, for example, was very much influenced by biology, especially Darwinism (see Crippen 2010, 2011), and his idea that experience is what remains after selective interests chisel away at sense impressions recalls Aristotle's claim that experience arises from accumulation of incidents with important regularities reinforced and combined. As Dewey said:

> [In] modern biology— experience concerned with life —[it is] impossible to think of life as primarily a way of knowing. [It] is primarily a mode of doing and suffering, i.e., passive, not from the standpoint of knowledge, but of doing, suffering disappointment, defeat. (Modern science tends in the same direction.) You get a certain principle of organization, you get away from the Humeian [notion]: every distinct perception [is] a distinct existence. Life has a certain structure, not a cognitive structure, but one which is a correlate of the function, i.e., continuity (Dewey 1910 – 1911, 134).

In *Reconstruction in Philosophy*, he added: "In order that life may persist, [its] activity has to be both continuous and adapted to the environment. This adaptive adjustment, moreover, is not wholly passive; is not a mere matter of the moulding of the organism by the environment" (Dewey 1920, 84).

Dewey, as a matter of course, nearly always insisted on establishing "working connections between old and new subject-matters" (Dewey 1929, viii). Dewey would do the same in developing his concept of experience. He urged that "[e] xperience itself has its retrospective, conservative, aspect," as the Greeks suggested, and which incidentally supplies a reason why Plato and Socrates, in their roles as social critics, downplayed its epistemic value. He also stressed the "projective, prospective aspect" of experience, in line with the Moderns. "Genuine intelligence is the function of adjustment of the two aspects to each other [...]" (Dewey 1910 – 1911, 135). According to Dewey's (1920) formulation, therefore, experience is "an affair primarily of doing" in which the living creature "does not wait passive and inert for something to impress itself upon it from without." Rather, it "acts in accordance with its own structure, simple or complex, upon its surroundings," and "changes produced in the environment react upon the organism and its activities" in turn. "The living creature undergoes, suffers, the consequences of its own behavior. This close connection between doing and suffering or undergoing forms what we call experience" (Dewey 1920, 86). Thus to use a simple example, a blind woman, in pushing her cane into the world, meets resistance and thereby undergoes consequences of her action. Out of this meaningful interplay of doing and undergoing, perceptual experience of the world emerges.

## 3 Going out to receive

Dewey's account of experience fits with current embodied accounts of cognition, and the corrections they suggest for older mainstream views. Many might assume, for example, that monarch butterflies—famous for their multigenerational migrations—possess complex internal cognitive maps when they only have pinhead sized brains, and interactions of bodies with complex environments explain much of their behavior, as when mountain ranges funnel them towards their destination. Noë (2009) and others such as Barrett (2011) have suggested we similarly err when locating the seat of our being inside our heads. As Noë indicated in the title of his 2009 book, this is wrong because we are in fact "Out of Our Heads." "[P]erception," he said elsewhere, "is not a process in the brain" (Noë 2004, 2), but "a mode of skillful exploration of the world" (Noë 2004, 194).

Noë thereby mirrored Dewey (1916), who a century earlier and building on ancient notions wrote that cognition "is not an event going on exclusively within the cortex [...]. It involves the explorations by which relevant data are procured and the physical analyses by which they are refined and made precise[.]" Seen thus, "[h]ands and feet, apparatus and appliances of all kinds are as much a part of it as changes in the brain" (Dewey 1916, 14). So without denying inner mechanisms of mind or brain, Dewey reasoned that bodily actions in the world already account for much of the structure of experience (see Dewey 1920, 84–92). Merging the language of rationalist and empiricist schools that had focused on the inner realm, Dewey declared that perception is "an act of the going-out of energy in order to receive" (Dewey 1934, 53). Thus when fingers go out to caress a lacquered woodcarving, smoothness and shape, not to mention the capacities of the hand, are realized as effects and in perception, whereas a cat's digging claws realize sinewy toughness, suggesting that the difference between the perceptual worlds of the two is no mere mental or neural variation. The carving, as Merleau-Ponty said, "utilizes the time occupied by our tactile exploration or modulates the movement of our hand" (Merleau-Ponty 1945, 315), thereby patterning a perceptual experience.

In the case of vision, Dewey similarly reasoned that "seeing involves the co-operation of motor elements" (Dewey 1934, 53), that is, coordination around things in the world. Of course, he granted we sometimes look without overtly acting; yet we also spend most of our waking life handling and ambulating, which means synchronizing actions around objects and settings. Here and elsewhere, said Dewey, "[m]otor and sensory structure form a single apparatus and effect a single function" (Dewey 1934, 255), indicating that even when there is no overt action, seeing remains "an affair of readiness on the part of motor equipment" (Dewey 1934, 98). Indeed, if we did not see things so, we would bump, bumble and be functionally blind. Thus we see contours of bottles as things we can handle, and hallways as spaces for movement. When we see shape and line, we see "ways in which things act upon one another and upon us; the ways in which, when objects act together, they reinforce and interfere" (Dewey 1934, 100–101). Dewey accordingly insisted that when we see an object "[i]t is not just the visual apparatus" that becomes active, "but the whole organism" (Dewey 1934, 122; also see 100).

It is for comparable but more specific and carefully elaborated reasons that O'Regan and Noë proposed in their landmark 2001 article that visual perception is "*the activity of exploring the environment in ways mediated by knowledge of the relevant sensorimotor contingencies*" (O'Regan & Noë 2001, 943), so that "seeing is a way of acting" and "exploring" (O'Regan & Noë 2001, 939). The point can be demonstrated by considering devices employed by individuals with sensory impairment. For the blind, as Merleau-Ponty observed, a cane ceases to be an object; through "exploration of things," which develops into skilled action, it becomes "an area of sensitivity, extending the scope and active radius of touch, and providing a parallel to sight" (Merleau-Ponty 1945, 143). Noë (2004) provided a more sophisticated illustration, citing Bach-y-Rita's (e.g., 1983; 1984; and Kercel 2002) tactile-vision substitution system. Here a camera mounted on the head transmits stimulation through vibrations on skin or electrical current on the tongue. Individuals who *actively explore* their surroundings come to develop quasi-vision in a relatively short time. They can identify positions and numbers of objects and grasp them. They even experience illusions normally associated with vision.

On such grounds O'Regan and Noë (2001) concluded that "[v]isual experience rests on know-how, the possession of skills" (O'Regan & Noë 2001, 946). Noë elaborated:

> Like touch, vision is *active*. [...] You and your eyes move around the scene the way you move your hands around the bottle. As in touch, the content of visual experience is not given all at once. We gain content by looking around just as we gain tactile content by moving our

> hands. You enact your perceptual content, through the activity of skillful looking (Noë 2004, 73).

Compare this to remarks made by Dewey:

> As we manipulate, we touch and feel, as we look, we see; as we listen, we hear. [...] The eye attends and reports the consequence of what is done. Because of this intimate connection, subsequent doing is cumulative and not a matter of caprice nor yet of routine. In an emphatic artistic-esthetic experience, [which means skilled, among other things,] the relation is so close that it controls simultaneously both the doing and the perception (Dewey 1934, 49–50).

Because of this, Dewey (see Dewey 1934, 175) and Noë (see Noë 2004, 73) rejected camera snapshot models of vision, arguing instead that perception develops in time with unfolding actions.

Noë's enactive sensorimotor theory of course "proposes not that perceiving is *for* acting, but rather that perceiving is constituted by the exercise of a range of sensorimotor skills" (Noë 2004, 90). Dewey, while arguing perceptual organs are vehicles for action, also held that there is no perception without joint action and coordination around things in the environment. He wrote:

> Nothing is perceived except when different senses work in relation with one another except when the energy of one "center" is communicated to others, and then new modes of motor responses are incited which in turn stir up new sensory activities. Unless these various sensory-motor energies are coordinated with one another there is no perceived scene or object. But equally there is none when—by a condition impossible to fulfill in fact—a single sense alone is operative. If the eye is the organ primarily active, then the color quality is affected by qualities of other senses overtly active in earlier experiences. In this way it is affected with a history; there is an object with a past. And the impulsion of the motor elements which are involved effects an extension into the future, since it gets ready for what is to come and in a way predicts what is to happen (Dewey 1934, 175).

Speaking in terms Dewey would have appreciated, O'Regan and Noë also observed that "seeing involves testing the changes that occur through eye, body, and attention movements" (O'Regan and Noë 2001, 947) and therefore "activity on the part of the animal as a whole" (Noë 2004, 2).

If perception—visual and otherwise—is, as Noë urged, "a mode of activity on the part of the *whole* animal" (Noë 2004, 111; emphasis added) whereby the organism coordinates with its surroundings; and if the manner in which "things look, smell, sound, or feel (etc.) depends, in complicated but systematic ways, on one's movement" (Noë 2004, 109), then it follows that perception occurs through modes other than but also including traditional categories of sense. Indeed, Dewey urged that motor, intellectual and emotional capacities are also involved (see Dewey 1934, 22, 53; also see Crippen 2014, 40–41). Consider the experience of eating popcorn, and how it mobilizes the tongue, jaw, hands, eyes and ears that hear that crunch and skin that feels its warmth, while also being part of cultural rituals and a socially integrative medium within movie theaters. This explains why a liquid popcorn stimulating the same excitations on taste receptors has little appeal. This also helps account for Zampini and Spence's (2004) finding that chips seem extra crispy when people hear crackling, for food is crispy not solely because it fragments easily, but also because it has a certain a look, sound and manner of mobilizing the jaw and tongue—an overall way of synchronizing sensitivities and actions. As Dewey (1926) wrote in an example about painting that could apply also to eating, jogging, skiing and numerous other activities: "[A]n integration is effected in the total set of organic responses; eye-activities arouse muscular activities which in turn do not merely harmonize with and support eye-activities, but […] evoke further experiences," for example, by directing attention, posture and gaze. "Moreover, as in every adequate union of sensory and motor actions, the background of visceral, circulatory, respiratory, functions is also consonantly called into action" (Dewey 1926, 122). Thus the experience of jogging or skiing, especially if carried to an aesthetic level, is highly integrated. Interactions with the trail fall into rhythmic unity and consequently experience too. More broadly, the experience hangs together like a narrative, with twists and turns in the trail introducing tension and repose, and the view reached after a strenuous climb involving increased heart rate, building pressure and then relaxation that follows from climax.

Dewey maintained that absent "interaction between the total organism and objects"—interaction in which doings and undergoings synchronize around objects and thereby become members of a "single act"—"[objects] are not perceived" (Dewey 1934, 54; also see 58–59). And, indeed, we experience food and ski trails through just such a joint mobilization of capacities. Here hands, eyes, tongue, heart and other organs are "instruments through which the entire live creature, moved and active throughout, operates" (Dewey 1934, 50); and here "an organized body of activities, including the motor elements [, is] necessary for full perception" (Dewey 1934, 256).

# 4 Brainless calculation and integration in AI, robotics and humans

For decades, psychologists have argued that "evidence indicates that as compared with consciously controlled cognition, the nonconscious information-acquisition processes are incomparably faster and structurally more sophisticated" (Lewicki, Hill & Czyewska 1992, 796). Part of the explanation seems to be that many complex processes are enacted through non-neural bodily structures, which perform "complex calculations by virtue of interacting with the environment" (Long 2011, 104). Or as Clark put it, many "computational ... operations are implemented" through bodily action, meaning "operations are not in the neural system alone but in the whole embodied system located in the world" (Clark 2008, 14).

In line with this idea and again attributing its seeds to the ancients while anticipating developments in AI, robotics and cognitive science, Dewey urged that before inner mental mechanisms show up, "[e]xperience carries principles of connection and organization within itself" by virtue of arising out of "adaptive courses of action, habits, active functions, connections of doing and undergoing" and "sensori-motor co-ordinations" (Dewey 1920, 91). Hence, he explained, "[s]ome degree of organization is indispensable to even [...] an amoeba" (Dewey 1920, 91). It must interact with its environment, else perish. Yet it cannot do so any way whatever. Its locomotive powers, its capacity to move materials in and out of itself, its shape and size all limit its possibilities of action. So too do the materials it encounters. Consequently its activity has "organization," "continuity in time" and "reference to its surroundings" (Dewey 1920, 91; cf. Lakoff & Johnson, 17–19), and therewith coherence and even the basic appearance of intentionality. Thus, as Noë put it, even with unicellular life "we find a subject and an environment, or an organism and a world. The animal, crucially, has a world; that is to say a relationship with its surroundings" (Noë 2009, 40). Hence "the life of the bacterium is not hidden within it." Its life "is a dynamic in which the bacterium, in its environmental situation, participates. And so it is for consciousness, more generally" (Noë 2009, 43; also see Thompson 2007, 74–75).

In recent decades, the artificial intelligence and robotics communities have made similar realizations, as with Herbert Simon's (see Simon 1996, 51) well-known ant example where the complexity of the ant's movement is a function of the complexity of the surface over which it strides. Dewey and Simon's observations about the amoeba and ant connect to recent work by John Long, who in a 2011 book recounted experiments involving simple light seeking robots he called

Tadros. Long varied the robots' tail stiffness, allowing them to compete in a kind of evolutionary game. He reported that they developed "better feeding behavior than their parents had—in a real sense, they got smarter. But [...] they did so by evolving their bodies, not their brains" (Long 2011, 95). Without claiming that his robots are going to win Nobel Prizes, Long insisted "that Tadros—by virtue of being goal directed, autonomous, and physically embodied—have intelligence" (Long 2011, 95–96); and Long, in line with Noë, who argued that "[m]eaningful thought arises only when the whole animal is dynamically engaged with the environment" (Noë 2009, 8), predicted further that if AI is ever to achieve human-like intelligence, "the AI has to be an embodied robot, and human-level intelligence is only achievable with a body and a brain" (Long 2011, 97).

In situations like this "part of the 'processing' is done by the dynamics of the agent-environment interaction, and only sparse neural control"—or the electronic CPU analogue—"needs to be exerted when the self-regulating and stabilizing properties of the natural dynamics can be exploited" (Pfeifer et al. 2007, 81). The body, Long explained, "[b]y virtue of being in the real world, interacting with real water, automatically solves [...] intensely complex" physical problems (Long 2011, 104), just as the human knee does when interacting with contours of terrain (Chemero 2009, 27). Long went on to say: "In response to the tail's coupled internal and external force computations, the body, to which the tail is attached, undergoes the yaw wobbles—recoil and turning maneuvers." Its body accordingly calculates and performs patterns of "acceleration that interact to produce the overall motion of the Tadro according to Newton's laws of motion (Long 2011, 104).

Of course, this is not to argue that amoebas or Tadros enjoy traces of conscious experience. This way of thinking, quipped Dewey, would be akin to concluding that because plows "originated from some pre-existing natural production, say a crooked root or forked branch, the latter was inherently and antecedently engaged in plowing" (Dewey 1925, 282). The point, rather, is that many of the integrative and calculative acts traditionally located in the inner mental realm are literally, which in this case means physically, carried out in the world. So while not denying either the existence of private consciousness or importance of the brain, Dewey—along with others such as Merleau-Ponty, Chemero, Thompson, Clark, Noë and Long—stressed that whatever else experience involves, "[t]he first great consideration is that life goes on in an environment; not merely *in* it but because of it, through interaction with it. No creature lives merely under its skin" (Dewey 1934, 13).

Though unfamiliar with Dewey and obviously Noë, Merleau-Ponty provided a nice summary of all this when he wrote that "[a]part from the probing of my eye or my hand, and before my body synchronizes with it"—the sensible object is a "beckoning" (Merleau-Ponty 1945, 214), which is to say, an invitation to actualize available possibilities. In the words of Dewey, "[i]t is clue in behavior, a directive factor in adaptation of life in its surroundings." "It is an invitation and inducement to act in a needed way" (Dewey 1920, 87)—"needed" because there are always limits; we cannot act in any way whatever. Thus while we "project" ourselves into the world by pushing into and otherwise probing it (see Dewey 1917, 7), we cannot perceive it in any manner whatever because we cannot act in any manner whatever. Whereas we can roll a bottle between our palms, the same action and hence experience is impossible with cinderblocks. Differences perceived here accordingly involve "a relation between organism and environment," but this, wrote Dewey, is no basis for "proclamations of the agnostic 'relativity' of knowledge"; it is "an argument for the ultimately practical value of distinctions." Distinctions "are *differences* made in what things would have been without organic behavior—differences made not by 'consciousness' or 'mind,' but by the organism as the active center of a system of activities" (Dewey 1908, 45; cf. Hurley 1998, 333). Thus we are, to repeat Noë's refrain, outside of our heads. Yet according to Dewey, this is inherent in Ancient Greek notions resuscitated by developments in biology, which is why Noë's variant of enactivism, like Dewey's account, appears to be a new name for an old way of thinking.

## 5 Experience, culture and history

From what has been said, it should be clear that Gallagher rightly noted that Dewey's "concept of cognition is not far removed from what today is called 'enactive cognition'." Gallagher added that "one thing that Dewey clearly emphasizes and that is all but missing in some discussions of the enactive approach [...] is the fact that cognition is always socially situated" (Gallagher 2009, 38). Dewey, for example, observed that

> [i]f we start with the traditional notion of mind as something complete in itself, then we may well be perplexed by the problem of how a common mind, common ways of feeling and believing and purposing, comes into existence [...]. The case is quite otherwise if we recognize that [...] we must start with grouped action, that is, with [...] interaction among individuals (Dewey 1922, 61).

Decades later and nearing the end of his life, Dewey (c. 1951) went on to lament that "Culture and Nature" might have been a better title for his *Experience and Nature* (1925) because so many had misunderstood what he meant by the term "experience." In these final sections, I will sketch Dewey's reasons for equating experience to culture, and discuss how it relates to and aptly summarizes his views regarding situated cognition, as well as connecting him to both recent and ancient thought, while also supplying future directions.

Drawing on Gibson's (1979) theory of affordances, Noë argued that the way objects limit and pattern movement reveals the shape of the object. Generalizing to vision, he further wrote that "[t]o learn how things are from how they look is to learn that the environment structures one's possibility of movement. It is to discover the structure of sensorimotor contingencies" (Noë 2004, 103). And in the same way activities of hands, mouths, tongues and eyes coordinate around things such as beer bottles and therewith form perceptual experiences, an entire weekend and therewith an experience, culture or world of revelry may organize around bottled beer. So too may the agricultural, industrial and commercial activities, not to mention the physical space, of a beer-producing town. Martin Heidegger noted that "the Old High German word *thing* means a gathering to deliberate on a matter under discussion, a contested matter" (Heidegger 1949, 172), and therefore a matter of concern; and in the just mentioned examples, life activity gathers around a concern for particular things. Heidegger thus suggested that worlds form around things. In *Grapes of Wrath*, John Steinbeck offered an instructive example of this. "In the daylight," he wrote, Great Depression migrants

> scuttled like bugs to the westward; and as the dark caught them, they clustered like bugs near to shelter and to water. [...] Thus it might be that one family camped near a spring, and another camped for the spring and for company, and a third because two families had pioneered the place and found it good. And when the sun went down, perhaps twenty families and twenty cars were there. [...]
>
> Every night a world created, complete with furniture—friends made and enemies established; a world complete with braggarts and with cowards, with quiet men, with humble men, with kindly men [and laws and rules on conduct, as Steinbeck later added]. Every night relationships that make a world, established (Steinbeck 1939, 264–265).

Steinbeck went on to say that "[a] certain physical pattern is needed for the building of a world" (Steinbeck 1939, 266). In his example, this included objects of concern such as "water, a river bank, a stream, a spring, or even a faucet unguarded. And there is needed enough flat land to pitch the tents, a little brush or wood to build the fires," and more besides (Steinbeck 1939, 266–267).

Steinbeck's description illustrates Dewey's notion that experience is equivalent to culture (c. 1951, 361–363; also see 1920, 92; 1922–1923, 351)—"culture" here understood as embodied patterns of activity within communities of interrelated agencies and things, patterns that constitute both our worlds and experiences, as when we talk about the "French experience," "culture" or "world," or the "world" or "experience of parenthood." This concept of experience is in fact very old, as Dewey well recognized. Plato, for instance, sometimes understood experience as sedimented customs (see *Laws* 720a; also see Dewey 1920, 92), and a similar notion can be drawn from Aristotle's view that experience comes from accumulated memories (see *An. Post.* 99b35–100a8; *Met.* 981a7–9).

Expanding on ancient views and echoing more recent scholars such as Hubert Dreyfus (2003), Dewey observed that the Greeks lacked introspective psychology and even conceptions of mental acts (Dewey 1910–1911, 52). For the ancients, Dewey elaborated, "the ultimate standard was custom—what had to be done. That meant that things were learned by apprenticeship; the things they learned were the things to do" (Dewey 1910–1911, 32). Their concept of experience fit with this. "Experience [...] signified a store of practical wisdom, a fund of insights useful in conducting the affairs of life." So conceived, "experience is exemplified in the discrimination and skill of [...] good [artisans]" (Dewey 1925, 354).[2] Given that we almost invariably become skilled or experienced in the context of communities, and there are cases in which we use the words "experience" and "culture" interchangeably, it is not difficult to see how Dewey came to identify the two. These connections are suggestive in the context of enactive theory, which follows Dewey—and to some extent the ancient Greeks—in holding that perceptual experience is constituted through skilled acting.

Dewey's notion that experience is equivalent to culture bears striking resemblance to ideas of the Russian developmental psychologist Lev Vygotsky, a contemporary of his commonly cited in today's embodiment literature, who emphasized the blurring of the individuals with their social worlds (see Barrett 2011, 62). In *Human Nature and Conduct* Dewey noted that "each person begins a helpless, dependent creature" (Dewey 1922, 62), and added in *Reconstruction in Philosophy* that

> [...] the contacts of the little child with nature are mediated by other persons. Mother and nurse, father and older children, determine what experiences the child shall have; they constantly instruct him as to the meaning of what he does and undergoes (Dewey 1920, 92).

Then, summing up, he wrote:

---

2 If Aristotle is to be trusted, then Dewey—while a sensitive reader of the ancient Greeks—oversimplified when he suggested that ancient Greeks equated experience (*empeiria*) to art (*tekhnē*), for Aristotle's *Metaphysics* (981a–9) clearly differentiates between the two. Yet Dewey (c. 1910–1911) also appears to have been aware of this since he said: "The Greek *techne* [...] covered Knowledge and Skill. It meant intelligent skill—i.e., skill directed by knowledge" (Dewey 1910–1911, 47–48).

> [t]here is doubtless a great mystery as to why any such thing as being conscious should exist at all. But [...] there is no mystery in its being connected with what it is connected with. [...] And by this is meant both that it will be shared by those who are implicated in the associative custom, or more or less alike in them all, and that it will be felt or thought to concern others as well as one's self (Dewey 1922, 62).

This view, which has received empirical support in the studies of Colwyn Trevarthen (e.g., 2011) showing that infant and parent are coupled units, was in turn echoed by Noë, who observed that "[c]hildren are not separate; they are not observers; they are regulated by their mothers' soothing or alerting tones, eye contact, gestures, and touch. A mother is literally one of the structures constituting a child's psychological landscape" (Noë 2009, 31).

He further observed that if "[l]anguage, tools, and collective practices make us what we are," then we have "no reason to suppose that the critical boundary is found in our brain or skin" (Noë 2009, 67–68). Indeed, "our cognitive powers require for the very exercises the existence of a sociolinguistic environment" to the extent that "language is itself socially manufactured and shared by linguistic communities." Seen accordingly, [o]ur minds cross out of the skull and get supported by shared sociolinguistic scaffolding" (Noë 2009, 88).

Dewey's thinking, as stated, developed in the context of ongoing efforts to circumnavigate the debate between *a priorist* and empiricist psychologies (e.g., 1906, 469–475; 1917, 16–27; 1920, 81–91)—the debate, to put it crudely, about whether the "inner subject" imposes form on the "outer world" or vice versa. Against rationalists, Dewey (1922) argued that our ways of cognizing worlds follow from our ways of inhabiting them, from habits of action. "Reason pure of all influence from prior habit is a fiction," he wrote. But so too are the "pure sensations" of the empiricists, for the ability "to single out a definitive sensory element in any field is evidence of a high degree of previous training, that is, of well-formed habits" (Dewey 1922, 31). In line with O'Regan and Noë's conclusion, therefore, that "visual sensation and visual perception are different aspects of a person's skillful exploratory activity"; and their additional conclusion that visual awareness depends on the "integration of these patterns of skillful exercise into ongoing planning, reasoning, decision-making, and linguistic activities" (O'Regan & Noë 2001, 970–971), Dewey observed that

> [a] moderate amount of observation of a child will suffice to reveal that even such gross discriminations as black, white, red, green, are the result of some years of active dealings with things in the course of which habits have been set up. It is not such a, simple matter to have a clear-cut sensation. The latter is a sign of training, skill, habit (Dewey 1922, 31).

Dewey thus chided rationalists for not being empiricists, that is, for not recognizing the priority of experience; yet this is, strange to say, also why he attacked empiricists. "Our ideas," he wrote, "truly depend on experience, but so do our sensations. And the experience upon which they both depend is the operation of habits" (Dewey 1922, 32)—in other words, skilled actions principally enacted not in our heads, but in shared cultures or worlds.

# 6 Future in the past

While this paper is not a critique, I am surprised at the extent to which Dewey—a leader in both philosophy and psychology—has been absent from enactive cognitive science until recently. I have heard some justify this on the basis that current researchers are doing science, and Dewey is "mere history." However, if science is an excuse for neglecting obvious predecessors and ignoring history, then so much the worse for science. Following Dewey's example, on the other hand, might progress enactivism into relatively unexplored areas.

Most obviously, Dewey has, in effect, written an enactive account of aesthetics that meshes with recent psychological theories describing aesthetic perception as a process whereby we rapidly assess what it is possible to do in a setting. So, for example, in the same way that some works of art have a quality that draws audiences into them, some settings entice people to penetrate deeper, as when we encounter ones with trails disappearing around bends (e. g. Kaplan & Kaplan 1989, 58). Dewey, moreover, has offered hints to how enactive theory might illuminate synesthesia, a weak point some scholars have noted (e. g., Fingerhut 2011). Dewey, with other historical figures such as Merleau-Ponty, has done this by noting the extent to which synthesis of perceptual modalities occurs through the body coordinating around things, as opposed to being a purely neurological occurrence. Because we habitually encounter things this way, Dewey urged that "[t]he eye, ear, or whatever, is only the channel *through* which a total [interaction] takes place" (Dewey 1934, 122). So while "[w]e see a painting *through* the eyes," it is a mistake to suppose that visual "qualities as such, are central if not exclusive." We can "perceive, by means of the eyes [...], the liquidity of water, the coldness of ice, the solidity of rocks" (Dewey 1934, 123).

Dewey, by virtue of drawing on the past, offers more besides. For instance, incorporating ancient ideas into an essentially modern psychology, he developed notions about self and morality. A civil rights activist, to consider an example, acts in but also against the world. Resisting "existing custom[s] is the beginning of individuality in mind" (Dewey 1922, 87–88), and it also shapes the direction of her life, giving it purpose and form. She cultivates a disposition by developing habits, for habits are dispositions towards or, as Dewey wrote, "demands for certain kinds of activity; and they constitute the self" (Dewey 1922, 25). "Were it not for the continued operation of all habits in every act, no such thing as character could exist" (Dewey 1922, 38). They even underlie "will," for they are inclinations to chase after and want certain things. Thereby the woman not only becomes an individuated self, but also an integrated one. Aristotelian and by extension Thomistic conceptions of humanity and morality are based on understandings emphasizing habits in just such ways. So in the work of Dewey and thinkers before him, arguably there are already well-developed seeds for an enactive theory of morality—another area relatively unexplored so far.

Seen as culture, experience is a historical phenomenon not merely in the sense that it unfolds in time, but also in the sense that it is a shared heritage and story of a community. While philosophers do not unanimously endorse this concept of experience, most grant that experience is a basic material upon which philosophical inquiry works, and all are centrally preoccupied with shared cultural experiences or stories. After all, the very words and concepts philosophers use and analyze only have meaning in the context of publically owned heritages. Above all, Dewey was a philosopher of experience and by extension a historian of philosophy. By virtue of being a historian and insisting on establishing "working connections between old and new subject-matters" (Dewey 1929, viii), Dewey developed ideas that remain cutting edge today, for instance, his notion that experience is culture and that it arises through skilled acting. We can learn from Dewey's example.

## References

Aristotle (1941). "Metaphysics." W. D. Ross (Trans.). In: R. McKeon (Ed.). *The basic works of Aristotle.* New York: Random House, 681–926.

Aristotle (1941). "Posterior analytics."G. R. G. Mure (Trans.). In: R. McKeon (Ed.). *The basic worksof Aristotle.* New York: Random House, 108–186.

Bach-y-Rita, P. (1983). "Tactile vision substitution: Past and future." In: *International Journal of Neuroscience* 19, 29–36.

Bach-y-Rita, P. (1984). The relationship between motor processes and cognition in tactile vision substitution. In *Cognition and Motor Processes*, ed. by A. F. Sanders and W. Prinz, 150–159. Berlin: Springer.

Bach-y-Rita, P. & Kercel, S. (2002). "Sensory substitution and augmentation: Incorporating humans-in-the-loop." In: *Intellectica* 2, 287–297.

Barrett, L. (2011). *Beyond the brain: How body and environment shape animal and human Minds.* Princeton: Princeton University Press.

Chemero, A. (2009). *Radical embodied cognitive science.* Cambridge, MA: MIT Press.

Chemero, A. (2013). "Radical embodied cognitive science." In: *Review of General Psychology* 17, 145–150.
Clark, A. 2008. *Supersizing the mind: Embodiment, action and cognitive extension.* Oxford: Oxford University Press.
Crippen, M. (2010). "William James on belief: Turning Darwinism against empiricistic skepticism." In: *Transactions of the Charles S. Peirce Society* 46, 477–502.
Crippen, M. (2011). "William James and his Darwinian defense of freewill." In: M. R. Wheeler (Ed.). *150 Years of Evolution: Darwin's Impact on Contemporary Thought and Culture.* San Diego: San Diego State University Press, 68–89.
Crippen, M. (2014). "Body phenomenology, somaesthetics and Nietzschean themes in medieval art." In: *Pragmatism Today* 5, 40–45.
Dewey, J. (1906). "Experience and objective idealism." In: *The Philosophical Review* 15, 465–481.
Dewey, J. (1908). "The practical character of reality." In: *Philosophy and civilization* [1931]. New York: Minton, Balch & Company, 35–56.
Dewey, J. (c. 1910–1911 [pub. 1993]). *Philosophy and education in their relation.* transcribed from lectures by E. R. Clapp & J. J. Chambliss (Eds.). Boulder: Westview Press.
Dewey, J. (1916). *Essays in Experimental Logic.* Chicago: University of Chicago Press.
Dewey, J. (1917). "The need for a recovery of philosophy." In: *Creative intelligence: essays in the pragmatic attitude.* New York: Henry Holt and Company, 3–69.
Dewey, J. (1920). *Reconstruction in philosophy.* New York: Henry Holt and Company.
Dewey, J. (1922). *Human nature and conduct. An introduction to social psychology.* New York: Henry Holt & Company.
Dewey, J. (1983/1922–1923). "Syllabus: types of philosophical thought." In: *The middle works, 1899–1922, vol. 13.* ed by J. A. Boydston. Carbondale and Edwardsville: Southern Illinois University Press, 352–395.
Dewey, J. (1925). *Experience and nature.* Chicago: Open Court Publishing Company.
Dewey, J. (1926). "Affective thought." In: *Philosophy and civilization* [1931]. New York: Minton, Balch & Company, 117–125.
Dewey, J. (1929 [1958 reprint]). *Experience and nature.* second edition. New York: Dover Publications.
Dewey, J. (1934). *Art as experience.* New York: Minton, Balch & Company.
Dewey, J. (1981/c. 1951). "Portion of unfinished introduction." In: *The later works, 1925–1953, vol. 1.* ed. by J. A. Boydsten. Carbondale and Edwardsville: Southern Illinois University, 361–364.
Dreyfus, H. (2002). "Existential phenomenology and the brave new world of the matrix." In: *The Harvard review of Philosophy* 11, 18–31.
Fingerhut, J. (2011). "Sensorimotor signature, skill, and synaesthesia: Two challenges to enactive theories of perception." In J. Fingerhut, S. Flach & J. Söffner (Eds.). *Habitus in habitat III: synaesthesia and kinaesthetics.* New York: Peter Lang, 101–120.
Gallagher, S. (2009). "Philosophical antecedents of situated cognition." In: P. Robbins & M. Aydede (Eds.). *The Cambridge companion of situated cognition.* Cambridge: Cambridge University Press, 35–52.
Gibson, J. J. (1979). *The ecological approach to visual perception.* Boston: Houghton-Mifflin.
Heidegger, M. (1949). "The thing". Albert Hofstadter (Trans.). In: *Poetry, Language, Thought* [1971], New York: Harper & Row, 163–180.

Hurley, S. (1998). *Consciousness in Action.* Cambridge, MA: Harvard University Press.
Kaplan, R. & Kaplan, S. (1989). *The experience of nature: A psychological perspective.* New York: Cambridge University Press.
Lakoff, G. & Johnson, M. (1999). *Philosophy in the flesh.* New York: Basic Books.
Lennon, J. (1960). "The notion of experience." In: *Thomist: A Speculatlve Quarterly Review 23*, 315–343.
Lewicki, P., Hill, T., & Czyzewska, M. (1992). "Nonconscious acquisition of information." In: *American Psychologist* 47, 796–801.
Long, J. (2011). *Darwin's devices: What evolving robots can teach us about the history of life and the future of technology.* New York: Basic Books.
Madzia, R. (2013). "Chicago pragmatism and the extended mind theory: Mead and Dewey on the nature of cognition." In: *European Journal of Pragmatism and American Philosophy* 5, 193–211.
Menary, R. (2007). *Cognitive integration: mind and cognition unbounded.* Basingstoke: Palgrave Macmillan.
Merleau-Ponty, M. (1945). *Phenomenology of perception.* Colin Smith (Trans.) [1962]. New York: Routledge & Kegan Paul Ltd.
Noë, A. (2004). *Action in perception.* Cambridge, MA: MIT Press.
Noë, A. (2009). *Out of our heads: Why you are not your brain, and other lessons from the biology of consciousness.* New York: Hill and Wang.
O'Regan, K. & Noë, A. (2001). "A sensorimotor account of vision and visual consciousness." In: *Behavior and Brain Sciences* 24, 939–973.
Pfeifer, R. et al. (2007). "On the information theoretic implications of embodiment—principles and methods." In: M. Lungarella et al. (Eds.). *50 Years of artificial intelligence: Essays dedicated to the 50th anniversary of artificial intelligence.* New York: Springer, 76–86.
Plato (1963). "Laws." A. E. Taylor (Trans.) In: E. Hamilton and H. Cairns (Eds.). *Plato: The collected dialogues.* Princeton: Princeton University Press, 1125–1518.
Rockwell, T. (2005). *Neither brain nor ghost: A nondualist alternative to the mind-brain identity theory.* Cambridge, MA: MIT Press.
Simon, H. (1996). *The sciences of the artificial.* Cambridge, MA: MIT Press.
Skorburg, J. A. (2013). "Beyond embodiment: John Dewey and the integrated mind." In: *The Pluralist* 8, 66–78.
Solymosi, T. (2011). "Neuropragmatism, old and new." In: *Phenomenology and the Cognitive Sciences* 10, 347–368.
Solymosi, T. (2013). "Neuropragmatism on the origins of conscious minding." In L. Swan (Ed.). *Origins of mind.* Biosemiotics 8. Dordrecht: Springer Publications, 273–287.
Steinbeck, J. (1939). *The grapes of wrath.* New York: Viking Press.
Thompson, E. (2007). *Mind in life: Biology, phenomenology, and the sciences of mind.* Cambridge, MA: Harvard University Press.
Trevarthen, C. (2011). "What is it like to be a person who knows nothing? Defining the Active intersubjective mind of a newborn human being." In: *Infant and Child Development* 20, 119–130.
Varela, F., Thompson, E. & Rosch, E. (1991). *The embodied mind: Cognitive science and human Experience.* Cambridge, MA: MIT Press.
Zampini, M. & Spence, C. (2004). "The role of auditory cues in modulating the perceived crispness and staleness of potato chips." In: *Journal of Sensory Studies* 19, 347–363.

Tullio Viola

# Peirce on Abduction and Embodiment

**Abstract:** The paper explores C. S. Peirce's relevance to embodied cognitive science, by discussing his notion of abduction, or hypothetical inference. It first maintains that the concept of abduction may be read as Peirce's anti-intuitionist attempt to explain the phenomenon of intellectual creativity. Then, it investigates the consequences of that concept for theories of enactive, embodied, and extended mind. Peirce is shown to rely on an enactive understanding of logical inference and to believe that our bodily constitution plays at least a minimal role in thought, as we need the cooperation of "the eyes and the hands" to bring about new cognitions. Yet the latter idea stops short of turning into a full-fledged commitment to embodiment, as Peirce also stresses the possibility for thought to undergo different material realizations. This may help us to recognize a difference between Peirce's account of cognition and phenomenologically oriented theories of embodiment. The paper suggests that this difference is related to pragmatism's and phenomenology's differing views as to whether we can have an intuitive grasp of reality.

## 1 The problem: inference versus intuition

In 1867—the "decisive year" of his early career as a philosopher (Fisch 1984)—Charles S. Peirce presented to the public for the first time his well-known tripartition of logical inferences. Alongside deductions and inductions, long recognized by logicians and philosophers as two distinct kinds of reasoning, he introduced a third and irreducible notion, namely "hypothesis", later to be renamed "abduction".[1]

Over time, the exact definition of abduction would undergo sensible changes. But the basic idea remains quite stable: given an observed fact, abduction suggests a conjectural explanation from which the fact may possibly follow. It is a highly fallible inference, in that it merely advances a hypothetical explan-

Many thanks to Alex Arteaga, Mark-Oliver Casper, Rossella Fabbrichesi, and Sascha Freyberg for their comments and suggestions.

1 "On the Natural Classification of Arguments", W2, 238–248 (1867). Peirce started using the term "abduction" in the late 1890s—first as a translation of Aristotelian *apagôgê* (CP 1.65, c. 1896), then, a bit later, in a systematic fashion (e.g., EP2, 95 [1901]).

ation of the fact observed, without giving any justification that the hypothesis actually holds. But as if providing a compensation for this weakness, abduction is the only kind of inference that introduces *thoroughly new content* into the argument. It thus differs from deduction and induction, which limit themselves to making explicit, generalizing, or confirming the information already contained in the premises.

The notion of abduction has proven one of Peirce's most momentous conceptual innovations. In its ambition to reshape our understanding of how new ideas come about, it wielded a decisive impact on philosophy and its neighbouring disciplines—including the cognitive sciences (see Magnani 2009; Queiroz/Atã 2013). In order to fully appreciate the scope of this impact, however, it is advisable to look at the notion's original context of formulation. As often happens with Peirce, in fact, behind the novelty of his terminology one may discern deep and telling links with the history of philosophy.

In particular, Peirce's treatment of abduction makes explicit its indebtedness to traditional debates on the scope and limitations of inferential reasoning. In a text from the early 1900s, we read that all "three radically different kinds of arguments which I signalized in 1867 [...] had been recognized by the logicians of the eighteenth century." What these logicians had "failed to recognize", however, is the "*inferential character* of one of them" (EP2, 205 [1903], my emphasis). In other words, they had already understood that *deductions* and *inductions* alone do not suffice to give a complete account of our reasoning faculties, but they mistakenly assumed that the only alternative to these two kinds of reasoning is a non-inferential or *intuitive* faculty, one that brings about new cognitions in an immediate fashion. Peirce, on the contrary, wished to defend the existence of abduction as an *outright logical inference*, which obeys precise logical rules and therefore does not partake of intuitive immediacy.[2]

The antithetical nature of abduction and intuition becomes even clearer if we look at the so-called "anti-Cartesian" papers that Peirce published just a few months after his 1867 classification of logical arguments.[3] In those celebrated essays, Peirce attacked Cartesian epistemology precisely on the ground that

2 See also CP 8.227–228 (c. 1910): "The general body of logicians had also at all times come very near recognizing the trichotomy [of inferences]. They only failed to do so by having so narrow and formalistic a conception of inference (as necessarily having formulated judgments for its premises) that they did not recognized Hypothesis [...] as an *inference*". Or CP 2.102 (c. 1902): "In what I [...] said about 'Hypothetic inference' I was an explorer upon an untrodden ground."
3 The papers are: "Questions Concerning Certain Faculties Claimed for Man" (1868), W2, 193–210; "Some Consequences of Four Incapacities" (1868), ibid., 211–242; "Grounds of Validity of the Laws of Logic: Further Consequences of Four Incapacities" (1869), ibid., 242–272.

humans have no faculty of intuition: a faculty he in turn defined as the ability to produce a "cognition not determined by a previous cognition of the same object" (EP2, 193 [1903]). In other words, all cognitions must be determined by previous ones; they necessarily have premises from which they derive inferentially.[4] The only way that we can produce new cognitions is through *inferences from signs*. But if that is true, it immediately follows that there has to be at least one kind of inference which not only fulfils a *transformative* function (i. e., which is able to recombine the content of the premises into a new conclusion), but can also let *genuinely new cognitions* enter the mind. Otherwise the entire phenomenon of creative intelligence would remain unexplained. And this inference is indeed hypothesis, or abduction (EP2, 209, 213, 226 [1903]).[5]

This argumentative line is particularly relevant to the question of Peirce's relation to the research programs on situated, embodied, or distributed cognition. The anti-intuitionist stance first formulated in Peirce's early papers, in effect, has been one of the main reasons why twentieth-century cognitive scientists have shown interest in the pragmatist tradition (Jung 2009, 190–192; Paolucci 2011, 71–72; Fingerhut/Hufendiek/Wild 2013, 32–43). Peirce's insistence on the impossibility for humans to have intuitive thoughts entails the idea that thought and cognition do not ultimately rest on subjective certitude, but are the outcome of a practical interplay between subjects and their environment. Moreover, Peirce equates the non-existence of intuitive cognitions with the thesis that all thought is of the nature of signs: a view that lets the boundary between what is "internal" and what is "external" to our head lose much of its importance. Thought is not (only) the product of our brain, but is *embedded in signs*.

So, once abduction is seen as *Peirce's anti-intuitionist answer to the problem of intellectual creativity*, it becomes clear that indicating its scope and purport is a crucial move in the study of the relations between pragmatism and embodied cognitive sciences. At the same time, precisely this reading of abduction highlights a major and well-known problem in Peirce's philosophy. To put it as concisely as possible, it is not altogether clear in which sense abduction may be said

4 Peirce avoids the infinite regression that seems to be lurking here by appealing to a property of continuous geometry, namely the fact that a segment of finite length contains an infinite number of points. Analogously, the finite span of time occupied by our trains of thought does not prevent cognitions from having an infinite number of premises. See ibid., 175–179, 209–211.

5 But compare also HP2, 899–900, 1901. To be more precise, at this moment the creative power of hypotheses is still partly shared with inductions (see 216, 238–239). But as we will see, later on induction would limit itself to confirming, strengthening or generalizing cognitions already formed. By 1903, for instance, we find the statement that abduction is "the only logical operation which introduces any new idea" (EP2, 216).

to *create new knowledge while remaining a full-fledged inference*. In other words, a tension seems to exist between the *creative* and the *transformative* nature of abduction: a tension that, if not solved, would jeopardize not only Peirce's epistemology as such, but also the whole conceptual scaffolding Peirce handed down to the cognitive sciences.[6]

In the following pages I will suggest a framework for the solution of this problem that, far from giving in to these pessimistic conclusions, actually strengthens the tie between Peirce's theory of knowledge and embodied cognitive sciences. I will do so by moving precisely from the antithetical nature of abduction and intuition, and by using that as a foil to discuss Peirce's relation to the three notions of *enactive*, *embodied*, and *extended* cognition. In this way, we will be able not only to qualify the particular notion of situated and distributed cognition to which Peirce's position leads, but also to contrast it with other positions—in particular those stemming from phenomenology.

## 2 Abductions enacted

Let me start by taking a closer look at what I have already referred to as the main puzzle of abduction. This puzzle has been often discussed with recourse to a well-known passage from a 1903 lecture on "Pragmatism as the Logic of Abduction", in which Peirce suggests a syllogistic account of abductive reasoning:

> C is observed;
> but if A were true, C would be a matter of course;
> *hence*, there is a reason to suspect that A is true. (EP2, 231 [1903])[7]

Here, Peirce follows an idea already resorted to in his 1860s papers (W2, 46 [1867]; W3, 325–326 [1878]), in that he describes abduction as the inverted form of a deductive syllogism. Instead of drawing the conclusion C from the major and minor premises, abduction infers the minor premise as a possible explanation of C, once the rule "if A, then C" is given.

The seeming perspicuity of this formal description, however, also highlights the very problematic logical status of abduction. If taken at face value, in fact, the syllogism reported above can hardly be said to be logically valid. It is a way of "affirming the consequent" (Magnani 2009, 28.) which we may only jus-

6 Among the classical discussions of this problem are Frankfurt 1958, Kapitan 1990, Pape 1994, Hintikka 1998.
7 Cf. Frankfurt 1958, 594; Paavola 2005, 132.

tify on the ground of its being the only way for the new hypothesis, A, to enter the mind. Yet how this is precisely supposed to happen is not clear. It will be noted, in fact, that A appears twice in the syllogism: first in its premise (as the antecedent to the rule "if A, then C"), then in its conclusion. At which point should we locate the genuinely abductive act? If we locate it in the major premise, then we are obliged to deny abduction any inferential character, for the premise is not derived inferentially, but posited. If, on the contrary, we locate it in the conclusion, then we may safeguard its inferential nature, but must discard its creative power, as the conclusion limits itself to *endorsing as valid* a hypothesis that has been already introduced.[8]

Peirce seems to have been well aware of this tension. Indeed, precisely because he saw it, he often felt the need to make clear that abduction, while having a similarity to the non-logical faculties of instinct and intuition, "*nonetheless* is a logical inference" (EP2, 231 [1903], my emphasis). Nor is this surprising: the tension of which I am speaking is the immediate consequence of abduction's origin in Peirce's attempt to overcome intuitionist epistemologies by moving a traditionally instinctive faculty within the realm of logic. The notion of abduction, in this sense, should be seen less as Peirce's "definitive solution" to the problem of inferential knowledge, and more as a challenge that he posed to himself, and that he tried to meet throughout his life by spelling out its traits with ever increasing precision.

In the course of these attempts, he did at times relapse into descriptions that cut the tie between the two souls of abduction, and took it to be either a simple instinct, or the endorsement of a previously formulated hypothesis. And in point of fact, some interpreters have maintained that a narrow reading of abduction as the endorsement of a hypothesis (or as the "recognition" of that hypothesis as valid) is the only way to avoid paradoxical formulations.[9] These readings, however, not only fly in the face of Peirce's insistence that abduction ought to encompass "the first starting of a hypothesis *and* the entertaining of it" (CP 6.525 [c. 1901]);[10] but they also risk relapsing into precisely the opposition between an intuitive and an inferential faculty which Peirce wanted to get rid of in 1867.

The only alternative to such pitfalls is to advance an interpretation that, instead of keeping apart the two souls of abduction, construes the creative moment

---

**8** In the latter case, abduction would come very close to an inference to the best explanation. See G. H. Harman's original formulation (1965, 89): "In making this inference one infers, from the fact that a certain hypothesis would explain the evidence, to the truth of that hypothesis".
**9** Frankfurt 1958.
**10** Cf. Anderson 1987, 45: "Peirce quite explicitly states that abduction is both an insight and an inference. This is a fact to be explained, not to be explained away."

as an integral part of the inferential or transformative one.[11] To achieve this task, the first thing we should do is to look at the inner development of Peirce's views on abduction. As has been often remarked, in fact, two distinct phases may be discerned (Tiercelin 2005; Paavola 2005).

The first phase encompasses Peirce's early years. In that period, Peirce regarded abduction as a form of inference clearly distinguished from the other two, and definable in syllogistic terms. "Abductive syllogism" infers the antecedent of the rule from the rule itself, plus the observation that the consequent holds (W2, 46–48, 58 [1867]). This definition also has the advantage of elegantly solving a crucial difficulty, namely the exact differentiation between the two poles of synthetic reasoning: abduction and induction. For induction, too, is now described as a syllogism, but one that proceeds in a complementary way, namely by inferring the general rule from a sample of both the rule's antecedent and its consequent.[12] So, while induction performs a *generalization* from a sample to the whole, abduction suggests "the existence of a fact quite different from anything observed", which explains the observation itself (W3, 332 [1878]).[13]

Later on, Peirce gradually stopped seeing his three classes of inferences as clear-cut syllogistic forms. Instead, he construed them as three stages of a broad inferential mechanism, which underlies all of our cognitive and perceptual processes. This inferential mechanism has an eminently practical nature, and runs in great part in a tacit way. Far from needing explicit beliefs to be formulated in the mind, it mainly hinges on the role of *habits*, which are the practical, embodied, and sometimes pre-conscious equivalent of the major premise of a syllogism, "if A then B". (That is, habits link a set of stimuli, or observations, to a set of "practical reactions".)

Correspondingly, the very notion of inference enlarges dramatically. There is a considerable set of cognitive and practical acts that are almost perfectly analogous to inferences, in that they share their logical structure, and only differ from inferences proper by the fact that they escape our ability to fully control them. In this broader sense, inferences do not even need to be discursive:

---

11 Cf. Tiercelin 2005, 397; Kapitan 1990, 503–505: "the possibility might be explored that the creativity belongs to an earlier phase of abductive thinking".

12 If we know, for instance, that *a*, *b* and *c* are samples of A, and that they share the property B, we may provisionally conclude by induction that all *A* is *B*.

13 In a number of texts, Peirce also advanced the idea that a hypothesis is an induction from qualities instead of things. So, if we know that the object *a* has the qualities *P*, *Q*, *R*, we may infer that it has all other properties which explain its belonging to a particular class. See in particular W8, 151 (1892).

they may constitute practical reactions to a general disposition (or habit), or other kinds of semiotic activity.[14]

The theory of logical inferences inaugurated in this second phase of Peirce's thinking may therefore be called an *enactive* one, in the sense that it is realized in (and undistinguishable from) the manipulative and explorative activities of subjects within their environment.

This idea helps us to give a first answer to the question with which we started off: namely, how to avoid tracing too sharp a line between the moment in which the hypothesis first enters the mind and that in which it is endorsed. Abduction is now no longer understood as a self-enclosed kind of reasoning. Rather, it is a creative moment within a much broader inferential process: a moment that is inextricably tied up with the two other forms of inference. Overall, abduction is responsible for first suggesting a relationship between a given observation and a possible explanation.[15] But once this first sparkle of a new idea is there, our task is far from complete. We will need to draw a set of practical consequences from the hypothesis we have started forming, in a way that is comparable to a *deduction*. In turn, these reactions will provide the basis for further *inductions*, which are tantamount to the experimental testing of a theory, and will feed back onto the pre-existing habit, gradually strengthening or correcting it.[16]

What is more, all three forms of inferences may be performed in more or less conscious ways. This means that what may appear to our consciousness as the product of a sudden abduction is often not entirely created at the moment in which the conscious inference is performed. Rather, it was already present in the mind in a more inchoate way.[17] Also in this sense, the creative and the transformative aspects of abduction are extremely difficult to disentangle.

---

**14** Cf. the famous passage in CP 6.286 (1893): "A decapitated frog almost reasons. The habit that is in his cerebellum serves as a major premiss. The excitation of a drop of acid is his minor premiss. And his conclusion is the act of wiping it away. [...] What he lacks is the power of preparatory meditation." See also EP2, 188 (1903): "There are, as I am prepared to maintain, operations of the mind which are logically exactly analogous to inferences excepting only that they are unconscious and therefore uncontrollable and therefore not subject to criticism."

**15** Pape 1998 insists on the *relational* logic underlying Peirce's understanding of creative intelligence.

**16** The whole mechanism is described with particular clarity in W8, 151–152 (1892); EP2, 85–114 (1901); ibid., 226–241 (1903). As for induction, see also ibid., 502 (1909): "[A]s the sample [of sensations] is enlarged, the inference becomes stronger and stronger."

**17** Cf., e. g., Peirce's review to James' *Principles of Psychology* (1891), CP 8.65: "This is hypothetic inference in form. The first premise is not actually thought, though it is in the mind habitually."

Indeed, the processual and partly sub-conscious functioning of abduction goes so far as to merge almost indistinguishably with perception.[18] As Peirce famously put it in 1903, the two phenomena "shade" into one another "without any sharp line of demarcation" (EP2, 227 [1903]). This is a significant aspect of the dynamic we are analyzing, because it means that abduction, even in its most explicit and conscious manifestations, *always* retains a perceptual element. Abduction is an intellectual operation that takes hold of a new idea by somehow "seeing" or "perceiving" it, the same way we grasp a general element in our perceptual experience.[19] Yet this does not amount to equating "seeing" with intuition, mainly because of what I have called the enactive nature of our inferential faculties. Perception is not a contemplative mirroring of reality, but an exploration of the environment.

It now becomes clear that the syllogistic form in which Peirce, as late as 1903, couches his description of abduction, is little more than a remnant of his early attempt to classify his three kinds of inference by means of a somewhat static modification of the traditional syllogisms. If taken too literally, in fact, that account may prove misleading.[20] Strictly speaking it is not correct to say that the hypothesis "A" is formed *either* in the argument's premise or in its conclusion. Rather, *A emerges in the course of the inferential process*, gradually surfacing to consciousness, but nonetheless being, from the outset, inferential.

## 3 "The Eyes and the Hand"

We have shown that the enactive theory of Peirce's mature years emphasizes the *process* through which abductions are formed—a process that cuts across any distinction between "the first starting of a hypothesis and the entertaining of it". What remains is to shed light on the mechanism that presides over the very first steps of an abductive act: those in which new cognitions are created,

**18** Here I am mainly following Tiercelin 2005.

**19** See Tiercelin 2005, 397: "The only way to break the circle is to provide an explanation which [...] analyzes abduction as emerging *in* the course of reasoning itself [...]. Peirce's claim is that this happens precisely as soon as one analyzes what goes on as a kind of *seeing* or *perceiving.*" (The problem here intertwines with another thorny issue of Peircean scholarship, namely whether perception is inferential or direct.)

**20** Scholars do not seem to have paid enough attention to the doubting note on which Peirce ends his 1903 syllogistic description ("Whether this be a correct account of the matter or not [...].", EP2, 231): a note we do not find in similar treatments from the 1860s and the 1870.

as it were, utterly from scratch. To do this, we will look at two important analogies of which Peirce avails himself when speaking of abduction.

The first analogy is closely related to what I have been saying about abduction as a kind of "seeing", and places the abductive inference in relation to the semiotic category of *iconic signs*. Faced with a set of uninterpreted facts, we are able to create an entirely fresh explanation of those facts, in that we put forth a hypothesis that is an *icon* of them. (For instance, Kepler's description of the elliptical orbit of planets was an icon of the observed longitudes of Mars.)[21]

To understand the full purport of this idea, we should briefly look at Peirce's reflections on the nature of icons. Although he did set off from a quite traditional understanding of the latter in terms of *similarity* to their object, Peirce gradually felt dissatisfied with this definition. He thus ended up carving a new, more operational definition of the icon, which emphasizes its being the *only kind of sign that is able to generate new content*, on account of its ability to reveal more "truths concerning its object than those which suffice to determine its construction" (CP 2.279 [1902]).[22] Icons are, in other words, *inherently creative and productive signs.*[23] As such, they are the equivalent of abductions in the domain of semiotics.

Take, for example, a geometrical diagram. We are able to draw the figure of a triangle if we follow the instruction to draw, say, "a closed plane figure with three straight sides". But once we have drawn the figure, we will be able to discover other properties of it, which were not contained in the original instructions. (For instance, that the figure has three angles, and that the sum of these angles is 180°.) Now, Peirce believes that this creative power belongs not only to diagrams, but to *all kinds* of iconic signs.[24] And this property is, in turn, related to an even more fundamental hallmark of icons, namely the fact that they denote their object by means of a quality they share with it, although their possessing this qual-

**21** EP2, 106 (1901); CP 2.96 (1902); EP2, 287 (1903).

**22** Stjernfelt 2007 (especially ch. 3) has first drawn attention to the central role of this operational notion of icon in Peirce. It is to him that I owe the term "operational definition". More recently, see Bellucci/Paolucci 2015.

**23** W5, 380 (1886): "No quality or character of any kind can be conveyed or made known, except by the means of an icon."

**24** A particularly revealing passage is CP 7.281 (c. 1895), in which Peirce speaks in very similar terms of the mathematician's and the artist's work. On the one hand, "[an] example of the use of a likeness is the design an artist draws of a statue, pictorial composition, architectural elevation, or piece of decoration, by the contemplation of which he can ascertain whether what he proposes will be beautiful and satisfactory." On the other hand, "[t]he utility of likenesses to mathematicians consists in their suggesting in a very precise way, new aspects of supposed states of things."

ity is not dependent on the existence of the object itself. In this way, they make it possible to investigate that quality from all possible aspects, as if it were a mathematical hypothesis, shorn of the constraints of reality.[25]

The analogy between abductions and iconic signs is also, for Peirce, an opportunity to further articulate the complementary character of abductions and inductions. As the former are related to icons, the latter depend on the other class of signs that establish an *immediate* connection to the world, namely *indices*. Indices do not signify by resemblance, but by *physical contiguity:* they hinge on a sheer "action-reaction" duality between the sign and its object.[26] Their mode of signification, therefore, corresponds to the definition of induction as that phase of the inferential process in which our hypotheses are put to a direct experimental test.

At the same time, the second analogy I want to briefly discuss slightly complicates the symmetry between induction and abduction, in that it confers on the latter, too, a physical and almost bodily tinge. As Peirce says, in fact, part of abduction's creative power consists in its ability to "[put] together what we had never before dreamed of putting together" (EP2, 227 [1903]); And at least once, he likens this operation to the way we acquire new motor "skills" by integrating different motor reactions into a new composite movement. The example he offers is that of somebody attempting to "mov[e] his two hands simultaneously and in opposite directions": What one does in that case is attend to the single, more elementary movements that compose the movement one is seeking to acquire, until "suddenly a general conception of the action springs up", acting as a force that "unites" the more elementary acts, and renders the performance of the new act "perfectly easy" (W8, 152 [1892]).[27] Although the analogy is far

---

**25** CP 2.304 (1902): "An *icon* is a sign which would possess the character which renders it significant, even though its object had no existence; such as a lead-pencil streak as representing a geometrical line." CP 4.447 (1903): "A pure icon can convey no positive or factual information; for it affords no assurance that there is any such thing in nature. But it is of the utmost value for enabling its interpreter to study what would be the character of such an object in case any such did exist. Geometry sufficiently illustrates that."

**26** W5, 379 (1886): "The index has no generality in itself. It does not depend on a mental association, but upon a real reaction between the mind and the external world at the moment when the index acts. The index asserts nothing; it only says "There!" It takes hold of our eyes, as it were, and forcibly directs them to a particular object, and there it stops."

**27** That this second analogy is intimately related to the first is made clear in CP 7.498 (c. 1898): "[T]here are several ideas dimly in my mind from time, none of which taken by itself has any particular analogy with my grand problem. But someday these ideas, all present in consciousness together but yet all very dim deep in the depths of subconscious thought, chance to get joined together in a particular way such that the combination does present a close analogy to from crystal clear, what we do get from it is the idea that the physical performance of the elementary acts is necessary for the birth of the new conception—though it is not sufficient, in that it must be accompanied by a "desire to perform the motion" (W8, 152 [1892]). In other words, there is a generative force inherent to the performance, which, if steered in the right direction by desire, triggers creativity.

We should read Peirce's two analogies together, as indicating two complementary sources of abduction's creative power. The first is related to a fundamental metaphysical and semiotic property of icons; the second, to our ability to perform physical acts. Abduction is, therefore, rooted in a duality between a motor and an iconic dimension.[28] This duality is a crucial hallmark of Peirce's epistemology: it not only structures his understanding of abduction as strictly construed, but permeates his understanding of scientific creativity *as a whole* as one which revolves on the double core of "*visual images and muscular imagination*".[29]

In particular, both Peirce's philosophy of mathematics and his diagrammatic logic are based on the idea that every act of intellectual creation relies on our ability to *carry out concrete manipulative experiments* with icons or diagrams that represent our situation of departure, and subsequently *to observe the results* of our own manipulation. Mathematics is, in this sense, a "science of the eye" (W2, 315 [1869]): one in which the most momentous discoveries are introduced by means of a concrete operation on a diagram. This operation reveals new properties of the object that we would not have been able to observe beforehand. Concurrently, Peirce developed a sophisticated graphical system of logic, which he held to be more powerful and complete than a linear or symbolic one.[30] Mainly as a result of these logical studies, he often insisted on the primacy of visual means of thinking over linguistic ones (see above, fn. 24). But also well beyond the boundaries of logic and mathematics, he contended that creative reasoning in general "is not done by the unaided mind, but needs the cooperation *of the eyes and the hand*" (Peirce to J. M. Hantz [1887], quoted in Houser 2000, xxix). Or as he puts it in a text on "Logical Machines": Although the mind, if left to itself, is similar to a machine in its inability to overcome the limits of

---

my difficulty. [...] It must be [...] its *analogy*, or resemblance in form, to the nodus of my problem which brings [that combination] into vividness."

**28** This duality has, of course, very ancient roots. See Rosen 1961.

**29** W8, 19 (1890): "It is not a historical fact that the best thinking has been done by words, or aural images. It has been performed by means of visual images and muscular imagination." Cf. ibid, 24; CP 2.778 (1902).

**30** On Peirce's logical and mathematical writings see CP, vol. 4; Peirce 2010.

its engineering, "the mind working with a pencil and plenty of paper has not such limitations" (W6, 71 [1887]).

The latter quote shows Peirce writing at the very onset of artificial intelligence, but already capable of recognizing one of its most pressing shortcomings—namely the inability to disentangle complex situations by dint of spontaneous decisions (Fingerhut/Hufendiek/Wild 2013, 51–52). And in his attempt to articulate the reason why human intelligence is not affected by this shortcoming, Peirce appears to commit himself, at least minimally, to an embodiment thesis in the proper sense of the word—that is, to the thesis that our very bodily constitution influences the way we think. Logical machines, in fact, cannot deploy the same intellectual faculties as we do, because they do not have the two specific organs that structure the human way of exploring the environment: "the eyes and the hand".

We will presently see that Peirce's theory of embodiment is actually countered by his insistence on the "external" character of cognition, and on the latter's ability to be realized in different material structures. Yet this should not prevent us from recognizing that Peirce did assign a role to the bodily constitution of human beings. Before spelling out this point in detail, however, let me recall that the duality of "the eyes and the hand" in scientific reasoning can also be read in semiotic terms, as yet another resumption of the two axes along which semiotic structures gain an immediate link to reality—iconicity and indexicality.[31] Taken together, these two classes of signs are the main instrument through which Peirce was indeed able to recover those aspects of traditional intuitionism that most mattered to him (that is, the possibility of an immediate contact with reality and the emphasis on the creative spontaneity of intelligence) without giving up his overall anti-intuitionist orientation.[32] Both icons and indices, in fact, establish an immediate link to their object (either physical or qualitative) *while remaining semiotic elements.* They take part in a broader mechanism that rules out, by definition, the existence of intuitive cognitions.[33]

---

**31** On Iconicity or Firstness as immediacy see, e.g., EP1, 248 (1887–88). On Secondness, EP2, 224 (1885); CP 5.487.

**32** See in particular Jung 2009's chapter on pragmatism.

**33** To be more precise, manipulation is also related to the other end of ampliative reasoning, namely *induction*, or the experimental testing of a theory. In the sense of the relation between abduction and iconicity, Peirce links induction to the semiotic category of *indexical signs.* (See, e.g., EP2, 106, [1901]). This parallel is explained by the fact that, within the process of inference, induction is the moment in which we attend to reality to confirm our hypothesis, in order to adjust or generalize it. In this sense, it is crucially related to what Peirce often calls the "Outward Clash" of external reality: its ability to respond to our hypotheses in a way that is absolutely independent from what we may think about them.

## 4 The "external" mind versus the lived body

If we take seriously the latter observation, we may also better understand some later statements in which Peirce might seem to go back to an intuitionist position. Speaking of the act of "observation" that accompanies all act of reasoning, and may be described as a "first acquaintance with an inference", Peirce says that "when it is not yet adopted, [it is] a perception of the world of ideas" (EP2, 233 [1903]).[34] But to understand these words correctly we should bear in mind that observation is *already* part of an inferential act: it is not a simple and ultimate psychological antecedent. Moreover, as a cognitive element, it is embedded in a semiotic chain, which is by definition mediated and non-intuitive.

Furthermore, abduction is unlike intuition in that it is never *unfailing*. "The abductive suggestion comes to us like a flash. It is an act of insight, although of *extremely fallible insight.*" (EP2, 227 [1903], my emphasis) For this reason, Peirce never maintains that the "perception of the world of ideas" represents an *alternative* to ordinary empirical inquiry. This is indeed one of the most important differences with epistemic intuitionism, as the latter typically presents intuition and dianoetic reason as two irreducible paths by which we gain access to reality. And it also has a particularly significant consequence with regard to the knowledge of ourselves. If we reject intuitive knowledge, in fact, then "our whole knowledge of the internal world is derived from the observation of the external facts" (W2, 205 [1868]). That is, it is not formally dissimilar to other kinds of empirical inquiry.

Peirce's theory of abduction may thus become the linchpin of a philosophy that does not limit mental phenomena to what happens within the mind (or to the "private" dimension of thought), but draws attention to the role of both the interplay with the environment and the *irreducibly intersubjective aspect of experience.* In particular, external elements such as artefacts or symbolic systems may be seen as the catalysts of a "manipulative" kind of abduction, thanks to their ability to store information, reshape the traits of a given situation, or suggest ways of categorizing the environment that transcend those we may have developed through natural selection (Magnani 2009, in particular ch. 3). From this perspective, Peirce's theory of abduction seems indeed to reveal many similarities to today's "extended mind" paradigm.

I would like to conclude my paper by arguing that we can better understand these similarities if we go back, once again, to Peirce's anti-intuitionist stance.

**34** Cf. Tiercelin 2005, 398.

This time, however, we will look not only at the potentialities of this position, but also at its limitations.

In particular, Peirce's interest for the embodied nature of experience and cognition never goes so far as to touch upon another important aspect of contemporary theories of embodiment, namely the peculiar status of the human body as a *lived body* that constitutes the primitive horizon of experience, and is present to us in a first-personal manner. This, indeed, marks a profound difference with another philosophical perspective from which cognitive scientists have recently been drawing, namely phenomenology. Think in particular of Merleau-Ponty, and his development of Husserlian intentionality into the notion of *corps propre*. For Merleau-Ponty, the body is not an object among others, but "that by which there are objects".[35] It is the locus of a primitive intentionality that makes experience possible in the first place, and represents the paramount way we get in touch with the world. Itself liable to be known in an immediate or *intuitive* way, this primitive intentionality manifests itself as a "spontaneity in the flesh" in the phenomenon of physical effort, and is therefore the pivot of Merleau-Ponty's first-personal analyses of the human body as the very condition of the possibility of perception and cognition.

To better understand the difference I am talking about, it is impossible not to recall a very famous text from 1902, in which Peirce seems to present a genuine "extended mind" thesis that pivots on a depreciation of inner experience (CP 7.364–367 [1902]).[36] As he argues therein, psychology is the study of *mind*, not of *consciousness*, and the two are by no means equivalent. Not only, in fact, do many mental phenomena take place unconsciously; but more generally, consciousness is a mere *feeling*, therefore "the inward aspect of things", while mind is "essentially an *external phenomenon*". Consequently, the real nature of mental or semiotic elements reveals itself as embodied in the external world. For example, it is more correct to say that the thoughts of a writer "reside [...] in the printed copies of his book" than in his brain; or that my faculty of language resides in my tongue, and in the inkstand I use for writing, as much as it resides in "a lobe of my brain". If a psychologist removed that lobe, in fact, I would certainly be bereft of language. But just the same would happen "if he [...] filched my inkstand", or if he cut off my tongue. The conclusion Peirce draws from these examples is that when we talk about the "localization" of mental faculties, we should

35 Merleau-Ponty 1945, 121: "Ce qui l'empêche d'être jamais un objet [...], c'est qu'il est ce par quoi il y a des objets. Il n'est ni tangible ni visible dans la mesure où il est ce qui voit et ce qui touche." Engl. transl., 105.

36 See in particular Paolucci 2011, to whom I owe the insistence on the notion of "external mind".

use this term in a *relational* sense: "the inkstand and the brain-lobe have the same general relation to the functions of the mind." Relations among elements have a clear priority over their material constitution (CP 7.364–367 [1902]).[37]

The relational—or *functionalist* (Chauviré 2004, 72–73)—position Peirce presents here is one of the most distinctive features of his philosophy. It also bears a clear similarity to so-called "functionalism" in the cognitive sciences, in particular with regard to its emphasis on the "multiple realizability" of mental processes.[38] We might, therefore, be tempted to follow a distinction that has become customary in today's debate, and suggest that Peirce's emphasis on the "external" nature of mind, as a precursor of the *extended mind* hypothesis, is indeed contrary to a strong *embodied* theory in Merleau-Ponty's sense.[39] Peirce, in effect, does not focus on our ability to "interiorize" the inkstand into our own bodily schema.[40] Rather, he highlights the role of the inkstand in writing and communication, and on external objects' abilities to realize semiotic processes. Our language ability could not be deployed without some kind of material realization. But whether this material realization overcomes the limits of our proper body is not particularly significant. This is just one particular case of a more general metaphysical and semiotic fact: In Peirce, no intellectual content can ever exist without some degree of "embodiment", as he himself says, or *material expression.* Nonetheless, very different material entities (for instance: "a daily newspaper, a great fortune, a social movement", EP2, 435 [1908]) can "embody" the same idea. Or to put the same point in slighty different terms, while embodiment theorists tend to confer mental properties on matter (i.e., on the body), Peirce insisted that "matter is effete mind";[41] that is, mind rendered dull and

---

**37** In the same text, Peirce makes another two interesting similitudes: he says that consciousness is "external" to the mind the same way electricity is external to the wires that transport it; or again, that a psychologist studying consciousness would be akin to an ichthyologist studying water. Scholars of extended mind would recognize in the latter simile an analogy to Andy Clark's discussion of the "swimming machine" (Clark 2001, 143. See also Hufendiek 2012, 444–447).

**38** CP 7.366: "So my faculty of discussion is equally localized in my inkstand. *It is localization in a sense in which a thing may be in two places at once.*" (My emphasis.)

**39** The opposition between functionalism and a stronger understanding of embodiment has been articulated by Clark 2008. Gallagher 2014 has already and rightly argued, however, that pragmatism escapes this dilemma (although he has focussed more on Dewey than on Peirce). On the crucial but ambivalent role of the term "external" in Peirce, cf. Paolucci 2011.

**40** Cf. Merleau-Ponty's examples (car driving, music playing, the feather on a lady's hat) in Merleau-Ponty 1945, ch. 3.

**41** W8, 22 (1890): "The only possible way of explaining the connection of body and soul is to make matter effete mind, or mind which has become thoroughly under the dominion of habit, till consciousness and almost extinct". See also ibid., 13–16 (1890), where Peirce's criticizes the "materialist" thesis that "we think with our muscles".

less spontaneous by a sort of "sedimentation" process. The difference is subtle, but crucial.

Still, it remains that equating Peirce's position to an outright extended-mind functionalism would be reductive. Peirce certainly thought that different material elements may realize the same idea. But it would be hard to prove that he held these material realizations to play no role whatsoever in the determination of the idea. In particular, we have seen in the previous section that his mature philosophy entails an appreciation of the peculiar structure and cognitive potentialities of the human body.[42] Furthermore, in his later years he laid particular emphasis on the metaphysical role of the two kinds of cognitive immediacy on which I have already insisted, namely the *qualitative immediacy* of iconicity and the *physical, action-reaction immediacy* of indexicality. Along these two axes, the specificities of the body as the locus of "qualitative feels" and "willed actions" (to use the terminology of Andy Clark) may be vindicated again (Clark 2008, 52, 55). Lastly, while extended mind functionalists typically emphasize the causal role of the environment for a central organism that processes information and outsources calculations, Peirce's paramount interest is to put forth a picture of Nature as embodying intellectual and semiotic values in a way that is not necessarily dependent on our own relation to them.

Rather, the gist of the difference between Peirce's version of embodiment and the phenomenological tradition may be found in something that is implied by my latter remark. Borrowing the words of Claudine Tiercelin, Peirce "absolute [ly] refus[es] to limit mind to the fields of consciousness or of the self",[43] and consequently tends to give up the focus on human subjectivity much more radically than phenomenology. In the end, this may be just another way of articulating the difference between the pragmatist conception of *sign* (essentially triadic and external) and phenomenological *intentionality* as two alternative bases for a theory of the mind. But as I have maintained throughout, it also depends on the uncompromising position Peirce decided to defend at the beginning of his philosophical career: We do not "have [...] an intuitive faculty".[44]

---

**42** See also EP2, 474 (1914), where Peirce describes the relation between mind and body as "intimate".

**43** Tiercelin (1998, 23, fn. 23) uses this expression to aptly characterize the difference between Peirce and another chapter of French philosophy, namely nineteenth-century spiritualism. Spiritualism and French phenomenology, however, are related at important junctures, among which, it seems to me, the problem of intuition and cognitive immediacy.

**44** See again "Questions Concerning Certain Faculties Claimed for Man", especially W2, 201.

# References


Anderson, D. R. (1987). *Creativity and the Philosophy of Charles S. Peirce.* Oxford: Clarendon Press.

Bellucci, F. & Paolucci, C. (2015). "Peirce e l'iconismo." In: *Versus. Quaderni di studi semiotici* 120, 3–13.

Chauviré, C. (2004). "Aux sources de la théorie de l'enquête. La logique de l'abduction chez Peirce." In : B. Karsenti/L. Quéré (eds.). *La Croyance et L'Enquête. Aux Sources Du Pragmatisme.* Paris: Éditions de l'EHESS.

Clark, A. (2001). *Mindware. An Introduction to the Philosophy of Cognitive Sciences.* Oxford: Oxford University Press.

Clark, A. (2008). "Pressing the Flesh: A Tension in the Study of the Embodied, Embedded Mind?" In: *Philosophy and Phenomenological Research* 76 (1), 37–59.

Cuccio, V. & Caruana, F. (2015). "Il corpo come icona. Abduzione, strumenti ed Embodied Simulation." In: *Versus. Quaderni di studi semiotici* 120, 93–104.

Fingerhut, J., Hufendiek, R. & Wild, M. (2013). "Einleitung". In: Idd. (Eds.). *Philosophie der Verkörperung.* Berlin: Suhrkamp.

Fisch, M. H. (1984). "The Decisive Year and Its Early Consequences." In: W2, xxi-xxxvi.

Frankfurt, H. G. (1958). "Peirce's Notion of Abduction." In: *The Journal of Philosophy* 55 (14), 593–597.

Gallagher, S. (2014). "Pragmatic Interventions into Enactive and Extended Conceptions of Cognition." In: *Philosophical Issues* 24, 110–126.

Harman, G. H. (1965). "Inference to the Best Explanation." In: *The Philosophical Review* 74 (1), 88–95.

Hintikka, J. (1998). "What Is Abduction? The Fundamental Problem of Contemporary Epistemology." In: *Transactions of the Charles S. Peirce Society* 34 (3), 503.

Hufendiek, R. (2012). "Draw a Distinction. Die vielfältigen Funktionen des Zeichnens als Formen des Extended Mind." In: U. Feist & M. Rath (Eds.). *Et in imagine ego.* Berlin: Akademie Verlag.

Jung, M. (2009). *Der bewusste Ausdruck. Anthropologie der Artikulation.* Berlin/New York: De Gruyter.

Magnani, L. (2009). *Abductive Cognition. The Epistemological and Eco-Cognitive Dimensions of Hypothetical Reasoning.* Berlin/Heidelberg: Springer.

Merleau-Ponty, M. (1945). *Phénoménologie de la perception*, Paris: Gallimard. Engl. transl. by C. Smith (1962). *Phenomenology of Perception.* London: Routledge & Kegan Paul.

Kapitan, T. (1990). "In What Way Is Abductive Inference Creative?" In: *Transactions of the Charles S. Peirce Society* 26 (4), 499–512.

Paavola, S. (2005). "Peircean Abduction: Instinct or Inference?" In: *Semiotica* 153/1, 131–154.

Paolucci, C. (2011). "The 'External Mind': Semiotics, Pragmatism, Extended Mind and Distributed Cognition." In: *Versus. Quaderni di Studi Semiotici* 112–113, 69–96.

Pape, H. (Ed.) (1994). *Kreativität und Logik: Charles S. Peirce und das philosophische Problem des Neuen.* Frankfurt/M.: Suhrkamp.

Peirce, Ch. S. (1931–35). *Collected Papers of Charles S. Peirce*, vol. 1–6 ed. by Ch. Hartshorne & P. Weiss, vol. 7–8 ed. by A. W. Burks. Cambridge, MA: The Belknap Press of the Harvard University Press. (Cited by volume and paragraph number.) [CP]

Peirce, Ch. S. (1984). *Writings of Charles S. Peirce. A Chronological Edition.* ed. by Peirce Edition Project. Bloomington: Indiana University Press. (Vols. 1–6 and 8 so far published.) [W]

Peirce, Ch. S. (1985). *Historical Perspectives on Peirce's Logic of Science. A History of Science.* ed. by C. Eisele, 2 vols. Berlin/New York/Amsterdam: Mouton Publishers. [HP]

Peirce, Ch. S. (1992–1998). *The Essential Peirce: Selected Philosophical Writings.* vol. 1 ed. by N. Houser and Ch. Kloesel, vol. 2 ed. by the Peirce Edition Project. Bloomington/Indianapolis: Indiana University Press. [EP]

Peirce, Ch. S. (2010). *Philosophy of Mathematics. Selected Writings.* ed. by M. E. Moore. Bloomington: Indiana University Press.

Queiroz, J., Atã, P. (2013). "Icon and Abduction: Situatedness in Peircean Cognitive Semiotics." In: L. Magnani (Ed.). *Model-Based Reasoning in Science and Technology.* Heidelberg: Springer.

Rosen, S. H. 1961. "Thought and Touch: A Note on Aristotle's 'De Anima.'" In: *Phronesis* 6 (2), 127–137.

Short, T.L. (2007). *Peirce's Theory of Signs.* Cambridge: Cambridge University Press.

Stjernfelt, F. (2007). *Diagrammatology. An Investigation on the Borderlines of Phenomenology, Ontology, and Semiotics.* Dordrecht: Springer.

Tiercelin, C. (1998). "Peirce's Objective Idealism: A Defense." In: *Transactions of the Charles S. Peirce Society* 34 (1), 1–28.

Tiercelin, C. (2005). "Abduction and the Semiotics of Perception." In: *Semiotica* 153 (1/4), 389–412.

Rebekka Hufendiek

# William James and John Dewey on Embodied Action-Oriented Emotions

**Abstract:** In this paper I present William James's theory of emotions and the central role it ascribes to the bodily reactions in emotions. I call the claim that there is a distinct pattern of bodily reactions that is constitutive for an emotion type the essential-ingredient view. I discuss some of the famous objections that cognitivists have raised against the feeling theory and characterize the position that cognitivist accounts take with regard to bodily reactions in emotions as the coincidental-byproduct view. With regard to current empirical research from psychophysiology I argue that the coincidental-byproduct view and the essential-ingredient view are both untenable. Yet the Jamesian view can be understood in a broader context as not aiming to find essential ingredients of emotions but rather taking the bodily arousal to be part of an organisms preparing for action. A reinterpretation of James in this manner has already been presented by John Dewey, who also provides a broader framework to think of emotion and perception as being co-constituted by action-tendencies.

## 1 James and the essential-ingredient view

In his classic essay *What is an Emotion?* (1884), William James argues that the feeling of bodily arousal is what constitutes an emotion. The commonsense view says that emotions simply cause bodily arousal; when we are sad, we therefore start crying or when we feel embarrassed, we blush. Yet James inverts the commonsense view and argues that in most cases we perceive something, like a dangerous predator or an offensive opponent, and this perception directly triggers bodily arousal, which is felt and evaluated only afterwards. James's central argument for this is of a phenomenological character: if you imagine a strong emotion and try to subtract the feeling of all involved bodily symptoms, nothing remains "but a cold and neutral mental state of intellectual perception" (James 1884, 193). We can refer to this argument as the "subtraction argument".

Upon a closer look, however, James makes two related claims. The first claim is that a mental state that does not involve the perception of bodily arousal would not be classified as an emotion at all. The second claim is that different kinds of emotions owe their special character to the different kinds of bodily arousal that constitute them. Both claims are nicely illustrated in the following quote: "What kind of an emotion of fear would be left, if the feelings neither of quickened heart-beats nor of shallow breathing, neither of trembling lips nor of weakened limbs, neither of goose-flesh nor of visceral stirrings, were present, it is impossible to think" (James 1884, 193 f.). These are some of the physiological elements involved in fear. Rage according to James is unimaginable with "no ebullition of it in the chest, no flushing of the face, no dilatation of the nostrils, no clenching of the teeth, no impulse vigorous to action, but in their stead limp muscles, calm breathing, and a placid face" (James 1884, 194). The bodily processes involved in emotions are their essential ingredients. Subtract them away and the mental state that remains is classifiable neither as a state of fear nor as an emotion at all. We can refer to this as the "essential-ingredient view." James's essential-ingredient view has been the target of many objections, which I discuss below.[1]

## 2 Nussbaum and the coincidental-byproduct view

Cognitivist approaches have criticized James view not only for ignoring the cognitive dimension of emotions but also for overestimating the role of bodily arousal in emotional episodes. Nussbaum (2001) argues that while emotions involve bodily sensations of many kinds, there are no constant correlations between specific bodily processes and certain kinds of feelings, thus bodily reactions appear to be a coincidental byproduct of emotions and not an essential ingredient. We can refer to this view as the "coincidental-byproduct view." Nussbaum's position represents a broad consensus among cognitivists. While cognitivists tend to argue that emotions are constituted or caused by cognitive states such as beliefs or judgments, they tend to downplay the role of the body in emotional processing.[2] With this view cognitivists such as Nussbaum stand in the tradition of the majority of authors in the second half of the 20th century in philosophy of mind and cognitive science, who take cognition to be disembodied.

1 Also, James' theory of emotions has been interpreted in various ways, some of which are perhaps more charitable than the present one. In this paper I focus on a systematic question concerning the standard reading of James' feeling theory. But see e. g. Ellsworth (2014) for the claim that James never aimed to present distinct emotion categories or Ratcliffe (2005) for the claim that James can account for the intentionality of emotions.

2 In this spirit Robert Solomon (with regard to his early works from the 70's) remarks, "In my original theory, it was by no means clear, that the body had any essential role in emotions" (Solomon 2003, 189).

Nussbaum puts forward a conceptual and an empirical argument in defense of the coincidental byproduct view. The conceptual argument, as Nussbaum articulates it, employs the possibility of bodiless creatures having emotions: "Whether we believe that bodiless substances exist or not, the reason it makes sense to imagine a bodiless substance having genuine emotions is that it makes sense to imagine that a thinking being, whether realized in matter or not, could care deeply about something in the world, and have thoughts and intentions associated with such attachments. And that's all we really require for emotion" (Nussbaum 2001, 60). Nussbaum argues here that it is conceptually possible to think of an emotion as an intense state of mind that is not realized in any kind of matter at all. Actually, James happily admits that it is no "contradiction in the nature of things, or that pure spirits are necessarily condemned to cold intellectual lives" (James 1884, 194). So, with regard to this particular argument there is no real disagreement between the coincidental byproduct and the essential ingredient view. What James instead wants to claim is that "*for us*, emotion dissociated from all bodily feeling is inconceivable" (James 1884, 194, emphasis added). After all, from a naturalist point of view, the mere conceptual possibility of disembodied emotions in possible worlds where thinking substances without a body exist, can be conceded. But what is of interest is how emotions are realized in living organisms in the actual world. And here, the hypothesis at stake is that to "care deeply" about something or someone involves specific bodily processing such as the release of oxytocin.

Therefore, it is the empirical argument that really is of importance for the evaluation of the Jamesian account. Walter Cannon (1927) did research to test the Jamesian hypothesis as early as the 1920 s. In his experiments, the spinal cord and vagus nerve in dogs and cats were cut to inhibit feedback from the body to the brain. The animals were tested with regard to their reactions to emotional stimuli like threatening shocks. Their ability to react emotionally to such stimuli appeared to be intact. According to Cannon, the animal tests prove that even if you cut the bodily feedback entirely, feelings do not seem to be reduced. Even today, Cannon's article ranks high on the list of the most quoted evidence for all variations of the coincidental-byproduct view.[3] It is often combined in one vein with a study by Chwalisz, Diener, and Gallagher (1988) who tested patients with spinal cord injuries. The outcome suggests that their emotional lives remain relatively stable. However, as William James already knew, cutting the vagus nerve and the spinal cord alone is not enough to prove him wrong:

**3** See for example Solomon 2001/2003, 26.

> A crucial test of the truth of the hypothesis is quite as hard to obtain as its decisive refutation. A case of complete internal and external corporeal anaesthesia, without motor-alteration or alteration of intelligence except emotional apathy would afford, if not a crucial test, at least a strong presumption, in favour of the truth of the view we have set forth; whilst the persistence of strong emotional feeling in such a case would completely overthrow our case (James 1884, 203).

More recently, Antonio Damasio (1999) carefully detailed the paths bodily feedback can take and which forms of damage could eliminate or reduce it. Damasio points out that an embodied account can include more than just the bodily reactions in the visceral organs. Musculoskeletal feedback from the body and the face is apparently quicker and contributes to interoception during emotions. That spinal cord injury patients do not lose the ability to feel emotions can be explained by the fact that in these patients the vagus nerve usually is still transporting information. Also, only a fraction of the enactment of emotions depends on the spinal cord. A large number of processes are mediated by cranial nerves at the brain-stem level that can act on the face and on the viscera, and by other brain-stem nuclei that can act directly on the parts of the brain above their own level. Also, a significant part of bodily feedback does not travel via the nerves at all, but by way of the bloodstream from where it directly enters the brainstem. Nevertheless, that the reduced amount of bodily feedback might also reduce emotional feelings is suggested by Hohmann (1966), who interviewed patients with a spinal cord lesion and found that they reported subdued experience for many emotions except sadness, and also found a correlation between the location of the spinal lesion and the degree to which their emotional experience was reduced. But even if the location of the lesion is high on the spine, this does not suggest that patients should entirely lack emotional feelings, since the body does not end above the neck. Face, skull, oral cavity, and tongue provide a massive input into the brain that enters at brain-stem level and cannot be inhibited by any spinal cord injury. Furthermore, most emotions express themselves in changes of the facial musculature, the musculature of the throat, and in autonomic changes of the skin in the face and scalp. The representation of these changes in the brain remains available even with spinal cord injuries (Damasio 1999, 290).

Similar arguments can be brought up against Cannon's interpretation of the animal experiments. Why should Cannon have expected his prediction, that cats or dogs with severed spinal cord and vagus nerve should have a complete loss of emotional display, to be true in the first place? A severed vagus nerve and spinal cord do not impede the pathways for the facial responses that come from the brain-stem, and are mediated by cranial nerves which were not compromised in Sherrington's or Cannon's experiments. It's therefore not surprising that the cats and dogs in the study showed anger expressions, even if they could not move their bodies (Damasio 1999, 291).

Another anti-Jamesian argument is that bodily reactions cannot be constitutive of emotions, since the reduction of bodily feedback does not always lead to reduced emotional experience. Yet the studies cited to prove this claim are, at the very least, highly ambivalent. Reisenzein and Stephan (2014, 43) argue, for example, that studies in which facial feedback has been partly blocked by Botox injections (Davis, Sengas, Brandt, Ochsner 2010) do not support the claim that facial feedback is necessary for emotions. Yet this is not what James or typical current views claim. The claim is that emotions (and thereby emotional feelings) are constituted through patterns of bodily arousal. The absence of one component of the bodily reaction should have an effect on the realization of the emotion and might also effect how the emotion feels but should not necessarily prevent its occurrence. This theoretical prediction fits with the results of the studies quite well. They show that subjects who received a Botox injection showed emotional muting in response to a mildly amusing video clip, while no such response was observed in strongly negative or positive clips. A plausible interpretation of this result is that in the case of strongly emotion-eliciting clips there is no remarkable difference in experience because there are many bodily reactions going on, while in the case of mild joy the feedback that stems from a smile is a central factor of the bodily reaction so that its absence makes a difference. Similar arguments can be made for other cases of reduced bodily feedback as well.[4]

With regard to the second claim, that particular kinds of emotions owe their special character to the pattern of bodily arousal that constitutes them, Nussbaum presents phenomenological and empirical arguments in favor of the coincidental-byproduct view. While, as we have seen, James considers the reactions of the nervous system in fear and rage to be distinct, according to Nussbaum the feeling of bodily arousal involved in an emotion can vary. While "many men report experiencing anger in connection with a boiling feeling," Nussbaum states, "my own experience of anger is that it is associated with tension at the back of the neck, or a headache that appears the next day" (Nussbaum 2001, 61). This phenomenological self-observation is meant to refute the idea that every type of emotion is associated with a specific kind of bodily arousal. Whether rage is associated with boiling blood, or tensed muscles around the neck, or another bodily reaction, or no bodily reaction at all can vary from person to person and is therefore no criterion by which individuate emotions. From merely the phenomenological point of view the disagreement between the essential-ingredient view

---

**4** See Laird and Lacasse 2014 for further arguments.

and the coincidental-byproduct view seems hard to decide, as reports in favor of either view can be found.

Nussbaum's argument appears to find support in the studies done by Stanley Schachter and Jerome Singer (1962), who injected subjects with adrenaline and exposed them to different scenarios eliciting mild anger or happiness. The result was that test subjects showed signs of anger when they were given an insulting questionnaire and signs of happiness when brought into a room where a stooge was doing funny things. The interesting point is that the reactions of the test group differed from those of the control group, who were not given an anterior adrenaline injection: members of the control group became less emotionally involved with the presented scenarios. Schachter and Singer concluded that autonomic arousal plays a role in emotion insofar as it is perceived as undifferentiated arousal and creates an evaluative need. The subject, feeling her heart beating faster and finding herself in front of an annoying questionnaire, labels her own state as anger and reacts accordingly. Thus Schachter's and Singer's experiment can be seen, as it has been for several decades, as evidence for the claim that emotions cannot be individuated according to a type of bodily arousal they alone involve.

Yet the Schachter/Singer study has been shown to be flawed in several ways. To repeat only one of the arguments brought up against it, it is highly doubtful whether the mere injection of adrenaline causes the same sort of physiological arousal in various subjects in the minutes that follow. The participants might be more sensitive to the happiness- or anger-inducing scenarios because of the adrenaline, but the perception of the anger-eliciting task together with the adrenaline in the blood might cause further bodily changes that are anger-specific, while the adrenaline combined with the funny situation causes bodily arousal that is specific to joy (see Prinz 2004, 70 f.).[5] So the frequent claim that empirical studies prove that emotion types cannot be distinguished by the bodily reactions they involve is not as well supported by evidence as it might seem at first.[6] This will become clear in the discussion of recent evidence in the next section. A closer view at the results of recent research render it highly likely that emotions involve something like core-patterns of bodily reactions that are evolutionary acquired and can be found cross-culturally.

---

5 For further arguments against the Schachter/Singer experiment, see Griffiths 1997 and McNaughton 1989.

6 Walter Cannon (1927) also argues that visceral feedback does not suffice to distinguish various kinds of emotions. However, this objection can be met with the same kind of argument that I have already mentioned, that bodily reactions involve much more than just visceral feedback from the spine and the vagus nerve.

## 3 Bringing in empirical evidence

The possible sources of evidence of the bodily reactions involved in emotions are numerous and far from having been tested and measured in detail. Even detailed overviews of recent studies are far from issuing clear and stable results. I present a brief and highly selective overview, which aims to show that there is clear evidence against the coincidental-byproduct view and that it is an open question how closely emotion types and discrete physiological responses are related. As has already been mentioned, the place to look for bodily reactions is not only the inner organs and the visceral feedback stemming from them, but also the endocrine system and the somatic nervous system that innervates skeletal muscles, including those of the face.

Indeed, research focused on facial expressions (and more recently on body posture) has been most successful. Paul Ekman has investigated the facial expressions accompanying emotions. By carrying out research on facial expressions for more than thirty years he has established a theory of basic emotions, which are expressed in ways that can be recognized panculturally. Ekman and his colleagues developed the facial action coding system (FACS), which codes observable facial muscle-movements, to test whether their configurations could be identified with the expression of specific emotions across cultures (Ekman 1997). The research on re-identifiable facial expressions in many cultures, including some which barely had any contact with the western world, resulted in the hypothesis that several facial expressions occur and can be recognized universally. The emotions for which Ekman found stable re-occurring patterns are fear, anger, joy, sadness, disgust, and surprise (Ekman 1971; see also Matsumoto et al. 2010 for a current synopsis).

Apart from the visual observation of facial expressions, which Ekman and colleagues performed, nowadays there is a growing amount of research using facial electromyography. Facial EMG directly measures muscle activity by detecting and amplifying the electrical impulses generated by muscle fibers when they contract. The lowering and contraction of the brows, for example, has been observed as part of the expressive reaction to unpleasant stimuli. It is involved in producing frowns. Therefore, the firing of motor units in this muscle region can be expected if a stimulus is judged to be unpleasant. But facial EMG studies have also found that activity of the corrugator muscle, which lowers and contracts the eyebrow, varies inversely with the valence of the observed stimuli and the reported emotional state, which means that the muscle, in neutral situations, is under slight tension and relaxes in the presence of pleasant stimuli (Schwartz et al. 1979).[7] The corrugator muscle can be described as a permanent bodily valence marker that sends proprioceptive feedback to the brain, which indicates the presence of pleasant, unpleasant, and neutral stimuli.

While the research on facial expressions alone only led to the discovery of the six aforementioned basic emotions (or rather seven, since Ekman later added contempt to the list), recent studies have extended the view to combinations of facial expression and body posture and found evidence for cross-culturally stable patterns of pride and shame as well (Tracy and Robbins 2007a; 2007b and Clark 2010; for discussion, see below). It is likely that further research will reveal evidence for even more emotion types. Facial expressions do not occur in isolation but have been shown to be systematically associated with physiological responses, such as skin conductance, cardiovascular activation, and somatic activity (Mauss et al. 2005), and with neuroendocrine activity (Lerner et al. 2005). Lerner et al. were able to show that fear expressions are associated with elevated heartbeat and cortisol levels, while anger and disgust show reduced levels.

With regard to the reactions of the autonomic nervous system, the evidence is ambiguous. Yet hardly anybody would claim today that there is no covariance between reactions of the nervous system and particular features of emotions at all. Even Lisa Feldman-Barrett (2006) who assumes that there are no emotion kinds with a unique and invariant autonomic signature, suggests that patterns of physiological responses can be divided into more general dimensions of threat and challenge, and positive and negative valence (see also Taylor 1991; Cacioppo et al. 2000). If we look at emotion *patterns* rather than *individual variables* of autonomic reactions, *no* current position would deny differences among coarse-grained emotional dimensions (Mauss and Robinson 2009; Harrison et al. 2013).

Recent overviews even suggest that emotion types can be much better individuated via patterns of autonomic nervous system (ANS) arousal than was previously thought. A main reason for this is its greater sensitivity to fine-grained differences and to the various sources of those differences. While most studies are, to a large degree, based on measures of cardiovascular, respiratory, and skin conductance responses, when it comes to individuation a lot depends on measuring more and more correct factors to individuate a certain emotion type. Sylvia Kreibig (2010) points out that, in the case of disgust, measurements of the increased heart rate and skin conductance tend to resemble other negative emotions like fear and anger. Surprisingly, there are few tests that look at gastrointestinal responses that might correspond to disgust, although lay association

7 See also Laird 2007 and Matsumoto et al. 2010 for general overviews of the feedback generated by facial expressions.

could draw such a connection and the few studies that tested for stomach electrical activity could actually find reactions.

Fear is among those emotions for which there are stable results for a unique ANS pattern already. Fear studies have used a number of different fear-induction paradigms and identified a broad pattern of sympathetic activation that includes increased heart rate, narrowing of the blood vessels and electrodermal activity, accompanied by an increase in respiratory activities associated with a decrease in blood carbon dioxide levels (Kreibig 2010; Harrison et al. 2013). Anger (elicited through harassment or personalized recall) is associated with a general increase in sympathetic activity as well; it also includes an increase in heart rate and systolic and diastolic blood pressure, but, unlike fear, anger is associated with total peripheral resistance (Harrison et al. 2013). Some studies also found that in anger there is an increase in forehead temperature, probably due to a general increase in facial circulation (Stemmler et al. 2001). Analogously, an increase in facial blood flow in anger was found in rhesus monkeys (Nakayama et al. 2005). Happiness, like many negative emotions, is associated with cardiac activation, increased blood pressure, and increased respiratory activity. However, in contrast to many negative emotions, it is associated with peripheral vasodilation (Harrison et al. 2013). When comparing different types of emotions there are similarities such that most positive and negative emotions involve some kind of arousal, which tends to be stronger in negative emotions such as fear and anger. However, taking a closer look at fear and anger in particular, it turns out that anger appears to be more strongly associated with vascular activity than fear, but less strongly associated with cardiac activity. These results render it highly likely that emotions involve something like core-patterns of bodily reactions that are evolutionary acquired and can be found cross-culturally (see also Colombetti 2014).

It must be noted, however, that there is a huge amount of data lacking in this field and a considerable amount of confusion and seemingly contradictory results brought forth in recent decades. For example, while Cacioppo et al. (2000) found fear to be associated with a greater heart rate increase than anger, Labouvie-Vief et al. (2003) found that imagery-induced fear and anger elicited comparable increases in heart rate. A plausible explanation for such differences is that many studies neglect the importance of context-dependent aspects of emotional reactions. Different studies compared, for example, the physiology of fear in situations as different as hearing a loud noise, looking at a picture of an amputated leg, giving a public speech, or imagining an intruder in one's house (Bradley 2007). Cardiac reactions also tend to differ when subjects process affectively similar events in different types of tasks. The results of the study by Labouvie-Vief et al., for example, might be explained by the fact that heart rate tends to decelerate during unpleasant picture-viewing, while it accelerates when unpleasant mental imagery is evoked (Lang et al. 1990). The context of perception or imagination has a strong influence on the heart rate. A more obvious case is that the psychophysiology of emotion in action can differ from vigilant anticipation. Even without further evidence, one can make good guess about the difference in heart rate measured when somebody meets a snake while she is jogging through the woods compared to somebody lying on the couch and seeing a snake suddenly appearing on TV. Therefore, the parameters of each task, which have implications for the motivational action, have to be carefully distinguished before comparing studies. While one could try to come up with bodily arousal patterns relative to the situations in which they occur, finding stable patterns is further complicated by the degree to which individual differences influence patterns of physiological reactions, including interpersonal differences in gender, age, temperament, and intellectual capacity, such as imagination and memory, just to name a few (Feldman-Barrett, Russell 2015).

Another complication stems from the fact that the presence of specific emotions is not only inferred on the basis of physiological reactions, but also on the basis of the behavioral actions they cause, and these behaviors tend to vary depending on contextual support. In response to an electric shock, even rats show a variety of different behaviors depending on their actual environment. If there is an exit, they escape; if there is another rat, they attack; and if there is neither, they freeze (Lang 1990). This tactical nature of emotional behavior, which is already present in animals far less intelligent than humans, constrains the efforts of psychophysiologists to make general inferences regarding the physiology of specific emotional states since freeze, fight, and flight reactions pose different metabolic demands and therefore also influence the emotional arousal evoked by emotional processes (Harrison et al. 2013).

The evidence above makes the Jamesian hypothesis, that a specific emotion has a specific physiological pattern, look too general and difficult to test since the physiological pattern of an emotion depends on the individual, on its concrete circumstances, and on the kind of task in which the emotion-eliciting cue appears. This not only calls for clean parameters to be defined in studies on the psychophysiology of emotions. It also raises the question of whether it makes sense to think of the bodily arousal involved in emotions in terms of essential ingredients, i.e., in terms of a pattern with stable features that can be re-identified in the various occurrences of an emotion-type in a subject over time as well as among different subjects.

## 4 The functionalist objection

We have seen that there is considerable evidence for re-occurring patterns of bodily reactions in emotions, but we have seen as well that the evidence does not suffice to identify sets of essential elements for each emotion as the essential-ingredient view would demand. The essential-ingredient view has no answer to a simple functionalist argument such as the one Nussbaum has constructed to support her view:

> All human experiences are embodied, and thus realized in some kind of material process. In that sense, human emotions are all bodily processes. But the question is, are there any bodily states or processes that are constantly correlated with our experiences of emotion, in such a way that we will want to put that particular bodily state into the definition of a given emotion-type? And here we run up against an issue well-known to biological researchers: the plasticity of the human organism, or, in other words, the multiple realizability of mental states (Nussbaum 2001, 58 f.).

Nussbaum here refers to the functionalist paradigm in philosophy of mind that was elaborated by Hilary Putnam (1960) and others in the context of computationalist theories of mind. At the heart of functionalism lies the idea that mental states like ours might be realized in several ways and should not be type-identified with reference to their particular neural underpinnings but, instead, with reference to their functional role.

Multiple realizability is a central concept in this context. It can be understood in two ways: first, with regard to an individual and, second, with regard to different kinds of living organisms. With regard to the individual, the findings on the plasticity of the brain suggest that the brain is a functional organ where, for example, after a lesion the functions realized by the damaged areas can come to be realized by other areas of the brain. Therefore, it does not make sense to claim that a particular arrangement of neurons in a particular individual realizes one particular type of mental state. Across species there are obviously functionally identical mental states, like pain or hunger, in most mammals, while at the same time we would not expect the neural machinery realizing a pain-state in a rat to be precisely the same as in humans. Even with regard to individuals from the same culture Nussbaum argues, “if we said that grief is always of necessity accompanied by the firing of so-and-so many neurons of such and such type, we would be likely to find hundreds of cases for which this just isn’t quite right” (Nussbaum 2001, 59). Nussbaum takes the central nervous system to be part of the bodily realization of an emotion, which in a broader sense is, of course, correct. But the Jamesian hypothesis in particular takes emotions to be perceptions of arousal in the peripheral nervous system. So the question is whether we find distinct patterns of bodily arousal there that can be perceived by certain parts of the brain (James thought these to be subcortical sensorimotor areas), and not whether the neurons in the brain form distinct patterns that can be identified.[8]

But Nussbaum's argument can be turned into an argument against James: it is not only the brain that is plastic, but the whole nervous system. Therefore one can have doubts about the possibilities of finding exactly the same reactions, e.g., in the endocrine system, in the same person over several years. And one can also have doubts about whether we find the same reactions in the endocrine system when observing fear in a female baboon compared with an old man. Put this way, it becomes easy to see what position Nussbaum is denying: she does not want to commit herself to the claim that there is a single pattern of neuronal activity or bodily arousal that is activated whenever somebody experiences a certain type of emotion. Instead, the functionalist idea Nussbaum refers to claims that a fear-state can be realized in various ways, and that emotions should therefore be type-identified with regard to the mental evaluations they involve.

Yet what remains unclear is what explanatory role is left for the body in such a functionalist account. On the one hand, Nussbaum claims, "all human experiences are embodied, and thus realized in some kind of material process. In that sense, human emotions are all bodily processes" (Nussbaum 2001, 58). So Nussbaum seems to commit herself to some version of materialism, as most functionalists do. But on the other hand, Nussbaum takes it to be a virtue of her definition that it includes bodiless beings such as gods and angels.

> Whether we believe that bodiless substances exist or not, the reason it makes sense to imagine a bodiless substance having genuine emotions is that it makes sense to imagine that a thinking being, whether realized in matter or not, could care deeply about something in the world, and have thoughts and intentions associated with such attachments. And that's all we really require for emotion (Nussbaum 2001, 60).

We have already seen that not even James would deny the mere conceptual possibility of such bodiless emotional creatures. Yet for Nussbaum, the reference to them is meant to support her view that emotions should be type-identified with regard to the evaluation that emotions involve.

---

**8** I have already mentioned that the view I want to defend here departs from the Jamesian hypothesis. But with respect to this particular point, the embodied-emotions hypothesis I am discussing here, just like the Jamesian hypothesis, does not depend on the question of whether there is a particular brain state that realizes an emotion, but rather on whether there is a particular set of bodily reactions that does so. However, Nussbaum denies both claims.

One may find it confusing, though, that Nussbaum refers to the functionalist argument of multiple realizability, but at the same time claims that thinking beings do not need to be realized in matter at all. Most functionalist theories that argue for multiple realizability do not deny the important causal role of matter in the making of mental states. They simply claim that several different material arrangements can fulfill the same causal roles. Multiple realizability is not equivalent to the complete arbitrariness of the realizing arrangement of matter. Yet what Nussbaum seems to have in mind is a position that claims that emotions are mental evaluations that can be realized in all kinds of ways. Coincidentally, in all animals, including humans, emotions are realized in their brains and bodies. But the ways in which they are realized vary enormously so that it does not make sense to refer to them when looking for features to classify emotions.

This appears to be a radical version of what Lawrence Shapiro (2004) calls the "separability thesis." According to the separability thesis, it is perfectly possible for a humanlike mind to exist in a non-humanlike body. Shapiro rejects this view, arguing that e.g. visual perception in humans depends fundamentally on the way we move our heads to gain information about the distance of objects. Shapiro makes similar points with regard to the other sensory modalities and concludes that "a description of various perceptual capacities cannot maintain body-neutrality and ...an organism with a nonhuman body will have non-visual and auditory psychologies" (Shapiro 2004, 190). What Shapiro means by body-neutrality is precisely what Nussbaum is arguing for: the view that the mind is a program that can be realized in several kinds of bodies and can be characterized in abstraction from the body. Like Shapiro, I think that the ways in which the body contributes to our special ways of perceiving the world have to be taken into account. There are a variety of ways in which this idea can be explained, by giving more or less weight to the precise role of the autonomic nervous system, several brain structures, and cognitive processes. But for any kind of naturalist theory, functionalist or not, Nussbaum's coincidental-byproduct view is a slightly-too-arbitrary description of the role of bodily reactions in emotions.

Furthermore, the evidence discussed so far clearly shows that the bodily reactions are far from arbitrary. Consider Nussbaum's example of rage again. Nussbaum reports that while many men say that they associate rage with a boiling feeling, she associates rage with a tense feeling around the neck and sometimes a headache occurring the next day. With regard to the studies discussed above, there might be several reasons for these differences: gender reasons might be important, but there is also evidence that people have different abilities when it comes to sensing their own heart rate (Barrett et al. 2004) and sensing their own respiratory frequency. Parts of this interoceptive awareness might also improve through training (Daubenmier et al. 2013; Bornemann et al. 2015). Yet tensed muscles, increased heart rate, and elevated body temperature all fit into the prototypical pattern of bodily arousal that is generally associated with rage. What would be disturbing is if Nussbaum reported feeling her heart beat slow down whenever she is angry, or if she associated it with a relaxation of her muscles from the corrugator all the way down to the belly and the toes. My guess is that in such a case we would doubt whether the feeling that Nussbaum describes could be labeled as a case of anger at all. This intuition, together with the evidence discussed above, should suffice to disqualify an approach that denies any significant functional role of the bodily changes involved in emotions altogether.

Still, there is a lesson to be learned from Nussbaum: if James's claim, that there are specific patterns of bodily arousal in humans that can be used to type-identify emotions, is taken literally, it seems hard to reconcile with the functionalist idea of multiple realizability. Yet the idea that all fear-states in all mammals under all circumstances should involve exactly the same pattern of arousal sounds unlikely. We need a better definition of what these patterns are and what it means to re-identify them in various organisms in various situations. The Jamesian question needs to be situated in a broader theoretical framework in order to be answered.

## 5 Bringing in Dewey

The essential-ingredient view and the coincidental byproduct view both seem to be inadequate with regard to current evidence. Bodily arousal plays an important role in emotions and there appear to be stable and evolutionarily acquired patterns at least for some coarse-grained emotion categories. But emotional episodes can differ strongly and are sensitive to many contextual factors. It is therefore unlikely that we will ever find crossculturally stable essential ingredients for each emotion type.

An essential feature of emotions is their motivational force. Emotions have been described as action-tendencies that respond to urgent situations in the environment (Frijda 1986). When it comes to explain this urgency bodily arousal obviously plays a constitutive role. We feel motivated to approach somebody in an aggressive way because of the increased heart rate, muscle tension, blood pressure, and so on. If emotions are described and individuated as action-tendencies the bodily arousal appears to be relevant and specific but not entirely stable over all possible contexts. Emotional episodes can differ strongly depending on the context in which they take place. Furthermore the bodily reactions involved in emotions are not hardwired but rather skillful. From birth on we adopt and train certain respiratory cycles, ways to tense and relax our muscles and show certain facial expressions and we also learn in which situations these emotional expressions are seen as appropriate reactions in the culture we live in. Therefore it is not only single emotional episodes that might differ, different histories of coping with different social environments might lead to slightly different patterns of bodily arousal in individuals and among cultures.

Such an action-oriented view has been articulated for perception and emotion by John Dewey (1895; 1896). In his writing on emotion Dewey aims to unite Darwin's and James's approaches to emotions. He criticizes James for his focus on emotional feelings as being constituted by bodily arousal. Instead he highlights that emotions are habituated actions and discusses this using the famous example of fear of a bear: "It is not the idea of the bear, or the bear as object, but a certain act of seeing, which by habit, whether inherited or acquired sets up other acts." (Dewey 1895, 17). Dewey here also claims that the emotional episode and the emotional object are inseparable and that the emotional object is constituted by the emotional episode. I disagree with this view and will leave it aside for now. The way Dewey puts the focus on action becomes clear when looking at his critique of the reflex arc concept in psychology and the way he takes sensation and motor activity to be closely intertwined and to likewise constitute perception. The *act of looking*, executed by the whole organism, is what is primary and both sensation (i.e., the processing of the input) and movement are parts of this act, which constitutes perception. Action and sensation are closely intertwined; they "have been so often bound together to reinforce each other, to help each other out, that each may be considered practically a subordinate member of a bigger coordination" (Dewey 1896, 359). While the hand permanently depends on the control and stimulation of the visual information, vision in turn must be stimulated and controlled by the body's movements. If the eye is not kept fixed on the goal of grasping, if there is no proprioceptive feedback about the body's posture, the task cannot be fulfilled.

Arguments for such an action-oriented approach with regard to emotions can be grounded on empirical data from various sources. From a psychophysiological point of view it can be doubted whether the input- and output-side of an emotion can be clearly separated. Several studies suggest that emotional behavior (e.g., bodily postures, facial and vocal expressions), which is commonly regarded as a mere output, can modulate emotions and sometimes even trigger them (see Niedenthal et al. 2005 for an overview). To test the influence of facial feedback on emotions and emotional feelings, for example, participants were given non-emotional tasks to guide the production of facial expressions without cueing the emotional meaning of the expression. Participants then had to evaluate certain stimuli, like cartoons, with regard to whether they were funny. Findings suggest that the intensity and quality of the participants' manipulated facial expression affected the intensity of their self-reported emotional feelings as well as their autonomic responses. Facial expressions modulate emotions and sometimes even trigger them (Laird 2007). When facial feedback is reduced through Botox injections, finer notes of emotional feelings tend to be reduced (Davis, Senghas, Brandt Ochsner 2010, chapter 2). Bodily postures seem to impact emotional experience in a similar way. In a study conducted by Stepper and Strack (1993), participants' bodily posture was manipulated by asking them to adopt one of two conventional working positions, an upright or a slumped posture. Participants then received positive feedback concerning their performance on an achievement task. Those who received success-feedback in the slumped posture felt less proud and reported being in a worse mood than participants in the upright position (for similar results see Duclos et al. 1989). A different study found that an emotion-specific tone of voice has been found to amplify emotional experience as well (Hatfield et al. 1995).

Taken together, expressive behaviors facilitate, modulate, and produce corresponding emotional reactions and feelings. This suggests that in emotional processing input and output processing pathways are not clearly separated. Rather, emotional processing works in feedback loops: the organism produces output that affects its input in a systematic way. It not only receives stimuli from the world but also is a source of stimulation to itself. Getting angry in a certain situation might not depend on the input received from the external world alone. The organism's entire situation including bodily posture and current facial expression plays a role as well.

The data from psychophysiology and neuroscience discussed so far suggest that emotions cannot be adequately described by the essential-ingredient-view that bodily ingredients constitute emotional feelings. What unifies a certain emotion type, according to an action-oriented-view is not its particular bodily profile alone but rather the bodily profile seen as constituting a certain action-tendency. We find evidence for bodily responses constituting broad and stable emotion categories like fear as a danger warning system that prepares for flight and anger as a response to an offense that prepares for aggressive behavior or guilt as a response to a rule violation that prepares to make amends. But the bodily profile of emotional episodes can vary strongly with regard to the particular context. Increase of heart rate differs when meeting a bear while already running to meeting a bear when sitting relaxed in a chair. Such live situations differ largely from seeing a bear on television. Furthermore we acquire different embodied skills throughout our lives and through our interactions with different social surroundings. When Aristotle describes anger as being accompanied by the boiling of the blood around the heart and Nussbaum describes anger as being accompanied by a slight tension at the back of the neck, they point to different embodied skills that both contribute to largely the same action tendency, namely aggressive behavior. Yet there are different ways of showing aggressive behavior. Some are more direct, some more subtle, some impulsive, some planned. These differences in how we are motivated to show aggression in anger is shaped by a learning process that takes place in a social setting. But these learning processes do not start with the body as a blank slate. Anger involves an increase of heart rate, blood pressure and muscle tension. So while the present approach can integrate variations in emotional action tendencies, this does not make the bodily constituents of these action-tendencies arbitrary. If Aristotle would describe his being angry as being accompanied by a pleasant relaxation of the muscles or Nussbaum would say that anger for her really feels as if the blood circulation would slow down in a way that puts her in a peaceful state of mind, I guess that we would rather doubt that they are describing that particular emotion that we call anger.

## 6 Conclusion and open questions

I have introduced the essential-ingredient view that is based on a certain reading of William James. I further discussed the objections that cognitivists such as Martha Nussbaum have raised against James and the coincidental-byproduct view that these authors defend. I argued that with regard to current empirical evidence from psychophysiology the coincidental-byproduct view is untenable. I also argued that James's view understood as an essential-ingredient view is untenable and needs to be put into a larger framework. Dewey already offers a rereading of James focusing on the action tendencies that are constituted by the bodily arousal rather than on the feeling. In this vein I suggest to think of emotions as being action-oriented and think of the bodily arousal as the functional equipment that realizes certain action-tendencies in various contexts in various ways. A dimension of emotions that I completely left out of the present paper is the cognitive or normative dimension of emotions.[9] Emotions are not just bodily responses they have intentional objects and can be described as being subject to semantic and social norms. A current embodied action-oriented approach should account for the way emotions represent the world and for the complex

9 But see Hufendiek (2016).

ways in which emotions relate to other emotions, perceptions and thoughts in a naturalist framework.

## References


Adelmann, P. & Zajonc, R. (1989). “Facial Efference and the Experience of Emotion.” In: *Annual Review of Psychology*, vol. 40, 249–280.

Barrett, L. F. et al. (2004). “Interoceptive Sensitivity and Self-Reports of Emotional Experience.” In: *Journal of Personality and Social Psychology*, vol. 87, no. 5, 684–697.

Barrett, L. F. et al. (2006): “Are Emotions Natural Kinds?” In: *Perspectives on Psychological Science* 1, 28–58.

Bornemann, B. et al. (2013). “Differential Changes in Self-Reported Aspects of Interoceptive Awareness through 3 Months of Contemplative Training.” In: *Frontiers in Psychology*, vol. 5, 1504.

Bradley, M. (2000/2007). “Emotion and Motivation.” In: J. Cacioppo, G. Tassinary & G. Bernston (Eds.). *Handbook of Psychophysiology.* Cambridge: Cambridge University Press. 602–642.

Cacioppo, J., Priester, J. & Bernston, G. (1993). “Rudimentary Determinants of Attitudes. II: Arm Flexion and Extension have Differential Effects on Attitudes.” In: *Journal of Personality and Social Psychology*, vol. 65, no. 1, 5–17.

Cacioppo, J., Tassinary, G. & Bernston, G. (2000). *Handbook of Psychophysiology.* Cambridge: Cambridge University Press.

Cannon, W. (1927). “The James-Lange Theory of Emotion: A Critical Examination and an Alternative Theory.” In: *American Journal of Philosophy*, vol. 39, 106–124.

Cannon, W. (1936). *Bodily Changes in Pain, Hunger, Fear, and Rage.* New York: Appleton-Century.

Cannon, W. et al. (1988). “Autonomic Arousal Feedback and Emotional Experience: Evidence From the Spinal Cored Injured.” In: *Journal of Personality and Social Psychology*, vol. 54, no. 5, 820–828.

Damasio, A. (1994). *Descartes’ Error. Emotion, Reason and the Human Brain.* London: Vintage Books.

Damasio, A. (1999). *The Feeling of What Happens: Body and Emotion in the Making of Consciousness.* New York: The Ecco Press.

Damasio, A. (2010): *Self Comes To Mind: Constructing the Conscious Brain.* New York: Pantheon Books.

Darwin, C.: (1872/2009). *The Expression of The Emotions in Man and Animals.* London: Penguin Books.

Daubenmier et al. (2013). “Follow Your Breath: Respiratory Interoceptive Accuracy in Experienced Meditators.” In: *Psychophysiology*, vol 50, no. 8, 777–789.

Davis, J. et al. (2010). “The Effects of BOTOX-Injections on Emotional Experience.” In: *Emotion*, vol. 10, no. 3, 433–440.

Dewey, J. (1895). “The Theory of Emotion 2. The Significance of Emotions.” In: *Psychological Review*, vol. 2., 13–32.

Dewey, J. (1896). “The Reflex Arc Concept in Psychology.” In: *The Psychological Review*, vol. 3, no. 4, 357–370.

Ekman, P. (1971). "Universals and Cultural Differences in Facial Expressions of Emotion." In: J. Cole (Ed.). *Nebraska Symposium on Motivation 4.* Lincoln: University Press of Nebraska, 207–282.

Ekman, P. (2003). *Emotions Revealed. Understanding Faces and Feelings.* London: Orion Books.

Ekman, P. & Rosenberg, E. (1997). *What the Face Reveals. Basic and Applied Studies of Spontaneous Expressions Using the Facial Action Coding System.* New York: Oxford University Press.

Ekman, P., Levenson, W. & Friesen, W. (1983). "Autonomic Nervous System Activity Distinguishes among Emotions." In: *Science*, vol. 221, no. 4616, 1208–1210.

Ellsworth, P. (2014). "Basic Emotions and the Rocks of New Hampshire." In: *Emotion Review*, vol. 6, 21–26.

Feldman-Barrett, L. (2006). "Are Emotions Natural Kinds?" In: *Perspectives on Psychological Science* 1, 28–58.

Feldman-Barrett, L. & Russell, J. (2015). "Ten Misconceptions About the Psychological Construction of Emotion." In: L. Feldman-Barrett & J. Russell (Eds.): *The Psychological Construction of Emotion.* New York: Gilbert, 45–79.

Frijda, N. (1986). *The Emotions.* Cambridge: Cambridge University Press.

Griffiths, P. (1997). *What Emotions Really Are. The Problem of Psychological Categories.* Chicago: University of Chicago Press.

Harrison, N., Kreibig, S. & Critchley, H. (2013). "A Two-Way Road: Efferent and Afferent Pathways of Autonomic Activity in Emotion." In: J. Armony & P. Vuilleumier (Eds.). *The Cambridge Handbook of Human Affective Neuroscience.* Cambridge: Cambridge University Press, 82–106.

Hufendiek, R. (2016). *Embodied Emotions. A Naturalist Account to a Normative Phenomenon.* New York: Routledge.

Hurley, S. (1998). *Consciousness in Action.* Cambridge, MA: Harvard University Press.

Hurley, S. (2001). "Perception and Action. Alternative Views." In: *Synthese*, vol. 129, no. 1. 3–40.

Kreibig, S. (2010). "Autonomic Nervous System Activity in Emotion: A Review." In: *Biological Psychology*, vol. 84, no. 3, 394–421.

Labouvie-Vief et al. (2003). "Age and Gender Differences in Cardiac Reactivity and Subjective Emotion Response to Emotional Autobiographical Memories." In: *Emotion*, vol. 3, no. 1, 115–126.

Laird, J. (2007). *Feelings. The Perception of Self.* Oxford: Oxford University Press.

Laird, J. (1974). "Self-Attribution of Emotion: The Effects of Expressive Behavior on The Quality of Emotional Experience." In: *Journal of Personality and Social Psychology*, vol. 29, no. 4, 475–486.

Laird. J. & Lacasse, K. (2014). "Bodily Influences on Emotional Feelings: Accumulating Evidence and Extensions of William James's Theory of Emotions." In: *Emotional Review*, vol. 6, 24–37.

Lang, P., Bradley, M. & Cuthbert, B. (1990). "Emotion, Attention and the Startle Reflex." In: *Psychological Review*, vol. 97, no. 3, 377–395.

Lang, P., Bradley, M. & Cuthbert, B. (1998). "Emotion, Motivation, and Anxiety: Brain Mechanisms and Psychophysiology." In: *Society of Biological Psychiatry*, vol. 44, no. 12, 1248–1263.

Lazarus, R. (1991). *Emotion and Adaptation.* New York: Oxford University Press.
Lerner et al. (2005). "Facial Expressions of Emotions Reveal Neuroendocrine and Cardiovascular Stress Responses." In: *Biological Psychiatry*, vol. 58, 743–750.
Mauss, B. et al. (2005). "The Tle That Binds? Coherence Among Emotion Experience, Behavior and Physiology." In: *Emotion*, vol. 5, no. 2, 175–190.
Mauss, B. & Robinson, M. (2009). "Measures of Emotion: A Review." In: *Cognition and Emotion*, vol. 23, no. 2, 209–237.
Niedenthal, P. et al. (2005). "Embodiment in the Acquisition and Use of Emotion Knowledge." In: L. Feldman Barrett, P. Niedenthal & P. Winkielman (Eds.). *Emotion and Consciousness.* New York: Guilford Press, 21–51.
O'Regan, K. & Noë, A. (2001). "A Sensorimotor Account of Vision and Visual Consciousness." In: *Behavioral and Brain Sciences*, vol. 24, no. 5, 939–1031.
Ratcliff, M. (2005). "William James on Emotions and Intentionality." In: *International Journal of Philosophical Studies*, vol 13, no. 2, 179–202.
Reisenzein, R. & Stephan, A. (2014). "More on James and the Physical Basis of Emotions." In: *Emotion Review*, no. 6, 35–46.
Schwartz, G., Ahern, G. & Brown, S. (1979). "Lateralized Facial Muscle Response to Positive and Negative Emotional Stimuli." In: *Psychophysiology*, vol. 16, 561–571.
Shapiro, L. (2004). *The Mind Incarnate.* Cambridge, MA: MIT Press.
Shapiro, L. (2010). *Embodied Cognition.* New York: Routledge.
Solomon, R. (1973/2003). "Emotions and Choice." In: R. Solomon. *Not Passion's Slave. Emotions and Choice.* Oxford: Oxford University Press. 3–24.
Stepper, S. & Strack, F. (1993). "Proprioceptive Determinants of Emotional and Nonemotional Feelings." In: *Journal of Personality and Social Psychology*, vol. 64, no. 2, 211–220.
Strack, F., Martin, L. & Stepper, S. (1988). "Inhibiting and Facilitating Conditions of the Human Smile: A Nonobtrusive Test of the Facial Feedback Hypothesis." In: *Journal of Personality and Social Psychology*, vol. 54, no. 5, 768–777.

Ling Zhu

# Feeling as the Force Dynamics of Thought. The Role of Feeling in the Jamesian Stream of Thought

**Abstract:** In *The Principles of Psychology*, William James famously distinguishes between 'the substansive parts' (or 'nucleus') and 'the transitive parts' (or 'fringe') of 'the stream of thought.' The transitive parts, which tend to be overlooked due to their elusive nature, consist of '*feelings* of relation' and '*feelings* of tendency'. What exactly does 'feeling' mean and what role does it play within the stream of thought? This paper discusses three different, and increasingly strong interpretations of the Jamesian 'feeling', culminating in the claim that feeling, as the force dynamics of thought, plays a central part in the constitution of *meaning*.

Relatively early in his book *The Principles of Psychology* (1890), when looking for a general term to designate 'all states of consciousness' (James 1890, 185) or 'mental states at large, irrespective of their kind' (James 1890, 187), William James oscillates between 'feeling' and 'thought'. Both terms seem to him to be suitable candidates, while both suffer from being equated with only specific kinds of mental states in their normal usage: while the word 'feeling' is hardly ever used to refer to thoughts and is even seen to be directly opposed to the latter, the word 'thought' is hardly ever used to refer to bodily sensations. James finally decides that he will use these two terms interchangeably, both in a wider sense than usual, to refer to consciousness in general.

Later in the book, James distinguishes between the 'substansive' and 'the transitive parts' of the flow of consciousness, which he calls 'the stream of thought' (James 1890, 243). By 'substansive parts' he means 'sensorial imaginations of some sort', which are stable enough to be 'held before the mind for an indefinite time,' while the 'transitive parts' consist of the *relations* between them, which are so fluid and elusive that they cannot be contemplated or reflected upon as such. James famously likens this internal differentiation of the flow of consciousness to the alteration between a bird's flights and perchings. Although James does occasionally speak of '*thoughts* of relations' (James 1890, my emphasis), he much more often speaks of '*feelings* of relations' (James 1890, 247, my emphasis), an indication that James actually conceives of feeling as being *part of* our consciousness—more precisely as belonging to the transitive part of consciousness—rather than standing for consciousness as a whole, as he had indicated earlier in the book. Given that this is indeed the case, what exactly do 'feelings of relation' mean and what role do they play within the stream of consciousness? This paper discusses three different answers to this question, which attribute a successively more important cognitive function to feeling, culminating in the claim that feeling, as the *force dynamics* of thought, is essential in the constitution of *meaning*.

# 1 Feeling as the evaluative mechanism of fringe consciousness

The first way to conceive of feeling is proposed by Bruce Mangan (1993) and elaborated by Russell Epstein (2000). Mangan and Epstein equate James' 'substansive'-'transitive' distinction of the stream of thought with the 'nucleus'-'fringe' division of consciousness. The substansive part, or the nucleus, is that portion of a mental activity that gains 'focal attention' and thus appears particularly clear-cut and vivid. The transitive part, or the fringe, is that part of a mental activity that goes on simultaneously in the background or periphery and is experienced as vague and elusive due to the fact that it stands *off* the focus of attention. According to Mangan, one of the major reasons that consciousness falls into this two-fold division is that it has limited processing capacity. The information it can explicitly represent at any given moment is of a very small scope, in contrast to a significantly larger amount of 'context information' (Mangan 1993, 99) or 'associative memory' (Epstein 2000, 564) that is equally active but in an implicit form. The latter is important because it constitutes the pool of potential resources consciousness can select from and put into focus in the next moment as it flows forward.

How does 'feeling' fit into the picture? For Mangan and Epstein, feeling belongs to the fringe consciousness and its function there is to 'tell' the consciousness which content to extract from the fringe section that would match with the preceding content in the nucleus section. As such, feeling serves to guide the forward movement of thought and to guarantee its consistency. Mangan's and Epstein's favorite examples in this connection are the so-called 'tip-of-the-tongue' (TOT) or 'feeling-of-knowing' (FOK) experiences, in which we have an anticipatory *feeling* of the target information we are looking for even before retrieving it completely from memory. Such experiences not only demonstrate that consciousness indeed has a fringe with a pool of latently active information, but more importantly this fringe possesses an *evaluative mechanism* that chooses which information is a fit for the progression of thought and which not. As Mangan puts it: “the heart of a TOT is the recognition of rightness” (Mangan 1993, 103). This recognition takes place in the form of feeling, most typically the feeling of familiarity, as in the TOT or FOK experiences.

In summary, Mangan and Epstein interpret James’ ‘feelings of relations’ as ‘feelings of the relations between the foregoing and the forthcoming content of the stream of thought.’ This interpretation finds support in James’ own remarks that ‘Relation [...] is constantly *felt* in the fringe, and particularly the relation of harmony and discord, of furtherance or hindrance of the topic’ (James 1890, 259, my emphasis). However, by being merely evaluative, feeling’s function within this interpretation is very limited. It is like the referee of a football match, who, despite his indispensable role, does not really partake in playing the game. It is an *accompanying phenomenon* to the thought process per se, which intervenes with the latter but does not actually possess a cognitive structure of its own.

# 2 Feeling as the force-dynamic pattern of consciousness

The second interpretation of Jamesian feeling derives from the psychologist Daniel N. Stern’s theory of vitality affects. Stern did not deal with or refer to James’ philosophy in his work. The reason for borrowing Stern’s idea to expound James’ is that it provides the most concrete way of understanding the *dynamics* of the stream of consciousness. The dynamic aspect of mental life is James’ primary concern in introducing the notion of the ‘stream’ and his critique of the psychology of his time is precisely to mistake the static elements of consciousness for all there is to it. Like a river, consciousness is characterized by its never-ending flow, its permanent motion. While it does contain relatively static tracts in the form of stable and definite ‘sensorial images’ (James 1890, 252)—that which James calls ‘the substansive parts’—, these are ‘steeped and dyed in the free water that flows round [them]’ (James 1890, 255). The dynamic aspect, which James calls ‘the transitive parts,’ of the stream of consciousness, tends to be overlooked because it is in the very nature of something moving that it cannot be arrested and examined for itself. It would be like ‘seizing a spinning top to catch its motion, or trying to turn up the gas quickly enough to see how the darkness looks’ (James 1890, 244). To put it in a somewhat Heideggerian way, the dynamics of consciousness escapes any attempt of objectifying representation (i.e. vergegenständlichende Vorstellung). Maybe herein lies the reason why James more often uses the expression '*feelings* of relation' than 'thoughts of relation' to designate the transitive parts of consciousness: we obviously cannot *think* of the dynamics of our mental motion—supposing that thinking always involves some form of objectifying representation, which in turn presupposes turning something moving into something static—but we can definitely 'sense' or 'feel' it.

On the basis of extensive empirical research that originally focused on the cognitive development of infants, Stern puts forward a two-fold structure of conscious experience that very much resembles James' substansive-transitive differentiation of the stream of thought. Stern distinguishes between the 'What'- and the 'How'-aspects of mental life (Stern 2010, 23). The 'What'-aspect refers to the *contents* of consciousness, such as a particular sensation, emotion, thought, memory or phantasy. The 'How'-aspect, on the other hand, refers to *the way the conscious experience itself unfolds through time*, for example:

> What does it mean and feel like to have a thought 'surge up,' 'fade in,' 'accelerate and crest,' or enter on the scene with a 'rapid attack'? What does it mean and feel like to have a 'feeling jump up and swell' or 'move down and settle deeper' (Stern 2010, 12f.)

'Surging,' 'fading away,' 'fleeting,' 'explosive,' 'crescendo,' 'decrescendo,' 'bursting' and 'drawn out' (Stern 1998, 54)—all these words and expressions can be used to describe the *dynamics* of mental processes. Stern calls these properties of conscious experience their 'temporal contour' (Stern 2010, 4), 'time profile' (Stern 2010, 4) or 'flow pattern' (Stern 2010, 8) on different occasions. These dynamic properties tend to be overlooked because we are usually more occupied with, if not even utterly absorbed, by the content (the "What") of our thought, for example the problem we are trying to solve or the object we desire to have; we tend not to attend to how the problem-solving or the desiring itself as a dynamic process develops over time, how it waxes and wanes, speeds up and slows down. According to Stern however, it is precisely these dynamics characters that enrich and *vitalize* our mental acts, make them a human experience rather than the mere mechanic processing of information, the reason why he coins the terms 'vitality affects' (1998) or 'forms of vitality' (2010) to refer to them. Although de facto, these dynamic properties are always attached to the contents of mental processes, they can, nevertheless, be abstracted from the latter and experienced on their own. They constitute a relatively independent layer of conscious experience that subtends the contential layer. In fact, Stern even believes that on the neural level, the dynamic 'strand of representation' (Stern 2010, 25) is the more fundamental and primary one, "along or around" (Stern 2010, 25) which the contential strand of representation is encoded. This two-fold structure of conscious experience is not unlike the structure of experience in the physical domain: *the prosodic contour* of someone's speech, for example, is dependent upon, but in principle also detachable from, the content of the speech; similarly, *the kinetic contour* of someone's movement may be detached from what the movement is about (whether strolling along the street or grabbing a chair) and observed for its own sake. These contours denote purely the movement patterns of an action, and if physical actions have movement patterns, then equally do 'mental actions,' i.e. 'the stream of thought' have them. Precisely these movement patterns are what Stern means by vitality affects and what we here call the dynamics of conscious experience.

It is important here to emphasize that the dynamics of consciousness does not mean the mere change of content in the sense of 'I am thinking of A and then the next moment thinking of B.' This would be a trivial understanding of Stern's concept of vitality affect. Rather, the dynamics of consciousness is defined by such properties as *intensity, directionality, speed, duration, rhythm* and by *the change of these properties over time* (cf. ibid.). For example, as we think, we might have a sense of the increasing or decreasing strength of our thoughts (intensity); we might feel how the thoughts rise, fall, twist and entangle (direction); or how the thought process accelerates and decelerates (speed); or how a thought lingers on or fleets away (duration); and when we look at a thought process over an extended period of time, we might even notice some form of distinct rhythmic or stress patterns. The dynamics of consciousness is a *Gestalt* formed out of the interplay of all these factors. What is unique about them is that they are all 'gradient features' (Stern 1998, 178), i.e. they each *change in degrees along a scale* (stronger and weaker, faster and slower, longer and shorter, etc.). As such they can build up a theoretically unlimited number of shades and nuances while maintaining a deep sense of continuity, because these different shades and nuances would all be values on the same scale. In contrast, the change of content during a thought process is discrete: the movement from one thought to another might follow some logical relation, but it is never a movement along the same graduated scale.

How does the above-mentioned remarks relate to James' ideas and to the question regarding the role of feeling in the stream of thought? In two respects: 1. It is very plausible to interpret James' substansive-transitive differentiation of the steam of thought as the distinction between the contential dimension (the 'What') and the dynamic dimension (the 'How') of consciousness; 2. It certainly is no coincidence that just as James refers to the transitive part of the consciousness as '*feelings* of relations' (my emphasis), Stern names the dynamics of consciousness 'vitality *affects*' (my emphasis). Their choice of word suggests that both thinkers agree that mental dynamics is experienced in the mode of feeling, rather than rational thought. The reason behind this is that—if we draw upon

Stern's analysis—mental dynamics consists of gradient properties that change in degrees, whereas the very nature of rational thought is *discursiveness* in the etymological sense of *discursus*, i.e. it moves from one piece of information to another in a discontinuous and discrete fashion.

Before we move on to the third interpretation of Jamesian feeling, there is one point worth expanding upon. By possessing both 'intensity' and 'directionality', the two properties that would define a physical force, mental dynamics can be more precisely characterized as *force dynamics*. This characterization conforms to our intuitive perception of our mental flow as being driven/impelled by some kind of force. According to Stern, this felt force is an 'inferred' (Stern 2010, 22) force in the sense that it is projected/transposed from our experience of the physical onto that of the psychological. The projection/transposal has its basis in our 'mental primitive (how the mind evolves to process certain events)' (Stern 2010, 22) of interpreting motion and action as the result of applied force. I will return to this point in the next section of the paper.

## 3 Feeling as a fundamental constituent of meaning

Stern's interpretation of feeling as the force-dynamic pattern of consciousness is a much stronger one than Mangan's and Epstein's, according to which feeling is a mere accompanying element to thought and without its own internal structure. Yet by separating the contential and the dynamic aspects of consciousness and attributing one to the realm of rational thought and the other to the realm of feeling, Stern subscribes to the 'thought/feeling dualism' (Johnson 2007, 216) deeply entrenched in the history of Western philosophy. Mark Johnson describes this dualism as 'align[ing] meaning with the cognitive and thus dismiss[ing] quality, feeling and emotion from any account of meaning' (Johnson 2007, 216). In his book *The Meaning of the Body*, Johnson combats this dichotomy by advancing the hitherto strongest interpretation of feeling. He argues that 'feelings of quality, connection and direction lie at the heart of logical reasoning' (Johnson 2007, 103) and 'of our capacity to experience meaning' (Johnson 2007, 53). In other words, feeling is not a mere concomitant to thought—which it is ultimately still within Stern's theory—but its essential component. It goes right into the core of the process by which we conceptualize and make sense of the world and exerts thus a genuinely cognitive function. Before I present Johnson's ideas in detail, I would like to quote the following passage from James in its full, unabridged form. It shows how, in an almost visionary way, James' thinking anticipated

Johnson's by including conceptual and logical relations in what he calls 'feelings of relations':

> If there be such things as feelings at all, then so surely as relations between objects exist in rerum natura, so surely, and more surely, do feelings exist to which these relations are known. There is not a conjunction or a preposition, and hardly an adverbial phrase, syntactic form, or inflection of voice, in human speech, that does not express some shading or other of relation which we at some moment actually feel to exist between the larger objects of our thought. If we speak objectively, it is the real relations that appear revealed, if we speak subjectively, it is the stream of consciousness that matches each of them by an inward coloring of its own. In either case the relations are numberless, and no existing language is capable of doing justice to all their shades.
>
> We ought to say a feeling of and, a feeling of if, a feeling of but, and a feeling of by, quite as readily as we say a feeling of blue or a feeling of cold. (James 1890, 245 f.)

Johnson's starting point is his theory of conceptual metaphor which he developed in collaboration with George Lakoff (2003). The central tenet of the theory is that our conceptual system employs structures that originate from our bodily being and interaction with the world. These structures, which Johnson calls 'image schemata' (Lakoff & Johnson 2003, 28), are metaphorically projected, by virtue of analogy, from experiences of the physical onto experiences of usually more abstract domains. It is important to note that conceptual metaphors do not merely 'describe' things and events that are already understood, say, in a literal way. Rather, they *are* the means by which we comprehend and make sense of these things and events in the first place. They are, in other words, *primary vehicles* by which our cognitive system structures and orders experience. An example of an image-schema is the 'container-schema.' We all know intimately what a container is by having a body whose boundary separates our inside from the outside world (cf. Lakoff & Johnson 2003, 29). We frequently project this knowledge of the physical inside-outside distinction onto experiences of non-physical nature, as can be seen in different linguistic expressions. We say for example 'X is still *in* the game, while Y is *out*,' treating the game as a container. Or on an even more abstract level, we say 'This argument has no substance *in* it,' treating in this case the argument as a container. Another prevalent image-schema is the 'up-down-schema.' Being constantly subject to gravity, we know all too well the opposition between up and down. Jumping up means the liberation from gravity, while falling down means succumbing to it. We map this value difference between up and down onto various states of affairs that exhibit an analog positive-negative polarity. We say for example 'I am feeling *up*' or 'I am in *high* spirits' when we feel positive, as against 'I am feeling *low/down*' or 'I am really *depressed*' when we feel negative (cf. Lakoff & Johnson 2003, 15). Even such utterly mundane and ordinary expressions entail the metaphorical use of the up-down schema as 'I want to *up*grade my membership' or 'I have to *down*-size my property.' Again, it should be emphasized that these expressions are not 'just a matter of language' (Lakoff & Johnson 2003, 6). Rather, they mirror the fact that 'the human conceptual system is metaphorically structured and defined' (ibid.).

While in his earlier work, Johnson focuses on what may be called 'static image-schemata' encompassing mainly spatial relations such as 'center-periphery,' 'front-back', 'near-far' and the above-mentioned 'inside-outside' and 'up-down', in a later book (1987), Johnson adds a deeper dimension to the theory of conceptual metaphor by introducing *force-dynamic* image schemata. Building on and further developing Leonard Talmy's seminal work (1985), Johnson points out that our conception, reasoning and logical inference are permeated by force-dynamic patterns. And just like all other image-schematic patterns, these force-dynamic patterns have their origin in our bodily existence, specifically in 'our forceful interactions in our world' (Johnson 1987, 48). We abstract these patterns from our physical, sensorimotor experience and transpose them onto the non-physical and mental realms in order to give structure and meaning to what we think and communicate.

Force dynamics includes several essential features. To begin with, force is a 'vectorial quantity,' meaning it always has a *direction* and a *degree of intensity* (or magnitude)—I have already mentioned these two major properties of force in relation to Stern's concept of 'vitality affect.' What does not play a role for Stern but is of utter importance for Johnson and also for Talmy is the thoroughly *relational* nature of force, i. e. the operation of force involves the interaction between at least two entities: the actor—the one who exerts the force—and the patient—the one who receives the force. Talmy calls the one 'antagonist' and the other 'agonist' (Talmy 1988, 53).[1] Now if we strictly adhere to the Newtonian law of physics that to every action there is an opposing reaction, the patient is also an actor by exerting a reacting force. So who is the actor/antagonist and who is the patient/agonist in a force-dynamic relationship is really a question of perspective. Depending on how the direction and the magnitude of force of the two entities stand *in relation to each other*, different *causative* effects will result from their forceful interaction. Suppose object B is stationary and object A is moving towards it, the following scenarios may ensue: A may *compel/force* B to move too; or, if B has an internal tendency to move and A merely puts this potential to realization, we are not talking about 'compulsion' but *facilitation* or *enable-*

1 Compare Jackendoff's translation of Talmy's antagonist-agonist dyad into a patient-actor relation (Jackendoff 1990, 130 f.).

*ment*; it is also possible that B *blocks* A's movement and becomes its barrier; A may also *overcome/remove* B's blockage and continue moving (cf. Johnson 1987, 45–48).

I will not enumerate all possible scenarios of the forceful encounter between objects here. What is significant is that physical force interactions generate relationships sophisticated and articulate enough to be metaphorically projected onto our thought and reasoning. Talmy (1988) has convincingly shown how much we think *in terms of* entities in force dynamic relationships, as most clearly reflected by the semantic structures of our language. For example, the following sentences are, upon closer look, each based on one of the four types of physical force dynamic relationships I have mentioned earlier on:

"Andy continued with his PhD studies because his parents wanted him to."

–*Compulsion* (Parents compel Andy to pursue PhD studies)

"Andy continued with his PhD studies upon receiving a scholarship."

–*Enablement* (Scholarship enables Andy to pursue PhD studies)

"Andy could not continue with his PhD studies because his application for a scholarship was rejected."

–*Blockage* (Lack of Scholarship blocks Andy's PhD studies)

"Andy worked part-time as a translator so that he could continue with his PhD studies."

–*Overcoming/removing hindrance* (Part-time work removes hindrance to pursue PhD studies)

In *The Body in the Mind*, Johnson (Johnson 1987, 48–57) argues that the meaning of the so-called 'modal verbs'—verbs which indicate possibility, actuality and necessity—is also built upon the analogy to physical force-dynamic relationships. The three most frequently used modal verbs 'must,' 'may' and 'can', for example, reveal as their underlying structures again the basic variants of force dynamic interactions 'compulsion,' 'enablement' and 'removal of hindrance':

"You must join the team, because we really need someone like you."

–*Compulsion* (Our need compels you to join the team)

"You can join the team, because you have the right qualifications."

–*Enablement* (Your qualifications enable you to join the team)

"You may join the team, because a vacancy has just come up."

–*Overcoming/Removing hindrance* (The vacancy removes the hindrance for you to join the team)

To push his argument yet another step further, Johnson (Johnson 1987, 63 f.) demonstrates that inference patterns of modal logic, which pertains to the possibility or necessity of truth, are equally grounded upon underlying force dynamic relations:

Logical necessity: $P\square \rightarrow P$ "If P is logically necessary, then P is true."

–*Compulsion* (The circumstances compel us to reach the conclusion)

Logical possibility: $\sim\Box\sim P \rightarrow \Diamond P$ "If it is not logically necessary that P is false, then it is logically possible for P to be true."

–*Removal of Hindrance* (Nothing prevents us from reaching the conclusion)
The above is a selective, but in no way exhaustive account of force dynamic patterns that underlie our language and thought. The following point, which will bring us back to our subject of 'feeling,' is lacking in Talmy's account and is only implied, but not really articulated in Johnson's account of force dynamics. Namely: Force is not something externally observable; observable are only the effects of force, e.g. static objects begin to move when being pushed. The only way force becomes directly accessible to our experience is through being *internally felt*. We are constantly exerting forces or being impacted by forces, even such elemental acts as breathing, looking at something and maintaining our bodily balance require the exertion of force, though due to the routine nature of these actions we tend to overlook this fact. There is no single moment in our life in which the application of force is absent. And feeling, which ranges from bodily sensations, specific emotions to more elusive affective states for which our language does not have names, is that mechanism that coordinates and optimizes our application of force. It signals to us how much effort we are applying and how much counterforce there is from the object we are acting upon. The sense of effort or effortlessness, of strain or relief, of hindrance or expedience all serves to provide feedback on our forceful actions and reactions in the world, so that we do not stamp on the ground in a way that would hurt our feet, nor shake someone's hand in a way that would make him scream out of pain. It would seem that with these remarks, I have turned feeling into a mere evaluative mechanism in the same way as have Mangan and Epstein have done. Yet it is important here to point out the difference: whereas Epstein's and Mangan's feeling lacks any internal structure, the feeling of force dynamics possesses very definite and complex structures, defined, as mentioned above, by *degree of intensity, direction* and very specific *causative relations between actor and patient*. As we feel force, we feel all these dimensions or properties of force in a highly differentiated way, the reason why we could fine-tune our movements and actions to suit so many different purposes.

To say, as do Talmy and Johnson, that fundamental meaning structures of our language, thought and reasoning are in essence force dynamic structures is tantamount to saying that feeling is a fundamental constituent of meaning, because force dynamics is only known to us by being subjectively felt. In other words: *it is by feeling that we grasp meaning*. Herein lies the essential difference between Stern's and Johnson's concept of feeling: While both interpret feeling as the force dynamics of thought, in Stern's account, force dynamics is external to the cognitive structures of thought, resulting in it being almost an epiphenomenon, whereas in Johnson's account, feeling becomes *embedded* in the meaning structure of thought.

# 4 Conclusion

This paper provided three different interpretations of James' notion of 'feelings of relations', which he suggests to be essential elements of the stream of thought. While James' own account accommodates and provides support for all three interpretations, they ascribe an increasingly important cognitive role to feeling. The crucial insight resulting from a synthesis of Stern's and Johnson's understanding of feeling is that it encompasses multiple factors including intensity, direction, temporal change and causative relations. The combination of these factors gives rise to complex force-dynamic structures which, as forcefully demonstrated by Johnson, grounds meaning structures of discourse and reasoning.

# References

Epstein, R. (2000). "The Neural-Cognitive Basis of the Jamesian Stream of Thought." In: *Consciousness and Cognition* 9, 550–575.

Jackendoff. R. (1990). *Semantic Structures.* Cambridge, MA: MIT Press.

James, W. (1890). *The Principles of Psychology.* New York: Henry Holt and Company.

Johnson, M. (1987). *The Body in the Mind: The Bodily Bases of Meaning, Imagination, and Reason.* Chicago/London: The University of Chicago Press.

Johnson, M. (2007). *The Meaning of the Body: Aesthetics of Human Understanding.* Chicago/London: The University of Chicago Press.

Lakoff, G., & Johnson, M. (2003). *Metaphors We Live by.* Chicago: The University of Chicago Press.

Mangan, B. (1993). "Taking Phenomenology Seriously: The 'Fringe' and Its Implications for Cognitive Research." In: *Consciousness and Cognition* 2, 89–108.

Stern, D. N. (1998). *The Interpersonal World of the Infant: A View from Psychoanalysis and Developmental Psychology.* London: Karnac Books.

Stern, D. N. (2010). *Forms of Vitality: Exploring Dynamic Experience in Psychology, the Arts, Psychotherapy, and Development.* Oxford: Oxford University Press.

Talmy, L. (1985). "Force Dynamics in Language and Thought." In: *Papers from the Regional Meetings, Chicago Linguistic Society* 21, 293–337.

Talmy, L. (1988). "Force Dynamics in Language and Cognition." In: *Cognitive Science* 12, 49–100.

# Index of persons

# Index of subjects

Bei Fragen zur Produktsicherheit wenden Sie sich bitte an:
If you have any questions regarding product safety,
please contact:

Walter de Gruyter GmbH
Genthiner Straße 13
10785 Berlin
productsafety@degruyterbrill.com